STATISTICS
FOR THE
TWENTY-FIRST CENTURY

STATISTICS
FOR THE
TWENTY-FIRST CENTURY

Florence Gordon
and
Sheldon Gordon
Editors

MAA Notes and Reports Series

The MAA Notes and Reports Series, started in 1982, addresses a broad range of topics and themes of interest to all who are involved with undergraduate mathematics. The volumes in this series are readable, informative, and useful, and help the mathematical community keep up with developments of importance to mathematics.

MAA Notes

1. Problem Solving in the Mathematics Curriculum,
    Committee on the Teaching of Undergraduate Mathematics,
    a subcommittee of the Committee on the Undergraduate Program in Mathematics, *Alan H. Schoenfeld,* Editor.

2. Recommendations on the Mathematical Preparation of Teachers,
    Committee on the Undergraduate Program in Mathematics, Panel on Teacher Training.

3. Undergraduate Mathematics Education in the People's Republic of China,
    Lynn A. Steen, Editor.

4. Notes on Primality Testing and Factoring,
    Carl Pomerance.

5. American Perspectives on the Fifth International Congress on Mathematical Education,
    Warren Page, Editor.

6. Toward a Lean and Lively Calculus,
    Ronald G. Douglas, Editor.

7. Undergraduate Programs in the Mathematical and Computer Sciences: 1985–86,
    D. J. Albers, R. D. Anderson, D. O. Loftsgaarden, Editors.

8. Calculus for a New Century,
    Lynn A. Steen, Editor.

9. Computers and Mathematics: The Use of Computers in Undergraduate Instruction,
    Committee on Computers in Mathematics Education, D. A. Smith, G. J. Porter, L. C. Leinbach, and R. H. Wenger, Editors.

10. Guidelines for the Continuing Mathematical Education of Teachers,
    Committee on the Mathematical Education of Teachers.

11. Keys to Improved Instruction by Teaching Assistants and Part-Time Instructors,
    Committee on Teaching Assistants and Part-Time Instructors, Bettye Anne Case, Editor.

12. The Use of Calculators in the Standardized Testing of Mathematics,
    John Kenelly, Editor, published jointly with The College Board.

13. Reshaping College Mathematics,
    Committee on the Undergraduate Program in Mathematics, Lynn A. Steen, Editor.

14. Mathematical Writing,
    by *Donald E. Knuth, Tracy Larrabee, and Paul M. Roberts.*

15. Discrete Mathematics in the First Two Years,
    Anthony Ralston, Editor.

16. Using Writing to Teach Mathematics,
    Andrew Sterrett, Editor.

17. Priming the Calculus Pump: Innovations and Resources,
    Committee on Calculus Reform and the First Two Years,
    a subcommittee of the Committee on the Undergraduate Program in Mathematics, *Thomas W. Tucker,* Editor.

18. Models for Undergraduate Research in Mathematics,
Lester Senechal, Editor.

19. Visualization in Teaching and Learning Mathematics,
Committee on Computers in Mathematics Education, Steve Cunningham and Walter S. Zimmermann, Editors.

20. The Laboratory Approach to Teaching Calculus,
L. Carl Leinbach et al., Editors.

21. Perspectives on Contemporary Statistics,
David C. Hoaglin and David S. Moore, Editors.

22. Heeding the Call for Change: Suggestions for Curricular Action,
Lynn A. Steen, Editor.

23. Statistical Abstract of Undergraduate Programs in the Mathematical Sciences and Computer Science in the United States: 1990–1991 CBMS Survey,
Donald J. Albers, Don O. Loftsgaarden, Donald C.Rung, and Ann E. Watkins.

24. Symbolic Computation in Undergraduate Mathematics Education,
Zaven A. Karian, Editor.

25. The Concept of Function: Aspects of Epistemology and Pedagogy,
Ed Dubinsky and Guershon Harel, Editors.

26. Statistics for the Twenty-First Century,
Florence and Sheldon Gordon, Editors.

MAA Reports

1. A Curriculum in Flux: Mathematics at Two-Year Colleges,
Subcommittee on Mathematics Curriculum at Two-Year Colleges, a joint committee of the MAA and the American Mathematical Association of Two-Year Colleges, *Ronald M. Davis,* Editor.

2. A Source Book for College Mathematics Teaching,
Committee on the Teaching of Undergraduate Mathematics, Alan H. Schoenfeld, Editor.

3. A Call for Change: Recommendations for the Mathematical Preparation of Teachers of Mathematics,
Committee on the Mathematical Education of Teachers, James R. C. Leitzel, Editor.

4. Library Recommendations for Undergraduate Mathematics,
CUPM ad hoc Subcommittee, Lynn A. Steen, Editor.

5. Two-Year College Mathematics Library Recommendations,
CUPM ad hoc Subcommittee, Lynn A. Steen, Editor.

First Printing
© 1992 by the Mathematical Association of America
ISBN 0-88385-078-8
Library of Congress Catalog Card Number 92-60194
Printed in the United States of America

Preface

We live in a society which is ever more dependent on statistical reasoning. Major political, social, economic and scientific decisions are made using statistical information. Statistical reports affecting virtually all aspects of our lives appear regularly in all the news media. Business and industry leaders constantly call for greater statistical understanding on the part of our workforce. Statistics has thus become the primary quantitative tool in most areas. In turn, statistics offerings have become one of the fastest growing segments of the undergraduate curriculum.

These trends are almost certain to continue with statistics assuming an ever greater role throughout our society. The students we teach today will spend virtually their entire productive lives in the coming century, a century that will require an even greater awareness of statistical ideas. We should therefore be offering statistics courses geared to the needs of the citizens of the twenty first century.

Yet, far too many of the students who take a statistics course describe it as the worst or most boring class they have experienced in college. While we constantly speak of the introductory statistics course as being the *first* course in statistics, it is actually the *last* course in statistics for the overwhelming majority of students. This is a particularly disturbing commentary when we consider that the material typically covered in an introductory statistics course forms the basis for the statistical reports and analyses that can be found all about us.

In the present volume, many of today's leading statistics educators address the problem of improving statistics education. All of their articles include suggestions for improving the introductory course, whether it is directed at a general liberal arts audience, at a narrower discipline-specific audience (such as social science, biological science or business majors), or at a more sophisticated mathematics/statistics audience. The volume presents an overview of the innovative and dynamic possibilities that can be used to bring any introductory statistics course alive: to make it more current, more realistic and more exciting to the students and to involve them directly in the subject as active participants. Our earnest hope is that all readers will find something here that will improve their courses. In turn, we hope that the first course in statistics will indeed be a *first* course and that our students will be ready and able to function statistically in the twenty first century.

Overview of the Contents

In the lead article, Robert Hogg provides a broad overview of many of the major problems with existing introductory courses in statistics and presents a variety of recommendations for improving such courses. These ideas are an outgrowth of a workshop that Bob organized involving 39 statistical educators. The following group of three articles address what the focus should be in the first course in statistics from three distinct perspectives: those of a statistician (David Moore), a group of users from different disciplines (Walter Chromiak, Jim Hoefler, Allan Rossman and Barry Tesman), and a mathematician (Gudmund Iversen). You will notice that many of the same themes pervade each of these three perspectives. There can be a single course that meets everybody's needs! But this course should focus on ideas and applications rather than on calculations or the probabilistic or theoretical underpinnings of the subject.

There are other important issues which may affect the first course in statistics and which must be considered. One is the implication for college courses of the increasing emphasis on statistics and probability in the elementary and secondary schools. Some of the leaders of the American Statistical Association's Quantitative Literacy Project (Ann Watkins, Gail Burrill,

James Landwehr and Richard Schaeffer) suggest that the day is not far off when colleges may have to offer remedial statistics courses. Further, all too often, mathematics departments concern themselves exclusively with student readiness for courses in the standard algebra through calculus track, but assume that everyone is ready for introductory statistics. Elizabeth Eltinge addresses this issue and discusses the need for appropriate placement into statistics courses.

The next two sections of this volume consist of a variety of ideas and suggestions for improving the curriculum in introductory statistics courses. Richard Schaeffer argues for the need to focus on data analysis as the key to understanding statistics. John Willett and Judith Singer show how any statistics course can come alive by using real-world data sets to involve the students in real situations using statistical arguments. Robin Lock and Tom Moore include a variety of fascinating "low tech" ideas for enlivening any statistics class. Harry Roberts describes the importance of having students conduct their own statistical projects to provide an understanding of *doing*, rather than *watching*, statistics in action. Martin Tanner and Robert Wardrop suggest some classroom activities for generating sets of data for statistical study. Gottfried Noether outlines the possibility of approaching the first course in statistics from a nonparametric viewpoint to avoid some of the complexities associated with the standard approach. Sandy Zabell shows that the search for randomness is not just a modern phenomenon associated with twentieth century statistics, but rather that it is a theme which has pervaded all human cultures; as such, it can provide an interesting new perspective to be interwoven into our courses. Many students enter introductory statistics courses with major misconceptions and problems with the underlying notions of probability. Psychologists Ruma Falk and Clifford Konold describe some of these common misconceptions which many mathematicians are likely to find very surprising.

Many of the efforts at improving statistical education focus on appropriate ways of using technology. Certainly, technology can eliminate the drudgery of performing complex statistical calculations by hand. However, it can also provide the tools for developing a much better understanding of statistical ideas. Laurie Snell and William Peterson address the question "Does the computer help us understand statistics?" with a wide variety of illustrations. Following that, we present a broad array of other perspectives and experiences on this subject using computer graphics simulations and demonstrations (Henk Tijms, Henry Krieger and James Pinter-Lucke, Florence and Sheldon Gordon, Elliot Tanis), spreadsheets (Deane Arganbright), systems such as HyperCard (Walter Chromiak and Allan Rossman) and graphing calculators (Iris Fetta). Based on all of these viewpoints, the answer to Snell and Peterson's question must be a resounding Yes!

The final section of this volume contains information on a variety of resources for statistical education. Thomas Moore and Jeffrey Witmer present an annotated bibliography of publications -- books, articles and periodicals -- that will be of great interest and use to instructors of statistics. Laurie Snell presents an annotated bibliography of articles of a statistical nature culled from newspapers and popular magazines; this ever-growing collection can provide a rich source of references to students to make the subject come alive as well as to instructors who might want to bring some current statistical discussions into the classroom. Finding appropriate sets of data to use for various statistical procedures can be a daunting task; Judith Singer and John Willett supplement their earlier article with an annotated bibliography of real-world datasets that are useful when teaching statistics. Finally, Tim Arnold introduces many of us to the wealth of resources for statistical education that are available through the growing electronic networks.

Acknowledgements

The editors would like to express their sincere appreciation to all of the authors who contributed articles to this volume and who cheerfully put up with a seemingly endless cycle of revisions and requests. Their work speaks profoundly of their intense interest in improving statistical education. It has been an honor and a pleasure to work with each of them.

Further, the development of a volume such as this involves the invaluable contributions of many people aside from those who actually wrote articles. We want to thank all of them. Warren Page, the editor of the MAA Notes series, initially invited us to organize such a volume; we want to thank him for having the vision to see the need for it. Many of the authors represented in the volume also served as formal reviewers at our request; many others served as informal reviewers as the authors themselves circulated preliminary versions for critique. They also provided us with extremely valuable suggestions in many long conversations and through correspondence. It was very welcome and helpful; we thank you all.

Our sincere appreciation goes also to the many people who provided assistance in the production of this volume. They include Don Albers, Eileen Pedreira Sullivan and Beverly Ruedi at the MAA offices in Washington. We also are particularly indebted to our son, Kenneth Gordon, who provided invaluable assistance throughout the typesetting and production process. Our sincere thanks also to Marilyn Campo for her expert retyping of some articles when the electronic processes broke down.

We want to acknowledge the support provided by the National Science Foundation under grants #USE-89-53923 and #USE-91-50440 and by New York Institute of Technology's Faculty Research Grant program. Their support made this project possible.

Finally, we want to thank you, the readers of this volume, for your interest in improving statistical education. It is your dedication that makes this volume necessary.

Florence S. Gordon
Sheldon P. Gordon

Old Westbury, NY
Selden, NY
July, 1992

Table of Contents

Preface vii

Part I: Issues in Statistical Education

Towards Lean and Lively Courses in Statistics
     Robert V. Hogg 3

Teaching Statistics as a Respectable Subject
     David S. Moore 14

A Multidisciplinary Conversation on the First Course in Statistics
     Walter Chromiak, Jim Hoefler, Allan Rossman and Barry Tesman 26

Mathematics and Statistics: An Uneasy Marriage
     Gudmund R. Iversen 37

Remedial Statistics?: The Implications for Colleges of the Changing
     Secondary School Curriculum
     Ann Watkins, Gail Burrill, James Landwehr, and Richard Schaeffer 45

Diagnostic Testing for Introductory Statistics Courses
     Elizabeth M. Eltinge 56

Part II: Innovative Curricula for Statistical Education

Data, Discernment and Decisions: An Empirical Approach to Introductory Statistics
     Richard L. Schaeffer 69

Providing a Statistical "Model": Teaching Applied Statistics using Real-World Data
     John B. Willett and Judith D. Singer 83

Low-Tech Ideas for Teaching Statistics
     Robin H. Lock and Thomas L. Moore 99

Student-Conducted Projects in Introductory Statistics Courses
     Harry V. Roberts 109

Hands-on Activities in Introductory Statistics
     M. A. Tanner and Robert Wardrop 122

An Introductory Statistics Course: The Nonparametric Way
     Gottfried E. Noether 129

The Quest for Randomness and its Statistical Applications
     Sandy L. Zabell 139

Part III: Technology in Statistical Education

Does the Computer Help Us Understand Statistics?
J. Laurie Snell and William P. Peterson 167

Exploring Probability and Statistics using Computer Graphics
Henk C. Tijms 189

Computer Graphics and Simulations in Teaching Statistics
Henry Krieger and James Pinter-Lucke 198

Sampling + Simulations = Statistical Understanding:
   Computer Graphics Simulations of Sampling Distributions
Florence S. Gordon and Sheldon P. Gordon 207

Computer Simulations to Motivate Understanding
Elliot A. Tanis 217

Using Spreadsheets in Teaching Statistics
Deane E. Arganbright 226

Using HyperCard to Teach Statistics
Walter Chromiak and Allan Rossman 243

Graphing Calculator Enhanced Introductory Statistics
Iris Brann Fetta 258

Part IV: Resources for Statistical Education

Bibliography of Materials for Teaching Statistics
Thomas L. Moore and Jeffrey A. Witmer 273

CHANCE: Case Studies of Current Chance Issues
J. Laurie Snell 281

Annotated Bibliography of Sources of Real-World Datasets Useful for Teaching
   Applied Statistics
Judith D. Singer and John B. Willett 298

Is There Anyone Out There?: Electronic Connections to the Statistics Community
Tim Arnold 314

Part I

Issues

in

Statistical Education

Towards Lean and Lively Courses in Statistics

Robert V. Hogg, et al[1]
University of Iowa

Background

Over the years, there has been considerable dissatisfaction expressed about the first course in statistics. Many instructors have been less than happy about the success of such courses. Students have usually been far more vehement about their dislike for the subject. Thus, far too often, introductory statistics has earned a reputation for being "the worst course that I took in college."

In response to these concerns, the author organized a special workshop on statistical education held at the University of Iowa in June, 1990. Thirty-nine individuals representing universities, colleges, business, industry and consulting firms accepted invitations to address the problems in statistical education and to propose possible solutions. The present paper represents a summary of the results of this workshop.

We make recommendations that the statistical community could and should adopt to improve the state of statistical education. Most of the participants believe that we, as statisticians, should reach out more to other groups, in particular to mathematicians who might be teaching courses in statistics. The introductory courses, to be most effective, can not be taught solely as mathematics, but must be data oriented, getting the students involved in projects having statistical components. That is, students and faculty must understand that statistics is a study of variation, and in it we try to find patterns despite the fact that the data points generally do not precisely fall on those "theoretical curves" due to "noise."

We believe that it is incumbent on the statistical profession to assist all, particularly mathematicians, who are teaching courses in statistics. In this report, we encourage workshops on teaching this subject and would like to see joint efforts between the American Statistical Association (ASA) and the Mathematical Association of America (MAA) to this end. The possibility of a series of workshops at regional and national MAA meetings is very appealing and many statisticians would like to be active participants in these. Such workshops would help us reinforce and fine tune the recommendations in this report.

As you will also note, the report does recommend that the academic profession in general should modify our present system of rewards. We can only achieve the goals of many recom-

---

[1]   Others participating in an invited workshop on statistical education in Iowa City, Iowa, June 18-20, 1990, were Mel Alexander, Jim Calvin, Patti Collings, Jon Cryer, Marilyn Dueker, Andrew Ehrenberg, Herman Friedman, Barry Griffen, Bert Gunter, Peter Hackl, Gudmund Iversen, Mark Johnson, Brian Joiner, Jim Landwehr, Bob Lochner, Richard Madsen, Bobby Mee, Bob Miller, Tom Moore, Peter Nelson, Ron Patterson, Harry Roberts, Tim Robertson, Ed Rothman, Ed Schilling, Jim Sconing, Elliot Tanis, Aaron Tennebein, Neil Ullman, Joe Voelkel, Ray Waller, Ann Watkins, Roy Welsch, Carl Wetzstein, B.F. Winkel, Jeff Witmer, Lianng Yuh, and Arnold Zellner.

mendations of groups in the mathematical sciences if there is a better balance among the teaching, research, and service aspects of the profession. To do this, we need champions from our profession to convince the leadership in universities, industries, and governments that this would be best in the long run for the students and the faculty.

In preparation for the workshop, most of the 39 participants, and several others not attending, wrote positions papers on some aspect of statistical education. The majority concerned a first course in statistics. As a group, we recognized several poor characteristics of science and mathematical education in general including statistical education in particular. These include:

(1) Improvements are needed in the K-12 curriculum as there is widespread science and mathematics illiteracy in the United States.

(2) There is not enough effort given to recruiting students to these areas.

(3) There is a lack of truly qualified teachers because the pool from which they come is drying up.

(4) College and university instructors are often required to teach large introductory classes allowing little or no interaction with students and severely limiting student involvement with others.

(5) Frequently students view courses as being "hard" because the class periods are dull and it is difficult to get good grades despite spending long hours doing homework.

(6) Sometimes there is a communication barrier. The instructors often forget to emphasize the "big picture" by concentrating on the solutions of narrow problems and eliminating any in-depth discussions of the basic ideas.

(7) In most universities and some colleges, there is no sense of community among most students in these areas. Accordingly, there is not much discussion during and after class periods. Somehow we should encourage teamwork and stress the importance of it.

Many of us in higher education use the excuse that students should be better prepared. They probably should be, and there are many efforts to improve education in K-12. However, the fact that we have these poorly prepared students does not excuse us from doing much better than we do. Besides, many of these "poor" students are really extremely good, and we, in colleges and universities, are turning them off because:

▶ it is not "fun" because we fail to communicate our enthusiasm and excitement about math and science (in particular, statistics);

▶ we do not encourage teamwork, which often results in the joy of discovery through lively discussion, observation and analysis;

▶ there are "grade wars" and students believe that they must beat others to get high grades;

▶ formulas are often not motivated and presented as an "end" rather than the "means";

▶ introductory courses have low priority among the faculty and administrators, and the students observe this;

▶ computer use is perfunctory or nonexistent.

Many of us believe that those of us in universities must "reach out" more to a number of groups, including:

(1) The high schools, in particular working with future and present high school teachers, strengthening math and science requirements and curricula, and often serving as role models.

(2) Colleges, with much more interaction between university and college professors and students.

(3) Graduate students, working with them on their teaching as well as their research.

(4) Junior faculty, serving as mentors in teaching as well as research.

(5) Colleagues in other fields, interacting and working on joint research projects.

(6) Business/industry/government, because many more joint efforts are needed here if America is going to continue to be in the "first class" category. Moreover, these persons can offer faculty and students glimpses of practical applications.

(7) The public, trying to improve numeracy and science literacy.

However, it is time consuming to do these things. Thus we must change the present system if our faculties are to get credit for these efforts. As it is now, most faculty members find it to their advantage to be "loners" doing research. None of us are against research; we support it strongly. On the other hand, we seek a better balance in our reward system recognizing that good teaching and valuable service are as important as research. We want to recruit strong, but possibly different, students to science and mathematics by improving our courses.

The preceding comments apply to statistical education as well as to science and mathematics education in general. We believe that some of the leaders in our profession must help in serious efforts to address these concerns, and we hope this report will encourage them to do so. Certainly ASA has taken some very positive steps with its Quantitative Literacy program in improving statistical presentations in K-12. Much more can be done, and it will take enormous efforts on the part of many to shake up the system enough to change science and math education for the better.

It is interesting to note that the "second Sputnik" (Japanese cars, electronics, etc.) has arrived for industry and business, but an equivalent impact on science education as followed Sputnik in the late 1950s and 1960s has not occurred. Some of us had hoped that the 1983 report, *A Nation at Risk*, would shock our leaders into action. Even though it was short and printed in big type, we are afraid few leaders read it -- or, if they did, they did not take it seriously enough to cause substantial action. Thus, we continue our decline to becoming a second-rate nation. Wake up America! Let us create demand for a return to quality education which is competitive with that of other industrialized countries.

Problems in Statistical Education

Clearly in a three-day workshop we could not address all of the problems listed in the preceding section even though they apply to statistics as well as to science and mathematics. We thought that we could do the most good by addressing the problems associated with our introductory courses. Inspiring courses early in the student's college career would help improve their numeracy and possibly encourage some to consider careers in science. That is, we believe the students should appreciate how statistics is used in the endless cycle associated with the scientific method: We observe Nature and ask questions, we collect data that shed light on these questions, we analyze the data and compare to what we previously thought, new questions are often raised, and on and on. It has proved to be an exciting process for many years; but, most of the time, the students do not experience this excitement in our classes. There is a consensus among statisticians that statistics offers a unique and useful way of looking at the world and helping to solve real problems -- or to exploit opportunities, but that statistics teaching in colleges and universities fails to communicate the potential, the utility, or the joy of discovery through investigation.

Unfortunately, statistics courses are seldom designed with any idea of what it is that students are supposed to be able to *do* as a result of having taken the course. Course objectives, if formulated at all, are expressed in terms of the specific topics covered in elementary texts rather than meeting the needs of the students. Statistics is seen as a "subject," rather than a problem solving tool to be used in the scientific method, or a useful way to look at the world around us.

Some of the problems associated with our introductory courses are:

1. Statistics teaching is often stagnant; statistics teachers resist change. The most popular elementary texts have evolved slowly over the decades. There is a tendency to present the same subjects, the same way, from the same books year after year. Meanwhile statistical methods are progressing rapidly.

2. Techniques are often taught in isolation, with inadequate motivation and with no connection to the philosophy that connects them to real events; students often fail to see the personal relevance of statistics because interesting and relevant applications are rare in many statistics courses. The applications, if any, are often contrived, even "phoney". Elementary courses are often taught by the youngest or least experienced teachers, who have limited, if any, personal contact with serious applications. The open-ended nature of statistical investigations and the sequential nature of statistical inquiry are not brought out. The students are not pushed to question their environment and seek answers through investigations.

3. Statistics is too often presented as a branch of mathematics, and good statistics is often equated with mathematical rigor or purity, rather than with careful thinking. (As George Box said: "Statistics has no reason for existence except as a catalyst for learning and discovery.")

4. Teachers are often unimaginative in their methods of delivery, relying almost exclusively on traditional lecture/discussion. They fail to take into account the different ways in which different students may learn, both individually and in groups, or the many possible modalities of teaching. They also fail to use the wide variety of simulations, experiments and individual or group projects which can make statistics come alive while simultaneously enhancing student understanding.

5. Teachers often construct the statistics course only for their satisfaction; not that of their students, not that of other areas in which the students study, and not that of the students' ultimate employers. There is little attempt to measure what statistics courses accomplish; statistics is too little *used*.

6. Many teachers have inadequate backgrounds: in knowledge of the subject, in experience applying the techniques, and in the ability to communicate in English. The word "statistics" has itself acquired bad connotations.

7. Statisticians may put their subject in a bad light for the students. They often fail to see any need to convey a sense of excitement.

8. Statistical notation is unnecessarily complex and inconsistent; the Greek alphabet is routinely used rather than reserved only for those occasions in which it is essential.

Although the problems of statistical teaching are severe, the picture should be seen in proper perspective. Innovative texts are available and achieving some degree of market penetration. More are needed. However, statistics teachers often face obstacles to good teaching that are imposed by the institutions for which they work, such as the following:

(a) There are the unpleasant realities in college and university teaching environments, like "megaclasses" which make it hard to do anything beyond lecture/discussion and seem almost to be designed to make students dislike the subject matter and to spur them into apathy. There are less extreme, but still annoying, constraints, such as the impossibility of changing short class periods into longer periods thus allowing for laboratory-type instruction and learning.

(b) University incentives do not encourage faculty to collect good examples, improve teaching methods, or spend time inspiring students. That is, there is no provision or incentive for faculty to acquaint themselves with new approaches or emphases in teaching statistics.

(c) The content of the courses (including the text) may be specified by others, leaving little freedom for an innovative statistics teacher to make improvements. The course may be required to attempt to cover more material than is reasonably possible.

(d) Location of statistics courses in mathematics departments is seldom ideal unless the mathematicians are sympathetic to a data analysis approach.

(e) Traditional grading systems may perform some necessary functions, but they often divert students from the fundamental objectives of learning statistics as well as other subjects. Examination requirements may be unnecessarily rigid.

(f) University logistical support for statistics teaching (and teaching in other areas as well) is often weak. Students may have no or, at best, limited access to computers or good interactive statistical software. Classroom amenities like good projection equipment may be lacking.

(g) Some obstacles are posed by the students themselves. While the instructors must try to motivate students, it is difficult to change the attitudes of many of them.

(h) Students often have a fear of formulas and mathematical reasoning. An even more serious problem is the desire for having problems with quick and clean answers that avoid the need for students to think, but make it easier for them to perform well on examinations. Some students may even like formulas because they are props for such problems. Students may be used to reading large volumes of materials relatively superficially, rather than a relatively few pages carefully, as is more suitable in statistical instruction. Unlike some courses, statistics is cumulative and does not lend itself to crash cramming sessions at the end of the term.

(i) Statistical thinking is different from that to which most students have been exposed. The ideas of uncertainty and variability are either ignored or dealt with poorly in everyday discourse and even academic study. Statistical thinking may even be counter-intuitive, as illustrated by the appearance of "hot streaks" and "momentum" or the notion that the "sophomore jinx" is a mysterious plague of nature.

(j) Much of the statistical reasoning students encounter in everyday life represents misuse rather than sound use of statistics. The bad reputation of statistics courses may become a self-fulfilling prophecy.

General Suggestions

The format of the workshop was the following. The participants were divided into four teams. Most of the time was spent in team meetings, with occasional plenary sessions to summarize the activities of each team. Although the four teams met separately and had lively discussions, there were several common suggestions:

1. State our goal(s)

Stating the aim of the course is an important, and often overlooked, step toward success-ful teaching. Participants spent much time debating the skills and knowledge they expected to impart to students in an introductory course, as opposed to the list of methods they wished to cover. One team came up with the following:

> "Our aim in a first course is to develop critical reasoning skills necessary to understand our quantitative world. The focus of the course is the process of learning how to ask appropriate questions, how to collect data effectively, how to summarize and interpret that information, and how to understand the limita-tions of statistical inferences. Statistical thinking is central to education."

Unfortunately, the typical introductory statistics course does not meet this goal, as it stresses mathematically precise statements and formulas applied to artificial data that are of little, if any, interest to most students. A typical textbook example begins with a question that has been formed to address one feature of data that has already been collected. Students gain little insight into how and why data are collected, how experiments are designed, and how analysis of one set of data leads to new questions, new experiments, and subsequent analyses in a contin-uing cycle of scientific inquiry. All too often, statistics is presented as a formal ritual, rather than as a dynamic study of processes.

Every team agreed that the typical introductory course should change by giving more atten-tion to graphical techniques, to simple topics in the design of experiments, and to the scientific method, and *much less time to hypothesis testing*. Most introductory textbooks devote a great deal of space to hypothesis testing, which workshop participants saw as being much less impor-tant than current coverage would indicate.

There was some agreement that this type of statistics course is valuable for students of all interests, although some participants favored the idea of tailoring courses to different aca-demic disciplines having students with different mathematical backgrounds. Another goal could be the identification of the interests of a group of customers and focus the course content and presentation on the needs of that group, possibly through joint efforts of statisticians and mem-bers of the other group.

There was also agreement that undergraduates interested in mathematical statistics, who are the largest pool of future graduate students in U.S. statistics programs, should depart from current standard practice and also take an additional introductory course in statistics that features data analysis and applications.

2. Analyze data and do projects

There was widespread agreement among workshop participants that students should work more with real data and with graphs. Many advocated projects in which students collect their own data and analyze them in written reports. Such projects combat apathy by allowing students to work with data that they find interesting. Moreover, projects give students experience in asking questions, defining problems, formulating hypotheses and operational definitions, designing experiments and surveys, collecting data and dealing with measurement error, sum-marizing data, analyzing data, communicating findings, and planning "follow-up" experiments that are suggested by their findings. It also has the further bonus of improving communication, writing, and organizational skills, all essential attributes for future careers.

3. **To compute or not to compute**

Properly used, the computer is a powerful and effective tool in teaching students about variability in data, particularly through statistical graphics. Most participants strongly support the use of the computer in introductory statistics courses. Having a computer available during class facilitates "on the spot" analyses of data, which teach students that there are often many analyses that can shed light on a problem and that simple graphics can tell one a great deal.

However, a variety of problems, including lack of equipment, lack of adequate projection systems, and lack of staff to operate statistical computing laboratories, severely limit current computer use in statistics courses. Indeed, a small group of participants saw such obstacles, coupled with the concern that teaching students to use the computer can shift attention away from statistical ideas, as sufficient reason to avoid computers in their teaching at the introductory level.

4. **Lecture less, teach more**

Much time was spent in team sessions exploring various modes of teaching and learning. Lecturing is but one way to communicate ideas to students. Classroom demonstrations and data collection add variety, as do experiments that are run in class. Much effective learning takes place when students work together in teams. Student projects can be done in small groups, but some experiments can be conducted in class. Of course, some teachers are faced with very large classes. Two ways to make a large class seem small are emphasizing project work and lecturing to a small sample of students who are brought to the front of the room at the beginning of each class period.

5. **What can the profession do to help?**

It was repeatedly suggested that a "First Course Corner" or "Activities Corner" be added to *The American Statistician*, in addition to the existing "Teacher's Corner" which deals mainly with issues that are only of interest in teaching graduate level courses. Such a new section in this publication would provide a place to share interesting sets of data, experience with projects and classroom experiments, and other ideas to improve the introductory course. It would also signify a commitment on the part of the profession to improve the teaching of introductory statistics courses.

Other suggestions included short courses and invited sessions at ASA and MAA meetings to teach teachers about such topics as process improvement (which few current statistics teachers studied in graduate school) or the effective use of student projects in the introductory course. Of course, it would be quite helpful if universities assigned their best teachers to introductory courses and reward appropriately excellent and innovative teaching.

In addition, some participants of the workshop are developing sample syllabi to share with other teachers. This idea has been used successfully in the calculus reform movement in the U.S., as it helps the hesitant teacher pare old lists of topics and methods down to the essential items, thus creating space in the curriculum for fresh ideas, such as student projects. For example, in statistical education, do many teachers at smaller schools recognize that most major statisticians now believe that we should not spend as much time on tests of statistical hypotheses as we have in the past? That is, more emphasis should be placed on estimation and prediction and less on hypothesis testing.

Detailed Suggestions

Clearly, with 39 individuals grouped into four teams, there were many suggestions and there certainly was not unanimous agreement on each of these. However, listed below are suggestions that most of the participants supported. The first list consists of topics that seem important for a first course.

Important topics to be included in an introductory course

(1-4 highest priority, 5-9 second, 10-13 third, 14-17 fourth)

1. Recognizing that statistics surround us in everyday living. Reported statistics are sometimes incorrect or misused; thus it is important for each of us to be a critical consumer of statistics given by the media. We must ask questions about the quality of the data and the reliability of the analysis.
2. Understanding variability: bias, sampling error, systematic error, measurement error, regression effect, etc. In particular, understanding: Actual observation = Fitted + Residual, and in statistics we try to detect the pattern (fitted) and describe the variation (residual) from that pattern.
3. Collection and summarization of data, including basic exploratory data analysis. (Some felt we should do more with writing up and explaining results.)
4. Graphs, including plotting data taken sequentially (that is, basic time-series concepts).
5. Sampling and surveys, including the importance of getting quality data.
6. Elementary designs of experiments, with some discussion about the ethics of experimentation and the distinction between observational and experimental investigating.
7. Formulation of problems and understanding the importance of operational definitions and the process of inquiry. That is, understanding the iterative nature of the scientific method, including: Plan-Do-Check-Act. We want the capability to make and understand predictions. That is, statistics represents a process concerned with gaining knowledge and solving problems and is not a collection of isolated tools.
8. Basic distributions (normal, binomial, etc.) as approximations to variability in data sets: modelling.
9. Correlation and regression and other measures of association. For example, there should be some illustrations of Simpson's paradox.
10. Elementary probability, including event trees and conditional probability (some included Bayes' Theorem).
11. Central limit theorem and law of averages.
12. Elementary inference from samples, recognizing there are not unique answers in statistics.
13. Ability to use at least one statistical software package.
14. Outliers and how statistical measures change with various changes in the data (that is, aspects of robustness).
15. Statistical significance vs. practical significance.
16. Categorical data and contingency tables.
17. Simulation.

   As well as these topics, there were other suggestions that could be used by the instructors and the students.

1. We need to produce syllabi so that less confident teachers can mimic more experienced teachers. Nondeterministic thinking often seems unnatural and we must help the inexperienced teacher and student.

2. Let us think more about processes rather than bowl models. Multiple sets of data should be analyzed, perhaps by multiple analyses and how these can be combined.

3. Students rely too much on formulas rather than on thinking. Do everything possible to improve student's self-esteem and his or her appreciation of and self-confidence in statistical studies.

4. Even with large classes, we need some interaction with students. Try using a small subsection as "your class" with the others as observers. Also select a few students each day for Mosteller's one-minute drill (list on a small sheet of paper the most important topic of the day, the muddiest, and the one about which the student would like to know more -- or other short comments).

5. Find appropriate articles, in the popular press and technical sources, for students to read. Think of projects (questions) for which students can collect and analyze their own data.

6. Statisticians are sometimes overconfident, even arrogant, about the intrinsic importance of their subject. This puts statistics in a bad light, and we should be aware of this and try to change this attitude.

7. Occasionally have a guest lecturer or colleague in another field or one from industry or government (we could consider retired statisticians). These might be held as informal sessions in the late afternoon or evening. The setting should encourage interactions between students and the visitor so as to expose the skills of an experienced statistician.

8. Work on improving our presentations (there was one suggestion that many of us should take acting lessons or use videotapes to help us improve). Some find it valuable to include some humor into courses.

9. Provide extra problems and old exams through the many "copy stores" that most colleges and universities attract.

Other suggestions were made that can not be accomplished by an individual instructor, but we would need an effort by a collection of people, such as members of the ASA.

Suggestions for the profession

1. Creating a journal on teaching statistics like the MAA's *The College Mathematics Journal* (at least have a "First Course Corner" in the *American Statistician*).
2. Newsletter on statistical education.
3. Providing information on teaching TAs to teach (MAA has one which was edited by Bettye Ann Case). Encourage extensive TA training and mentoring.
4. Developing teachers' networks.
5. Workshops on teaching statistics at MAA, AMATYC, NCTM, and ASA meetings.
6. Short courses on teaching statistics.
7. Poster sessions on teaching, student projects, and data sets.
8. Supporting faculty efforts to change the system.
9. Diagnostic tests for students.
10. Encouraging position papers, each prepared by two or three like-minded persons.
11. Funding for future conferences or workshops on statistical education. In particular, challenge NSF to provide appropriate funding for "reform" in statistics as it has in calculus, including faculty training activities.

12. Collection of "tidbits", namely good ideas for classroom teaching.
13. Writing a "Point of View" article for last page of *The Chronicle of Higher Education*.
14. Supporting efforts to modify the academic system, in particular, the reward structure and grading procedure.
15. Becoming aware of difference of jargon. For example, an engineer might call the statistician's independent variable (or factor) a "parameter."
16. Center for collecting materials like projects, videos and case studies.

Final Thoughts

Clearly, with 39 statisticians involved, we could not agree on the topics to be included in that first course. There will be a continuing need for several first courses. For example, one person clearly thought chi-square tests must be in such a course because his colleagues at his college expected it (and he thought his students appreciated it). Most, however, would decrease the role of tests of hypotheses; some eliminating that topic altogether.

We should ask ourselves what it is that we want the students to remember about that one statistics course that they took 10 years ago. We hope that it is not "The worst course I took in college" although that is frequently the answer that we get today. In thinking about ways to improve the answer, we should reread the goal of the course as suggested by the one team. Whether we agree with this or not, it will make us think hard about the process of continually trying to improve our introductory course.

While we did not prepare a syllabus for an introductory course, a group headed by

Professor Thomas L. Moore
Department of Mathematics
Grinnell College
Grinnell, Iowa 50112
(515) 269-4206

did start thinking about one for students in liberal arts. Moore has agreed to continue to collect ideas concerning this type of first course in case anyone would like to correspond with him. For first courses in other areas, it is clear that we must work with those clients; for example, we must find out what is important to the faculty in engineering before writing texts in engineering statistics. This requires time and energy on the part of the statisticians, and somehow these efforts must be rewarded properly.

We certainly have not solved everything and we must continue our efforts to improve statistical education. Many suggestions have been made in this report on how to do so. Certainly we should seek appropriate funding to get statisticians together to produce tangible materials. If we have future workshops, each team probably should focus on a very special and important topic (at this one, all teams worked on "the first course").

At professional meetings, we must have sessions devoted to statistical education, particularly what improvements are actually being made in these early courses. However, one good sign is that ASA is doing something: in addition to the QL program and the Center for Statistical Education, the topic of the 1992 Winter Meeting was Statistical Education, and a huge effort was made to get graduate students to attend. One way to do this is to get many employers (academic, business, industry, government) there to interview students. This Winter Meeting is an excellent time for these interviews. But, beside these employment opportunities, we wanted to provide a program that students and those interested in statistical education can enjoy. This includes statisticians from industry, business, and government too, because much of the program was devoted to education beyond that in the traditional academic community. Moreover, there should be comparable outreach activities in conjunction with professional meetings of other societies, most notably those in mathematics.

So, as we left our workshop, there was a great deal of determination that we would *not* let this topic drop. Education is too important to our profession.

Special Thanks

We appreciate the sponsorship of The University of Iowa and the American Statistical Association. In addition, we truly needed the financial assistance provided by the National Science Foundation, the ALCOA Foundation, the Ott Foundation, the Statistics Division of the American Society of Quality Control, and the Quality and Productivity Section of ASA.

In addition, there were the leaders of the four teams: Harry Roberts, Bob Miller, Tom Moore, and Ed Rothman/Bobby Mee. There were some excellent note-takers in Mel Alexander, Jim Sconing, and Jim Calvin. Three persons stayed an extra day to help organize all the notes and thoughts, namely Harry Roberts, Jeff Witmer, and Peter Hackl; Bob Hogg would have had a most difficult time preparing this article without their help.

Finally, it is fair to say that all participants took this workshop very seriously. While there were disagreements (a few wanted more revolutionary actions suggested), every one of the 39 statisticians made some contributions and their work before, during, and after the meeting is truly appreciated.

Robert Hogg is Professor of Statistics and Chair of the department at the University of Iowa. He has served as president of the American Statistical Association and is a Fellow of both the ASA and the Institute of Mathematical Statistics. His research interests involve statistics in quality improvement and robust methods and he is the author of five books and over 50 journal articles. He also maintains a very strong interest in the importance of statistical education.

Teaching Statistics as a Respectable Subject

David S. Moore
Purdue University

In referring to statistics as a respectable subject, I mean first that it is a subject in its own right. Statistics, though a mathematical science, is not a subfield of mathematics. And although statistics is a methodological discipline, it is also not merely a collection of methods that can be understood as ancillary to a substantive discipline such as psychology, business, or engineering. Statistics has its own substance, its own distinctive concepts and modes of reasoning. These should be the heart of the teaching of statistics to beginners at any level of mathematical sophistication.

In discussing the teaching of statistics as statistics, I will first show why statistics is not properly considered as a field of mathematics. Then I will illustrate the inadequacy of a theory-driven mode of teaching through discussion of a simple setting that is considered in every first course. Finally, I will make some comments on useful principles and content for teaching statistics as statistics.

Most statistics teaching is service teaching, in the sense that our students are concentrating in some field other than mathematics or statistics. My remarks about teaching are most immediately aimed at teaching elementary statistics to relatively non-mathematical students. But the principles apply to teaching statistics to engineers and even to mathematics majors. When we teach students of business or psychology, we are not strongly tempted to offer mathematics in place of statistics. The spirit of mathematics may nonetheless dominate our teaching, so that the distinctive concepts and reasoning of statistics are hidden in mathematical garb. When we teach mathematically strong students, the temptation to teach them mathematics rather than statistics is acute and must be resisted. I hope to persuade readers that the traditional math major sequence of probability followed by statistical theory is not a suitable introduction to statistics.

Probability is, of course, important for many reasons other than its applications in statistics. The teaching of probability is not my topic here. Our challenge is to improve or replace the statistics portion of the probability-statistics sequence. In a perfect world, the first statistics course for mathematics majors would offer an harmonious blend of theory and practice, the practice motivating the theory and in turn informed by theoretical findings. In our imperfect world, we might consider the model of the sciences: a separate required laboratory period devoted to actually working with data, preferably via interactive computing. Alternatively, why not admit that theory is not the proper starting place and that most of our students would better master the theory after some acquaintance with practice? Even mathematics majors, I think, are better served by studying statistics before mathematical statistics. There is certainly much to study.

Statistics is not Mathematics

Mathematics is an eminently respectable subject, honored at least since Plato. Statistics considered as a field of mathematics is not. The mathematics employed by statisticians is shallow by the standards of contemporary mathematics. Mathematicians therefore often consider statistics a shallow discipline. It is more just to see statistics as a field like economics or physics that makes heavy and essential use of mathematics, but is not part of mathematics and should not be taught to beginners as if it were mathematics. There are strong reasons to take this view. Reviewing these reasons will also give some sense of the nature of statistics as a distinct discipline.

Statistics has its own subject matter. Statistics is the science of data. More precisely, the subject matter of statistics is reasoning from uncertain empirical data. Data are more than numbers; they are numbers with a context. For example, 12.5 and 20 are numbers; as such, they carry no information. But if you hear that a wine contains 12.5% alcohol by volume, this context engages your background knowledge and does convey information. If you hear that the alcohol content is 20%, you at once note that this is too high to achieve by natural fermentation. Either the datum is incorrect or the wine has been fortified.

Calculating the mean of five numbers is an exercise in arithmetic. Calculating the mean alcohol content of five vats of wine as they complete fermentation is statistics, particularly because we look also at the variation among the vats and compare their mean with the mean alcohol content in other years and at other vintners. The context, and the emphasis on variation both within our sample and between samples, allow calculation to serve as the basis for statistical reasoning.

Statistics does not originate within mathematics. Historians have recently given a great deal of attention to the origins of statistics. Their work demonstrates that statistics is indeed a distinct discipline. Stigler [17] presents an authoritative treatment of the development of statistical ideas before 1900, while in Gigerenzer et al. [10], six randomly ordered authors offer a narrative that pays more attention to the broader social and intellectual setting. The development begins with methods for combining imperfectly precise observations in surveying and astronomy. These methods were later imported to deal with variable data in the social and life sciences, undergoing constant revision and expansion. Some of the early work was done by mathematicians (Gauss, Laplace); other equally significant advances were made by people who were certainly not mathematicians (Galton).

The struggle of better minds than ours to understand and quantify uncertainty is fascinating, full of interplay with social patterns and with the ideas of determinism and free will. Differing concepts of chance led to disagreement over the validity of statistical reasoning. The early growth of statistics was also marked by what can only be called confusion as statistical methods that were deeply entangled with the specific application that gave them birth were transported to other settings. It was only gradually apparent that we can apply common methods and common reasoning -- and now, common computer software -- to data from many fields. As Stigler [17, p. 361] says,

> The conceptual triumphs of the nineteenth century had been the product of many minds working on many problems in many fields, and one of the most striking of their accomplishments was the creation of a new discipline. Before 1900 we see many scientists of different fields developing and using techniques we now

recognize as belonging to modern statistics. After 1900 we begin to see identifiable statisticians developing such techniques into a unified logic of empirical science that goes far beyond its component parts.

Statistics, this "*unified logic of empirical science*," has never been a child of mathematics.

The practice of statistics is not mathematical. Statistics in practice is characterized by a dialogue between data and mathematical models. A model is used to analyze the data, but the data are invited to criticize and even falsify the model. Diagnostic methods, which detect various types of disagreement between data and model, are a major field of statistical research. Belsley, Kuh and Welsch [3] and Cook and Weisberg [7] present book-length treatments of now-standard diagnostic methods in the regression setting.

Derivation of the consequences of a model, comparison of properties of competing procedures based on different models, and the search for optimal procedures in specified circumstances are important, but these mathematical aspects of statistics are far from the full story. In recent years, the impact of ever-faster, ever-cheaper computing has turned statistical research as well as statistical practice somewhat away from the purely mathematical aspects of the subject. Diagnostic methods are generally computer-intensive, and the most effective produce graphic displays. The use of computer graphics for the study of complex data (see Becker, Cleveland, and Wilks [2]) is an example of an important aspect of statistics that is not yet mathematized. Graphics research is strong on bright ideas and on data description, and produces striking insight for at least some data. It is in search of a theory, much as new concepts in economics or physics may at first lack an adequate mathematical expression.)

Statistics has its own foundational controversies. Reasoning from uncertain empirical data raises knotty philosophical issues. It is not surprising that statisticians take differing points of view on these issues. Foundational controversies in statistics are entirely unrelated to the largely dormant controversies concerning the foundations of mathematics. What is most striking is that the position that statisticians take on foundational issues has an immediate impact on their statistical practice. Barnett [1] presents an exposition and comparison of the main contending positions. Even elementary text cannot avoid taking sides, even if only implicitly, on questions such as the role of prior information in inference and whether statistical tests are decision procedures or merely assess the weight of evidence.

There is a one-way traffic between statistics and mathematics. Statistics does not participate in the close relations among subfields that characterize contemporary mathematics. Like economics, statistics imports mathematical tools and ideas but does not export its own tools and ideas into mathematics. This clearly distinguishes statistics from probability theory, which *is* a field of mathematics. Probabilistic ideas such as Brownian motion and martingales are familiar to mathematicians and are put to use in other areas of mathematics -- and of course in economics, physics, and statistics. Central concepts of statistical theory such as sufficient statistics, maximum likelihood, and prior and posterior distributions are not familiar to or used by mathematicians. This, in itself, is enough to convince most mathematicians that statistics is not properly part of mathematics.

Teaching should not be Driven by Theory

Because statistics is not a subfield of mathematics, introductory instruction that presents statistics as if it were mathematics will give an inadequate picture of the field. More importantly, elementary instruction that is almost entirely non-mathematical but is guided by the theory lurking in the background is similarly misleading. I do not suggest that the mathematical theory of statistics is unimportant: you can never be too rich or too thin or know too much mathematics. But statistical theory is inadequate as a guide to either teaching or practice.

To see why, consider a simple setting discussed in all beginning statistics courses, *two-sample problems*. This setting is important in practice, and most of the comments I make about it apply to analysis of variance problems in general. The mathematical model for the two-sample setting states that the data are independent random samples from two normal populations:

$$X_1, X_2, \ldots, X_n \quad \text{iid} \quad N(\mu_1, \sigma_1)$$

$$Y_1, Y_2, \ldots, Y_m \quad \text{iid} \quad N(\mu_2, \sigma_2)$$

Statistical theory allows us to obtain tests and confidence intervals in this setting, and guarantees that they have some desirable properties. This is valuable. Why aren't the fruits of theory the whole story?

First, **the mathematical model is incomplete**. It does not capture the distinction between observational and experimental studies, one of the most important distinctions in statistics.

An observational study based on the yields obtained by random samples of farmers who planted two corn varieties, for example, provides little information about the relative merits of the varieties. The observed differences in yield may be due to systematic differences between the two groups of farmers. Perhaps they farm in different regions, or perhaps one variety is favored by better-informed farmers whose other good practices exaggerate the difference in yields. A *randomized, comparative experiment*, on the other hand, is designed to permit causal

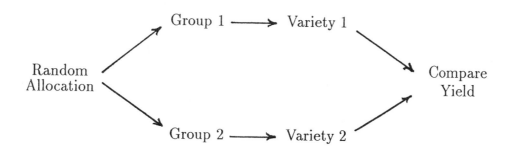

Figure 1: A randomized comparative experimental design to compare the yields of two crop varieties.

conclusions. Figure 1 outlines the simplest randomized comparative experiment. Plots of ground are allocated at random into two groups. All plots receive identical treatment (tilling, fertilizer, pesticides, etc.), except that Variety 1 is planted on one group of plots and Variety 2 on the other.

Experimentation is active data production in which treatments are actually imposed on experimental units. The random allocation creates groups of experimental units (the plots of ground) that are identical except for random variation before the treatments are applied. The comparative design assures that all factors other than the treatments (weather, for example) affect both groups. An observed difference greater than chance variation can reasonably explain must be due to the effect of the different corn varieties. The mathematical model is the same for both experiment and observation, but the statistical distinction between the two in fundamental.

Second, **the theoretical merit of a statistical procedure is not the same as its practical merit**. In the two-sample setting, we may wish to compare the centers of the two populations as described by the means μ_1 and μ_2. Or we may wish to compare their variability, described by the standard deviations σ_1 and σ_2. Evidence of unequal means would be assessed by the (two-sample) t-test, evidence of unequal σ's by the F-ratio test. In theory, these tests are of equal merit. Both are likelihood-ratio tests. That is, they are products of an honored organizing principle of statistical theory and share certain large-sample optimality properties (see for example Wilks [18]). Beginning instruction that is driven by theory often presents the t and F tests on the same level. Because the theoretically cleanest version of the two-sample t requires the assumption that $\sigma_1 = \sigma_2$, some texts even suggest that an F-test be conducted as a preliminary to the t-test.

In practical merit, however, these two tests are poles apart. The two-sample t is a most useful tool, while the F-ratio should almost never be used. (Warning: The F-tests encountered in analysis of variance compare two or more *means*; they are analogues of the two-sample t-test and share its usefulness. It is formal tests for equality of two or more *variances* that join the F-ratio in lack of practical value.) The reason for this parting of the ways is that no data are exactly normal. A procedure useful in practice must therefore be somewhat *robust*, that is, not sensitive to mild departures from normality. The two-sample t is quite robust, while the F-ratio is so sensitive to even small departures from normality that it is almost useless.

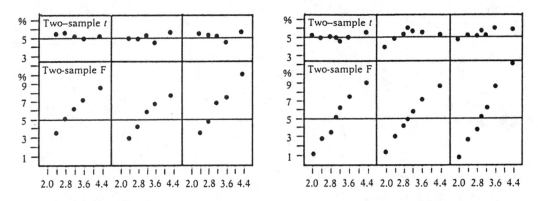

Figure 2: The actual significance levels of nominal 5%-level significance tests under several conditions. Reprinted with permission of W. H. Freeman & Co.

Figure 2, based on one of the many similar figures in Pearson and Please [15], presents some of the abundant evidence distinguishing t from F in robustness. The figure displays the actual significance level, found by simulation, of tests conducted at the nominal 5% level with samples of size 25. The left-hand panel shows the results for populations that are nonnormal and skewed to the right. The right-hand panel concerns symmetric populations, both normal and nonnormal. The horizontal scale in both cases is *kurtosis*, a standardized fourth moment that roughly measures how heavy the tails of the population distribution are. Normal distributions lie at 3 on the kurtosis scale in the right-hand figure, and symmetric but nonnormal distributions lie on either side of 3. Each panel in Figure 2 reports results for one-sided tests ("above" and "below" the null hypothesis of no difference) and two-sided tests ("beyond") for both the t and F statistics.

If a test is robust against nonnormality, the actual level will stay close to 5% even when the population is not normal. The results for normal populations appear in the right-hand panel of Figure 2 above 3. These points are all close to 5% on the vertical scale, as they should be. The other points show what happens when the tests are used on nonnormal data. The robustness of the two-sample t is remarkable. The true significance level remains between 4% and 6% for the entire range of distributions studied. The lack of robustness of the test for variances is equally remarkable. This test conducted at the 5% significance level can have true levels anywhere between 1% and 11%, even for symmetric distributions with no outliers. Such a test is hardly useful. As a check on the equal-variances assumption of the usual test for equality of means, it is silly. As Box [6] remarks, "To make a preliminary test on variances is rather like putting to sea in a rowing boat to find out whether conditions are sufficiently calm for an ocean liner to leave port."

My primary point is that theory is an imperfect guide to practice. But in light of the failure of the F-test, what ought we to teach our students about inference on spreads? In beginning instruction, nothing. There are no simple formal inference procedures that can be recommended in practice. (Nonparametric tests for spread have fatal problems also.) At an advanced level, more can be said. Modern jackknife and bootstrap methods (Boos and Brownie [5]) work well in moderately large samples. For smaller samples, Conover et al. [8] discover via a large simulation study some practically useful methods.

Finally, **theoretically-based procedures may require unrealistic assumptions.** Of course, not all unrealistic assumptions are barriers to practical use, as the robustness against non-normality of the t procedures reminds us. But the version of the two-sample t-test most often found in texts on statistical theory requires another assumption: not only do the two populations being compared have normal distributions, but their standard deviations σ_1 and σ_2 are equal. The test statistic derived in theoretical treatments of this setting is

$$T_p = \frac{\overline{X} - \overline{Y}}{s_p \sqrt{\dfrac{1}{n} + \dfrac{1}{m}}}$$

Here s_p is a pooled estimate of the common standard deviation. The pooled-sample t-test is a likelihood ratio test and the statistic T_p has a t distribution under the null hypothesis of equal population means. This test is somewhat robust against unequal σ's if the sample sizes are equal, but not otherwise. One study of the actual significance level of T_p using 5% t critical points found for $n = 5$ and $m = 15$ that the true significance level ranged from 0.24% to 31.7% for normal populations as σ_1/σ_2 varied. Moreover, we have just seen that the assumption $\sigma_1 = \sigma_2$ is difficult to verify.

The "natural" two-sample t statistic is obtained by standardizing the difference of sample means and then replacing the unknown and possibly unequal σ's by the sample standard deviations s_1 and s_2. The result is

$$T = \frac{\overline{X} - \overline{Y}}{\sqrt{\dfrac{s_1^2}{n} + \dfrac{s_2^2}{m}}}$$

Unfortunately, the statistic t does not have a t distribution under the null hypothesis and is not supported by any theoretical optimality properties. But its critical points can be very closely approximated by those of the $t(df)$ distribution, where the degrees-of-freedom parameter df is determined *from the data* by a somewhat messy recipe and is usually not an integer. A substantial literature (e.g. Leaverton and Birch [11], Scheffé [16], and Best and Rayner [4]) demonstrates the accuracy of this approximation for even quite small samples, and demonstrates in addition that when in fact $\sigma_1 = \sigma_2$, using t sacrifices very little power relative to T_p.

Statistical software makes the degrees of freedom df and tail probabilities for t distributions with non-integer degrees of freedom easy to compute. Procedures based on the statistic t are available in all standard software packages, and often are the default two-sample t procedures. The rather special pooled-sample procedures are less often used in practice. They ought not to prevail in teaching simply because they are rooted in theory.

This example of the shortcomings of a theory-driven approach to teaching basic statistics concentrates on formal *inference* designed to answer well-posed questions. That is the setting in which theory offers the most help, so the example in fact overestimates the usefulness of statistical theory in guiding instruction. The example is a counterexample to an unwise teaching style; it does not reflect the emphasis on data that wise teaching of statistics should display. We noted the importance of the *data production* design in drawing practical conclusions. We omitted the preliminary *data analysis* that would certainly be carried out: Do the data contain outliers that will distort the means and variances used as summaries? Do normal quantile plots suggest that the distributions are reasonably normal? What do, for example, side-by-side boxplots show about the comparisons of both center and spread in the two groups? Data analysis and data production must take their place with inference when statistics is taught as statistics rather than as mathematics.

Teaching Statistics as Statistics

Here is a first principle for improving statistics instruction:

> **Almost any statistics course can be improved by more emphasis on data and on concepts at the expense of less theory and fewer recipes.**

The "less theory" has already been justified. The "fewer recipes" reminds us that the goal of teaching is not to cover as many specific methods as possible. The "cookbook" approach is, after all, just another style of teaching mathematics. Beginning instruction in statistics should emphasize experience with data and variation, and the concepts and reasoning that statisticians use to understand data and variation. Some aspects of statistical reasoning are formal, even mathematical, while others are informal and not mathematical. Strategies for exploring data and principles of experimental design are examples of statistical fundamentals that don't lend themselves to either theory or recipes.

Statistics taught according to this first principle will offer enough experience working with data that it resembles a laboratory science. A second principle, less important but still very helpful, then follows:

Automate calculation and graphics as much as possible.

If the often-forbidding calculations required by statistical procedures are automated, instruction can offer experience with realistic data and can emphasize reasoning. Students can be expected to carry on a dialogue with the data, asking repeated questions that would otherwise be hopelessly time-consuming. If, for example, a scatterplot displays outlying points, good software makes it easy to repeat a regression analysis without these points to check their influence on our original conclusions. The teaching of statistics should in this respect mirror statistical practice, where computation and graphics are almost entirely automated.

The *content* of basic statistics can conveniently be divided into three headings: Organizing and summarizing data; producing data; drawing conclusions from data. A few brief remarks here will indicate the place of each in beginning instruction. These ideas are worked out in more detail (as far as the limitations of the authors and the intended audiences permit) in the texts Moore [12] for liberal arts students and Moore and McCabe [14] for students of statistical methods.

The first of these content divisions is often called *data analysis*. Data analysis seeks to discover the information in the data, without restricting this information to answers to questions we had posed in advance. Even when data are carefully produced to answer specific questions through formal inference, data analysis is an essential preliminary step. Unexpected effects or "obviously" erroneous observations may lurk in the data ready to distort standard procedures unthinkingly applied. Students should learn to *look* at data in their first exposure to statistics.

Because data analysis relies on simple graphical and numerical tools for displaying data, it is easy to regard it as a collection of elementary techniques that deserve only cursory treatment in college instruction. This attitude overlooks the importance of giving students actual experience working with realistic data. What is more, students must learn to "read" data as they learn to read words. Data analysis involves not only a collection of tools, but strategies for examining data intelligently.

Here are some data-analysis strategies. First, move from simple to complex, from describing the distribution of a single variable to examining relations between two variables and finally the complex connections possible among many variables. Second, look first for an overall pattern and then for striking deviations from the pattern. Third, use a natural progression of tools, graphical display followed by numerical summaries and then perhaps by a mathematical model that gives a compact description of the overall pattern. For example, data on two measured variables can be displayed in a scatterplot. We look for the overall pattern, perhaps a generally linear relationship, and for outlying points that require individual examination. Numerical measures help describe specific aspects of the data; for example, the correlation coefficient describes the strength and direction of the linear relationship. If there is an overall linear pattern, we may fit a line to the data so that pattern and deviations take the form of fit and residuals. The line is a compact mathematical model for the pattern of the data.

An essential part of learning data analysis is the gradual development of judgment in choosing tools and forming informal conclusions. Stemplots and histograms, for example, are alternative graphical tools for describing the distribution of a single variable, just as the five-number summary and mean plus standard deviation are alternative numerical descriptions. Students should be able to choose a graphical description based on the size and spread of the data set, then to select a numerical description using the pattern revealed by their graph. (There is not always a single "correct" choice.) Students should be able to describe the main features of the data in words backed by their graphs and calculations. An emphasis on working with data fits the renewed insistence that effective learning must be active. Lecturing, we hear, should be replaced or supplemented by group projects and by a classroom emphasis on problem-solving, discussion, and communicating results that go beyond a number or a graph. Thinking about real data will pervade a modern introduction to statistics; beginning the course with a thorough introduction to data analysis sets the proper tone for both content and learning style.

Good data are as much a human product as hybrid corn and compact disc players. The *design of data production* through sampling and experimentation is perhaps the most important role of working statisticians. In particular, the randomized comparative experiment has revolutionized the practice of many applied sciences and can claim to be the most influential contribution of statistics to science as a whole. The deliberate use of chance in data production is a striking statistical idea, originally controversial but now widely accepted. Other ideas that any introduction to statistics should present include: Experiment versus observation; the role of comparison or control in data production; the importance of nonsampling errors in practice, and the fact that formal inference ignores nonsampling errors.

Even this very brief look at data analysis and data production reveals a wealth of essentially statistical concepts and modes of reasoning that are not mathematical in nature. Until recently, it has been all too common to slight these topics in beginning instruction. A few days spent on "descriptive statistics" is not an adequate introduction to data analysis. Students receive too few data-analytic tools, no presentation of strategies for examining data, and most of all no significant experience actually working with data. A quick mention of random sampling as justifying the "iid model" is not an adequate introduction to data production. Designs for data production are not only very important in themselves, but help justify formal inference and clarify its place in statistical practice.

The final content division is *formal probability-based inference*. The first thing to be said is that this is a difficult subject. The reasoning of inference, particularly of tests of significance, is subtle. Worse, the reasoning rests on probability concepts, which are among the hardest to grasp in elementary mathematics. Teachers of statistics would do well to read Garfield and Ahlgren [9] where the unusual difficulty of probability ideas is substantiated by a review of research on learning. Garfield and Ahlgren go so far as to suggest that we should explore "how useful ideas of statistical inference can be taught independently of technically correct probability." The difficulty of probability and of the reasoning of inference is well-known to teachers of statistics. One unfortunate reaction is to attack the difficulty by spending much of a first statistics course trying to teach probability. Another, opposite in spirit, is to avoid difficulties by reducing inference to recipes.

A proper emphasis on data analysis and data production will reduce the time available for study of formal inference. Our response should be to present less probability (that's the heart of "less theory" in my first principle) and cover fewer specific inference procedures ("fewer recipes").

Probability is important in its own right, but its difficulty is a barrier to learning statistics. Statistical ideas are challenging enough in themselves. Here, then, is a third principle for improving statistics instruction:

A basic statistics course should cover no more probability than is actually needed to grasp the statistical concepts that will be presented.

We need surprisingly little probability. We should certainly omit all traces of combinatorics, a hard subject in itself. I omit even conditional probability unless the students are mathematically strong. There is little point in wrestling with conditional probability and Bayes' theorem if we will teach no statistical methods that require these topics. We ought to apply with rigor the principle that, within a beginning statistics course, probability exists only to introduce the statistical ideas and methods that are our focus.

Elementary statistics relies mainly on basic facts about random variables, their distributions, and their means and variances. Even this little can be presented less formally. Might we avoid the mathematical language of sample space and probability measure? Simulation can help convey both the hard idea that random variation has a pattern in the long run and specific facts such as the central limit theorem. The tools of data analysis, by now familiar to students, apply to the distributions produced empirically by simulation. Normal distributions, for example, can appear as compact descriptions of the overall pattern of data rather than as formal probability distributions. We need more imagination and more daring here.

The reasoning of inference, though difficult, belongs to the heart of statistics and cannot be evaded. It is not necessary to cloak the statistical reasoning in excessively mathematical garb. Happy the student who understands that "95% confidence" means "I got this result by a method that is correct 95% of the time in the long run." Happy even if she did not quite follow the derivation of the recipe from the sampling distribution of $\bar{x}$. Happier still, of course, if she understands that repeated random samples will give different values of $\bar{x}$, knows that the distribution of these values is close to normal in many settings, and sees that the 95% confidence interval arises from catching the middle 95% of this normal distribution. These ideas deserve careful treatment, including demonstration by simulation if possible. We should shed no tears if the list of specific confidence intervals presented is rather short. A student who grasps the reasoning can easily learn more recipes later. Besides, the recipes can be automated.

Which recipes for formal inference are essential in a first course? Tastes vary. Some teachers effectively convey the reasoning of inference while presenting only nonparametric methods. I prefer to emphasize the standard z and t normal-theory procedures for means and proportions, along with inference for least-squares regression. These are the most-used elementary methods in practice, have an accessible theory yet are relatively robust, and lead into advanced methods (analysis of variance and least-squares multiple regression) that have as yet no effective competitors for formal inference in complex settings.

The teaching of inference both completes and draws upon the teaching of data analysis and data production. Randomized data production motivates asking "What would happen in repeated sampling?" and reminds us that inference recipes can't compensate for blunders in producing the data. Data analysis shows how to look at the sampling distribution of $\bar{x}$ (shape, center, spread) as well as how to inspect the sample data before calculating the confidence interval. With skill (and perhaps luck) statistics will be seen as a unified discipline, the science of data.

Acknowledgment

The writing of this essay was supported in part by National Science Foundation grant DMS-8901922.

References

1.   Barnett, V. (1982) *Comparative Statistical Inference*, 2nd Ed. Wiley, New York.

2.   Becker, R. A., W. S. Cleveland, and A. W. Wilks, (1987) Dynamic graphics for data analysis. *Statistical Science* **2** 355-383.

3.   Belsley, D. A., E. Kuh, and R. E. Welsch, (1980) *Regression Diagnostics: Identifying Influential Data and Sources of Collinearity*. Wiley, New York.

4.   Best, D. J. and J. C. W. Rayner, (1987) Welch's approximate solution for the Behrens-Fisher problem. *Technometrics* **29** 205-210.

5.   Boos, D. D. and C. Brownie, (1989) Bootstrap methods for testing homogeneity of variances. *Technometrics* **31** 69-82.

6.   Box, G. E. P. (1953) Non-normality and tests on variances. *Biometrika* **40** 318-335.

7.   Cook, R. D. and S. Weisberg, (1982) *Residuals and Influence in Regression*. Chapman and Hall, New York.

8.   Conover, W. J., M. E. Johnson, and M. M. Johnson, (1981) A comparative study of tests for homogeneity of variances, with applications to outer continental shelf bidding data. *Technometrics* **23** 351-361.

9.   Garfield, J. and A. Ahlgren, A. (1988) Difficulties in learning basic concepts in probability and statistics: implications for research. *J. for Research in Mathematics Education* **19** 44-63.

10   Gigerenzer, G., Z. Swijtink, T. Porter, L. Daston, J. Beatty, and L. Krüger, (1989) *The Empire of Chance*. Cambridge Univ. Press, Cambridge.

11.   Leaverton, P. and J. J. Birch, (1969) Small sample power curves for the two sample location problem. *Technometrics* **11** 299-307.

12. Moore, D. S. (1991) *Statistics: Concepts and Controversies*, 3rd ed. Freeman, New York.

13. Moore, D. S. (1988) Should mathematicians teach statistics? (with discussion). *College Math. J.* **19** 3-35.

14. Moore, D. S. and G. McCabe, (1989) *Introduction to the Practice of Statistics*. Freeman, New York.

15. Pearson, E. S. and N. W. Please, (1975) Relation between the shape of population distribution and the robustness of four simple test statistics. *Biometrika* **62** 223-241

16. Scheffé, H. (1970) Practical solutions of the Behrens-Fisher Problem. *Journal of the American Statistical Association* **65** 1501-1508.

17. Stigler, S. M. (1986) *The History of Statistics: The Measurement of Uncertainty Before 1900*. Belknap, Cambridge, MA.

18. Wilks, S. S. (1962) *Mathematical Statistics*. Wiley, New York.

David S. Moore is professor of statistics at Purdue University. He is a Fellow of both the American Statistical Association and the Institute of Mathematical Statistics, as well as a member of the International Statistical Institute. Professor Moore served on the National Research Council's Committee on Mathematics in the Year 2000 and authored a chapter in the NRC's volume *On the Shoulders of Giants: New Approaches to Numeracy*. He was content developer for the Annenberg/Corporation for Public Broadcasting telecourse *Against All Odds: Inside Statistics* and for the new series of secondary school video modules *Statistics: Decisions Through Data*. Professor Moore is also the author of two well-known statistics textbooks.

A Multidisciplinary Conversation
on the First Course in Statistics

Walter Chromiak, Jim Hoefler, Allan Rossman and Barry Tesman
Dickinson College

The authors of this paper represent a variety of disciplines and levels of experience in teaching statistics. Walter Chromiak is a cognitive psychologist who teaches a statistics and research design course in the psychology department that he chairs; Jim Hoefler is a political scientist and public policy researcher who teaches a polimetrics course in the political science department; Allan Rossman is a statistician who teaches various levels of courses in statistics and probability in the mathematics department; Barry Tesman is a mathematician specializing in graph theory who has recently taught an elementary statistics course in the mathematics department for the first time. We and others at Dickinson College, a small (approximately 2000 students) liberal arts institution, have formed a multidisciplinary group to better coordinate statistics courses offered at the college and to develop new quantitative courses that better meet students' needs and the college's aims.

We have chosen to write this article as a conversation among the four participants: statistician (S), mathematician (M), political scientist (POL), and psychologist (PSY). Naturally, what follows is an accurate representation of how we actually speak -- no hesitations and always in complete grammatical sentences. More seriously, we express our own views, which are not necessarily those of other members of our fields or even departments, for that matter. We also focus on the situation that we know best, that of small liberal arts colleges. We hope that our conversation proves useful to those who are involved with statistics courses in various capacities.

Before proceeding to the conversation, however, here are some of the common themes that have emerged during the course of our discussions:

● The first course in statistics should be a **foundational** one, focusing on fundamental concepts of statistics as the science of reasoning from data. The course should not attempt to cover a set package of methods (*t*-tests, regression, ANOVA, chi-square tests, etc.); rather, it should **prepare** students for follow-up courses that would build on its foundation by teaching particular methods. At the same time, we do not mean to suggest that such techniques should not be taught. The point is that the course should provide students with a thorough grounding in the fundamentals.

● The course should strongly stress **communication** skills. Students should learn to be intelligent, critical consumers of quantitative information; they should also be taught to communicate numerical arguments clearly themselves. Students should learn to communicate both in writing and with graphs and other visual displays. The course should also allow students to develop their oral presentation skills.

● The course should emphasize student **involvement** with **data**. In order to help students understand data more clearly and to motivate their interest in studying statistics, the first course in statistics should provide students with many opportunities to analyze genuine data. These data sets should include both student-generated data on topics of their choice and data available from a variety of fields and concerning a range of issues.

Place in the Curriculum

S: Let's start by discussing the place of the first statistics course in the undergraduate curriculum. Small colleges typically do not have several professional statisticians on their faculty, so who should teach introductory statistics courses? Should each department teach its own version? Should the math department teach them all?

PSY: Most of the students in my course on research design and statistical evaluation have had no previous experience with statistics. Consequently, I spend much more time than I would like covering elementary statistical concepts, such as descriptive statistics, z-scores, confidence intervals and so on. Don't get me wrong; students need to know all those things, but it's very hard to cover that material properly and spend the time we need to learn analysis of variance. I would like my students to have had a course that established a foundation of statistical ideas on which I could build.

POL: I agree. I'd also like to be able to build on a foundation, to know that students weren't hearing ideas of sampling, say, for the first time in my polimetrics course.

PSY: But could we all agree on what this foundational course would entail? Over the years, a number of departments have talked about requiring a common statistics course. On the one hand, some have argued that such a course would eliminate the need for follow-up courses in individual disciplines. On the other hand, many of us have pointed out that it would be impossible to cover all of the techniques that various disciplines would want to be taught.

POL: It doesn't get us very far if every discipline were allowed to make parochial demands on the people teaching "intro stat". Basic statistical concepts can be taught quite effectively without having to deal with statistical routines that would have application only for a narrow subset of disciplinary majors, such as multiple regression analysis for economists. The objectives of a first course in statistics should be generic to the liberal arts.

M: I agree with the model of having a foundational course that's followed up by more specialized courses as needed. One of my frustrations as a mathematician teaching elementary statistics is that I'm never sure I've covered what students need for their major. I'm also never really comfortable pretending that I'm a psychologist or an economist for a day as I do examples from those disciplines.

S: If we're agreed that the first course should be foundational rather than trying to serve too many masters, who will teach such a course?

POL: My department could not afford to teach two statistics courses, so I'd like to see the foundation laid by someone else.

M: But not in the math department. I don't think that, as a mathematician, I am automatically the person most qualified to teach elementary statistics. I would imagine that there are a dozen or more faculty members, from the social and natural sciences, better prepared than I to teach the course.

S: I think that's actually an enlightened opinion. My impression is that many mathematicians who teach statistics do not realize that their training might have been insufficient to prepare them for the experience. I'm reminded of David Moore's paper, *Should Mathematicians Teach Statistics?*, which he answered with a resounding **No!** [3]

PSY: Come on, a mathematician who can't teach statistics?! I can hear people saying, "Statistics is all about numbers, isn't it? Who knows numbers better than a mathematician?" Right?

POL: I would imagine that most mathematicians have had more statistics than I've had.

M: I never had a single course in statistics as either an undergraduate math major or as a Ph.D. student in mathematics. In fact, I can't think of any of my fellow graduate students who took a statistics course while in graduate school. And I don't think my experience is unusual among mathematicians.

S: As I recall, many of the respondents to the Moore article pointed out that mathematicians are often pressed into service as statistics instructors, simply because there aren't enough statisticians to go around.

PSY: We want students to have a common framework from which to explore more advanced topics in statistics, but it sounds as if we're arguing that we shouldn't expect math departments to take sole responsibility for statistics courses. In fact, we seem to be saying that a variety of disciplines should have a hand in organizing and possibly even teaching a first course in statistics.

S: But it certainly makes sense to have a statistician oversee the course content and organization. After all, each discipline tends to focus on a handful of techniques, while the statistician has the broadest training and expertise.

POL: But what about those schools that are not graced with the luxury of having a trained statistician on staff?

M: Maybe some serious thought should be given to hiring one. Just think about it. If this were a history course, they would certainly hire an historian.

Skills to be Acquired

S: Let's move on to talk about what students should be able to do after their first statistics course.

POL: The most important skill to me is being able to communicate. Students should walk away from this course able to convey their thoughts about numerical arguments clearly and concisely, using both words and pictures as the vehicles. We should have students write sentences, at least, or better, paragraphs describing what they've found.

M: That's a good point! Students don't expect to write more than a phrase or two ("the results were significant") in what they think of as "just another math course". I agree that we shouldn't let them just report naked numbers, with statistical jargon thrown in here and there for good measure.

PSY: We should remember that communication is a two-way street, of course. Students should also be able to examine numerical information critically. For example, they should be able to assess whether a reported conclusion follows from the information presented or whether the study described suffers from obvious flaws. They should also be able to identify misleading graphs and tables.

M: In particular, students should be able to read and criticize popular articles containing numerical arguments, be they from *U.S.A. Today, Consumer Reports*, or the campus newspaper for that matter.

PSY: An important skill, related to the issue of communication, is the ability to prepare **useful** displays, both for analysis and presentation (and there is a distinction there). For example, EDA (Exploratory Data Analysis) techniques like stem-and-leaf plots, boxplots, etc. are quick, easy-to-do, and very informative. I particularly like them because you don't need a computer to construct them. We shouldn't assume that every student will always have a computer handy to do graphs and calculations.

POL: Another important skill is that of being appropriately skeptical and wary of numerical arguments, the ability to look for and recognize misleading presentations and conclusions. It's especially important to break through the popular conception that **any** argument which contains numbers must be inherently true. We need to rid students of that notion without letting them drift to the other extreme of cynically disregarding ALL numerical arguments.

S: Right. We want students to appreciate that statistics can be a very powerful and illuminating intellectual tool, but that statistical analysis needs to be applied with thoughtfulness and judgment.

M: A related point that has struck me is the necessity (and difficulty) of getting students to believe that there need not be a single right answer to all questions. They tend to think of statistics as a math course where there are right-or-wrong solutions to narrowly defined problems. Getting them to acknowledge and accept uncertainty is an important attitude to foster. Critical thinking, rather than learning how to plug values into formulas, should be our focus.

S: Wait a minute. I have to confess that I expected you to say that students must be able to conduct a chi-square test or perform an ANOVA or construct regression models or something like that. Are such skills not important to you?

PSY: I think that they're more or less important depending on one's field. The essential point, though, is that they may be too specialized for a strong foundational course in statistics. Besides, it's clearly too much to ask that a single course teach the basic ideas well and cover many specific techniques.

Concepts to be Understood

S: Let's consider, then, what the essential concepts to be covered in a foundational statistics course are. What concepts should students see and understand?

PSY: One key concept is simply that of measurement. Not everything has a natural yardstick; how can one measure unemployment, or intelligence, or scholastic aptitude, or teaching effectiveness? We should also point out that measurement difficulties and errors arise in the natural sciences, as well.

S: The concept of variability is certainly a fundamental one. This entails an understanding of distributions of data, which in turn leads to considering measures of center and of spread, skewness, outliers, and so on.

POL: Of course, it's important to ask where the data came from, so sampling is a central concept as well. Understanding strategies for sampling (simple random samples, stratified samples, etc.) and the strengths and limitations of sampling should be stressed.

PSY: Which leads to the basic principles of experimental design; such ideas as having a comparison or control group and the manipulation of independent variables should at least be mentioned. No, more than mentioned...

POL: Right. If we don't discuss the circumstances under which data are collected, the students see statistics as just an exercise in playing with numbers.

M: In other words, students will perceive statistics as just another math course.

PSY: To eliminate that possibility, we should make clear that someone (perhaps even the students themselves) has collected the data for a reason. That means that the data are not just a bunch of numbers; hidden somewhere inside the data are insights into some issue.

POL: And that issue might be whether a relationship exists between two variables, so the idea of association is also fundamental. Measures of association, such as the correlation coefficient, and principles like the distinction between association and causation are important.

PSY: That last point needs to be emphasized. It seems that one can't turn on the television without hearing an anchorperson tell us that scientists have discovered a connection between, say, church attendance and wealth. Lots of people probably infer that going to church will make you wealthy.

S: No one has mentioned probability yet; can a first statistics course be taught without it?

M: You have to teach probability, but not the way it's presented in most textbooks. I think that it's important to deal with it in the context of practical statistical problems.

POL: As opposed to...?

M: As opposed to suddenly shifting the course into math mode, treating probability as a separate subject by stating the axioms, deriving some propositions, and then solving generic "balls in urns" problems.

S: What about more formal inference procedures, confidence intervals and significance tests being the most common examples? And what about the Central Limit Theorem; its very name seems to anoint it as a fundamental concept.

M: Of course one must cover the Central Limit Theorem, but again being careful not to slip into math mode. Most of these students couldn't care less about its intrinsic mathematical beauty; it should be presented very intuitively and with practical demonstrations.

POL: The issue of confidence is a critical one, especially in the context of sampling. The ability to make strong confidence statements based on what might be a very small sample compared to the size of the population is something that students should understand. Who knows? They may even come to appreciate it for the startling intellectual achievement that it is.

PSY: Statistical significance falls into that same category. If students can understand the idea of significance in this course, my job in a follow-up course will be much easier when I teach analysis of variance and other techniques which involve testing and significance.

S: If one is to cover confidence and significance, though, it's especially important to stress a thoughtful understanding of these concepts. For instance, one must emphasize issues such as the 5% significance level not being sacred and the distinction between practical and statistical significance. In other words, the limitations of these ideas must be presented hand-in-hand with their applications.

Course Principles

S: Let's talk about the principles that you think should guide the first course in statistics. How should the course be taught? What experiences do you want the students to have encountered? For instance, many statisticians are advocating a "data-driven" approach which emphasizes DATA as the focus of the course and gives students lots of opportunities to analyze genuine data sets, to critique the data collection, and to draw appropriate conclusions [4]. Is this the right way to go?

M: I favor using real data mainly because I think that's the best way to engage the students, to get them motivated to learn some statistics. If you don't try to come up with glitzy examples that the students can relate to and genuinely care about, it seems to me that you've lost the opportunity to grab their attention. For example, have your students design an experiment to investigate whether Smart Food popcorn really makes one smarter, or have them do a taste testing experiment of bottled water vs. tap water. Do whatever it takes to capture their interest, in the process of teaching them the appropriate statistical concepts.

S: I agree completely with the importance of getting students involved analyzing data that they collect themselves. One of my favorite examples is having students count the lengths of sentences in newspaper editorials and compare the distributions of these sentence lengths for various newspapers (e.g., *New York Times, U.S.A. Today*, and the campus newspaper). I also think, however, that we shouldn't confine ourselves to data that students collect themselves. I also like to use available data that address important societal issues; two examples that come to mind are the infamous 1970 draft lottery [2] and the Berkeley admissions data (which concern alleged sex discrimination and illustrate Simpson's paradox [1]). Such genuine data can help to convince students that statistics does have an important role in society.

PSY: An anecdote comes to mind of the antithesis of using genuine data. When I was a student taking a statistics course, I thought that if I simply memorized the squares and square roots of the integers from 1 to 10, I'd be able to do any statistics problem, because the only examples we saw contained the numbers between 1 and 10.

POL: An interesting way to incorporate genuine data into the course is through the use of projects. These can vary considerably in scope. For instance, my polimetrics students conduct polls of registered voters prior to and following elections. The students are involved throughout the process -- from writing and rewriting the questions to conducting the interviews to coding and analyzing the data to writing a report of the results for the local newspaper. Such an extensive project is probably not suited to an introductory course, but I would regard smaller exercises such as the water tasting experiment and the sentence lengths comparison as useful smaller projects.

M: I think that mathematicians are often hesitant to use projects in their courses, first because it goes against the traditional method of teaching, but second because we doubt whether students have enough knowledge to do a good project until perhaps at the end of the course, when it's too late.

POL: But conducting projects can be a valuable learning experience in itself. There are all kinds of problems with the opinion polling project that my students work on - - we only call on two evenings, we miss a lot of people who work evening shifts, our source lists are incomplete, etc. -- but I think the students learn by thinking about and discussing all of the problems associated with the project.

S: We seem to be agreed that it's essential to involve the students actively with the material and to use genuine data as a means of doing that. What other principles are important?

PSY: It's important to make calculating easy for the students. For example, I think that having students use the computer as a tool for analyzing data is crucial. You can't do serious data analysis without using a computer to perform the calculations and to display graphs and tables.

POL: Doesn't your comment about the importance of computers contradict what you said earlier about not assuming that students will always have computers?

PSY: I don't think so. The point I was trying to make earlier was that we shouldn't teach statistics as if students will always have a computer and their favorite software available after they leave our course. We need to provide students with tools that will allow them to make sense of data even if there isn't a computer available. However, I don't think we should ignore computers. Let's use them whenever they help make life easier!

S: Another use of the computer is that it allows students to "play with" data, to explore and perhaps discover ideas for themselves. For instance, one can ask students to change one or two of the numbers to see the effect on the value of the correlation coefficient, leading them to discover some of its properties for themselves.

PSY: Besides, for many students, the calculations get in the way of understanding the ideas we're trying to teach. They're used to getting the "right answer" to a problem rather than thinking about the problem itself.

POL: I've already mentioned that I think projects are useful, so it seems to me that having students learn statistical ideas in the context of a particular problem or project really makes a difference. For example, teaching about correlations without connecting the idea of association with why someone would even ask the question of some particular set of data is silly.

M: Right. When my class did their Smart Food popcorn project, the students discovered the need for randomization and control groups as we worked with the problem. They could see why those ideas were important because they weren't presented in the abstract, but in the context of a question the students thought was interesting. They cared about what we would discover and I think that really helped.

S: So we're arguing that you shouldn't teach students about variability, for example, just because it's in the textbook. Rather, teach them about variability because they need to understand it to help them work on some problem that they care about.

PSY: Come on, I can just hear people saying "Right! Now I need to find problems students care about."

M: Just ask them! The students will tell you what they care about. If they're only interested in grades, analyze grades. If they're only interested in sex...

PSY: Having them work on projects and problems they care about is a good idea, but if you only present the idea once, you can't expect them to remember it. I'm guilty of this myself and I have to remember to keep bringing up things we've covered as we deal with new problems. For instance, after you've gone over descriptive statistics, don't assume that students will automatically calculate measures of center and spread and look at distributions for new data sets. You've got to keep driving home those points at every opportunity and practice what you preach.

POL: Practice does make perfect. You can apply the same principle to strengthening students' communication skills. If they only have to write once or twice during the course, they won't learn much. For example, I've had students review one another's papers and this let's them see that everyone doesn't produce a perfect paper the first time. It also gives them the chance to consider other opinions and use that information as they revise their own work.

M: But the important point is that they get more than one chance to write, just as they should get more than one chance to deal with centers and spreads.

S: We've talked about lots of principles we think instructors should espouse; is there anything in particular that you think instructors should avoid?

M: Don't teach formulas, teach **ideas**! Certainly don't waste everyone's time deriving formulas or results. It's more important for them to understand the properties of the t-distribution than it is to be able to derive it mathematically.

S: You're right! A first course in statistics is neither the time nor the place for such derivations. They add nothing to the students' understanding of the ideas and, in fact, would probably detract from it.

M: The problem with formulas is that students are accustomed to just plugging values into them and grinding out the answer without knowing what's going on. I think you have to work very hard to not let students work **mindlessly**.

Conclusions

S: Well, we seem to have covered quite a bit of ground in this conversation. If there was one thing you wanted to be sure to emphasize, what would it be?

M: Engage the students! Use problems and data that will capture their attention and leave them with the idea that statistics is useful and worth studying.

POL: I think being able to communicate what they've discovered is very important. Instead of simply testing for right answers, instructors should guide students in the area of numerical communication, then expect those students to convey thoughts about numerical arguments clearly and concisely.

PSY: The notion that the first course in statistics serves as a foundation for further work strikes me as important. We all agree that expecting this course to serve the needs of many different disciplines doesn't make sense. Even if students didn't take any other course in statistics, if they took the type of course we've described, they would have learned important skills that will serve them well in and out of school.

S: I'd like to reemphasize my appeal for the use of genuine data in the course, both in examples and in homework and exam problems. I think that my point ties together the three that have preceded it: genuine data can make the material more engaging for students, can make communication relevant since a real question is being addressed, and makes clear the need for a firm foundational understanding of statistics as the study of data.

References

1. Bickel, P. J. and J. W. O'Connell, (1975). Is there sex bias in graduate admissions? *Science*, **187**, 398-404.

2. Fienberg, S. E., (1971), Randomization and social affairs: the 1970 draft lottery. *Science*, **171**, 255-261.

3. Moore, D. S., (1988), Should mathematicians teach statistics?, (with discussion). *The College Mathematics Journal*, **19**, 3-35.

4. Moore, T. L. and R. A. Roberts, (1989). Statistics at liberal arts colleges. *The American Statistician*, **43**, 80-85.

Walter Chromiak is an Associate Professor of Psychology and chair of the department at Dickinson College where he teaches courses in research methods, statistical analysis, and human cognitive processes. His research interests focus on the differences between novices and experts in a variety of domains.

Jim Hoefler is an Assistant Professor of Political Science at Dickinson College who teaches a course in polimetrics and research. He also teaches in the Policy and Management Studies program, where he puts to good use his interests in public policy analysis (job training, health care, and education), state and local government, and public administration.

Allan Rossman is a statistician and Assistant Professor of Mathematics at Dickinson College whose primary teaching interests are probability and statistics, and their application to the social sciences. His research area is Bayesian statistical inference and decision theory.

Barry Tesman is an Assistant Professor of Mathematics at Dickinson College who, like many mathematicians before him, has been pressed into service to teach a statistics course -- which he found both challenging and enjoyable. His scholarship involves graph theory, combinatorics, and measurement theory.

Mathematics and Statistics:

An Uneasy Marriage

Gudmund R. Iversen
Swarthmore College

Looking Back

The history of statistics shows how broadly based statistics is, both in theory and practice. Most of the original demands that created statistics as we know it today were practical in nature. There came a time when individuals realized that it was important to know how many inhabitants there were and how many people were born and died. This led to the development of techniques for counting and organizing counts into tables as well as methods to interpret the tables. Astronomers measured the same star several times and obtained different results each time. This led to the notion of error and the distribution of errors, which, in turn, led to the development of the idea of the normal distribution. Questions of whether variables such as the heights of parents and offspring were related led to the development of correlation and regression analyses. As a latecomer in the early part of this century, it became desirable to see how the results from a sample could be used to say something about an entire population, and so statistical inference was born.

As the field of statistics grew and matured, it became possible to formalize these practical methods into a mathematical notation and framework. With statistical methods formalized this way in mathematical terms, it also became possible to manipulate the mathematics and derive new results that had not been obvious from purely practical needs. That way statistics became mathematical, and it was necessary to know mathematics to get an understanding of theoretical statistics.

Initially, the formal training of people interested in theoretical statistics took place in departments of mathematics. This went on until a large enough critical mass of statisticians existed to create departments of statistics. Most such departments came into existence after World War II. Even though we now have separate departments of statistics, statistics as a discipline rests firmly on a solid base of mathematics.

Teaching Statistics

Statistical education today encompasses a broad array of introductory offerings at many different levels of sophistication. A "first course" in statistics is now taken by large numbers of students with very different backgrounds, different levels of mathematical preparation and different academic needs and emphases. In the most elementary statistics courses, the level of mathematics is limited to a presentation of formulas for the computation of a variety of descriptive statistics and computations of values of the normal, t, chi-square and F distributions. If the entering students have a higher level of sophistication, the introductory course can go beyond simply using mathematics to express formulas. We start using mathematics for the derivation of new results. On the most advanced level, a text in statistics looks like any text in mathematics with its emphasis on theorems and proofs.

It is this double track of practical needs and mathematical sophistication that has brought statistics to where it is today. This double track of practice and theory also exists in the teaching of statistics. Perhaps it is also responsible for the widely acknowledged problems in statistical education today as different groups (statisticians, mathematicians, and practitioners such as biologists, economists, sociologists, etc.) with different viewpoints and backgrounds are teaching statistics.

The practitioners claim that they are the ones who know the subject matter for which statistical methods are being used. To them, this means they should be the ones to teach statistics to their students. They are correct about knowing the subject matter, particularly as it is applied in their own specialties, but their own training in statistics is often limited. One common consequence of limited training is that they do not see ways in which various statistical methods are related. For example, students seem to benefit and delight in the simple fact that the t-test for the difference between two means is only a special case of analysis of variance. Even many textbooks leave out a simple fact like this. Also, the statistical training of the practitioners is not necessarily up-to-date. This means that statistics is often being taught in a cookbook fashion directly from the textbook. Furthermore, recent advances in statistical theory are not being passed on to the students. Even the most simple aspects of exploratory data analysis have been slow to enter the common statistics curriculum.

Because of the intrinsic mathematical nature of statistics, some mathematicians also claim to be the ones who should teach statistics. At the other extreme, statistics courses at many institutions are given by the mathematics department because the dean sees statistics as being most closely allied to mathematics or because there are no statistically trained faculty available elsewhere. According to the CBMS study of the undergraduate programs in mathematics reported by Albers [1], 180,000 students enrolled in introductory statistics courses offered through mathematics departments in the fall of 1985, up 36% from 1980. An additional 66,000 students took advanced statistics courses that semester, up 52% from 1980. These numbers have likely grown considerably in the years since. As a result, we see that mathematics departments have assumed responsibility for teaching statistics to very large numbers of students.

Unfortunately, the typical member of a mathematics department has not had any formal training in statistics. It is more common for mathematics faculty to have had a graduate course in probability theory than in mathematical statistics, let alone in applied statistics. Consequently, many of the mathematicians who are called upon to teach statistics are neither qualified to do it nor comfortable with it. For instance, McKelvey et al. [3] report that in a survey of two-year college mathematics instructors, the question "I am entirely secure about my qualifications to teach statistics" elicited the following responses: 74% of PhD holders and 46% of masters holders responded Yes. In comparison, the responses to the comparable question regarding calculus were 91% and 86%, respectively. The situation at many four-year schools is likely not much better.

As a consequence, it is natural for such individuals to react to teaching a statistics course by falling back on their mathematical experience. Even without training in statistics, a mathematician immediately recognizes many aspects of statistics. For example, integrals used to find expected values represent ways of finding moments in general. Similarly, the matrix approach to least squares methods is only linear algebra in a bit of disguise. Beyond that, since statistical theory is formalized in mathematical terms, a mathematician can easily absorb it. This makes it possible for a mathematician to teach statistical theory. However, the more applied a statistics course is, the harder it is for a mathematician to teach.

Statistics in a mathematical context becomes a set of theorems and proofs together with formulas for the computations of parameter estimates. A person versed in this type of language can read and see directly the implications of the formulas. For example, it is not very difficult for a mathematician to see how an outlier will have a large effect on the value of a variance. In spite of the dampening effect of the square root, the mathematician can also see that the outlier will have a large effect on the standard deviation. For the typical student in an introductory statistics course who is ignorant of this mathematical language and lacks mathematical insights, the formula for the standard deviation has no meaning at all. It is possible to learn the meaning of each symbol in the formula, but the student still does not see how the various symbols relate to each other and what the formula means for the final result. The student's lack of mathematical understanding becomes a formidable barrier for the understanding of statistics when statistics is simply taught as a way to substitute data into the right formula at the right time.

Three Levels of Courses

How does the ability of a mathematician to teach statistics match up with the needs of the students who take a first course in statistics? Students come to statistics with varied backgrounds in mathematics. The more mathematics they have had, the easier it is for a mathematics department to accommodate them in their statistics courses. With three semesters of calculus and a course in linear algebra, the students are well prepared for a course in mathematical statistics. A student with some background in calculus can take a course that uses formulas to find expected values and derives the normal equations in simple regression. The third kind of student comes with only high school mathematics, and it is often considerably more difficult for a mathematician to teach an introductory statistics course for such students. This is also the course that statistical practitioners typically teach all over our campuses. This course is often taught with little success, whether it is taught by mathematicians or practitioners.

The objective in any introductory statistics course, regardless of what level of mathematics is used, is to expose the students to the central aspects of statistics. This includes an understanding of the role of randomness and variability, the desire to condense data down to a few characteristics in order to gain an understanding of a phenomenon without too much loss of information, the need for properly collected data in order to generalize beyond the sample, and a realization that statistical relationships are not necessarily causal. Beyond that, the objectives vary with the type of course. In a course in mathematical statistics, we want the students to use their mathematical knowledge to understand the theorems and proofs. Knowing these theorems will then enable the students to construct their own statistical models and study their properties. For example, they should be able to decide whether it is better to use maximum likelihood methods or least squares methods for the estimation of the model parameters.

As we move down on the mathematical scale, the statistics courses tend to become more applied. One possibility in these courses is that we may want the students simply to understand the kinds of statistical analyses they see in their readings of journals and books in other disciplines. When they see a contingency table with phi, chi-square, degrees of freedom and a *p*-value, they should be able to conclude how strong the relationship between the two variables is and to what extent the relationship is statistically significant.

Another common objective of such a course is for the students to be able to do their own statistical analyses when they leave the course. Students often collect data in science laboratories and in field work, and students who write senior theses often do empirical studies of one kind or another. They need to know how to use the statistical methods they have learned. The only way for the students to learn the methods is to have them do data analysis within the statistics course they take.

Either way, we want the students to understand something about the ways in which data are properly collected. We want the students to understand some of the choices we make in data analysis, even on such a simple level as making the proper choice between a median and a mean. Most of the time we also want the students to have some sense of statistical inference, even though this is becoming an increasingly controversial issue. Such understandings require some knowledge of statistical methods and an understanding of underlying statistical principles. The majority of textbooks in statistics reflect this type of goal. Judging from the reactions many students have to their introductory statistics course, the success of such a course is typically less than outstanding.

These three types of introductory courses affect a mathematics department differently. A mathematician is able to offer a first course in mathematical statistics. Many mathematicians are also able to offer the kind of first course that uses some calculus. But they often have a problem with the low-level introductory course that does not make much use of mathematics. This is also the type of course where the enrollment is much larger than for any other type of first course in statistics. As such, it requires special attention as we discuss in the next section.

Statistics as a Liberal Art

There is a new trend in many disciplines cutting across the more traditional approaches to teaching. This trend reflects a wish of many instructors to present their material in such a way that the students become as excited about the topic as the instructor is, without getting bogged down in trivial details. Currently, not many students come out of their introductory statistics courses with this kind of an attitude towards the subject. Mostly, they feel they have spent a semester learning how to plug data into a set of unrelated formulas. It is not surprising that they do not have much of a sense of what statistics is all about.

Instead, what if we could manage to share with the students the central focus of what statistics really is all about? What if we could convey why statistics is so interesting that it was worth spending the effort to get a PhD in the subject, and that statisticians would not want to trade their occupation for anything else? What if we could share with our students the beauty of the subject and the importance of statistics in their personal lives as well as for the rest of the world?

If we could do this, then we may be able to leave the students with more than facts, methods and formulas that are forgotten the moment the final exam is over. Instead, we may be able to leave the students with a sense that they have learned something they want to carry with them and something that is important for how they live their lives as good citizens in today's society.

It is hard to convey such a sense of excitement about mathematics. Students do not begin to see the essentials of mathematics until they are halfway through a college mathematics major. The barrier seems to be that a person has to have a great deal of technical, mathematical knowledge before he or she can begin to understand what mathematics really is all about. There have been attempts at such an approach to the teaching of mathematics, but these attempts have perhaps not been as successful as people had hoped.

Statistics is different enough from mathematics that it is possible to teach such a course in statistics. The course needs to be taught by an individual who has had considerable experience with data and with the uses of statistical methods. Such a course should try to convey the nature of uncertainty and how seemingly random events have patterns to them; patterns that statistics is created to look for. Indeed, it may be possible to define statistics as the set of methods used to study regularities in the face of variability.

Such a course should stress the prevalence of biased results that come from badly formulated questions in surveys and polls and the biased results that come from samples that are not properly random. The course should stress the need for good tabular, graphical and computational summaries of data. The course should also stress the study of relationships between variables and the role of other variables in our attempts at distinguishing between causal and spurious relationships. Finally, the course should stress that there is uncertainty associated with the results of any statistical investigation, and that this uncertainty affects how we use the results of the analysis.

We need go no further than today's newspaper, evening television news or the weekly newsmagazine to see why it is important for an educated citizen to have some knowledge of statistics and some sense of the uses of statistics. These sources educate us about what is happening in the world we are part of and help us understand this world better, but we need to be able to critically assess what we are told.

This means that an educated person should be able to evaluate the information that comes our way. It is important to be able to look at a graph and see ways in which the graph may be misleading. It is important to know that even if we are given the median income, such a number is only an estimate and does not deserve the implied accuracy that comes from reporting the number down to the last dollar. Because the number is an estimate, we should know that there is uncertainty attached to it.

A good deal of public policy in this country is determined by surveys and polls. It is a step in the right direction that reports from such polls now often include a 'margin of error' of plus and minus a few percentage points. Some newspapers even include information in small print on how the poll was conducted and what the meaning is of such an error term. A reader who has had a course in statistics containing material on the role of question formulation, sampling, and some statistical inference, is much better equipped to evaluate the results of such a poll.

Statistical reasoning will also help a person make a more critical evaluation of reports that two or more variables are related. We know that correlation between two variables does not necessarily mean that they are causally related. But it is often hard to distinguish between spurious and causal relationships. Anyone who has had a course in statistics containing a discussion of the role of control variables, is much more tuned in to the notion that we have to examine the role of possible control variables before we can draw any conclusions about cause and effect.

This way of looking at statistics means we should think of statistics more as part of a liberal arts education than as a set of methods and techniques. It may be easier to teach such a liberal-arts-type course now than some time ago because of the advances in computer hardware and software. The existence of good statistical software and the easy access to computers on our campuses has eliminated the need to teach much of the material we used to include in introductory statistics courses. There is no longer any need to struggle with things like the formulas for the computation of the variance for grouped data or the t-statistic for the comparison of two means using the pooled variance. It was never important how this was done, but it was necessary to teach these formulas so that the students could compute these quantities when they were required. Unfortunately, such formulas too often became the primary focus of statistics courses, at least in the students' minds. Now that we need not emphasize such computations, we are given free time for other and hopefully more relevant topics.

Liberal arts should be more than the original seven classical subjects; it should be a learning process that helps a person develop into an understanding and contributing member of society. An important part of that learning is the ability to understand how the empirical world functions with its social, biological and physical variables. For a textbook that takes this liberal arts approach to the teaching of statistics, see Moore [4]. For a further discussion of statistics as a subject in a liberal arts curriculum, see Iversen [2].

Those who intend to become statisticians need to learn statistics in terms of specific methods and techniques the way any trade or profession is taught. But only a small minority of students become statisticians, and many of the large numbers of students taking introductory statistics may be better served by being exposed to statistics as a liberal arts course than the more usual course with its emphasis on the nuts and bolts of statistics. We must remember that for most students, their first course in statistics is also their last course in statistics. A good liberal arts type course can awaken an interest in statistics that may lead some students to consider it as a career. A dreary nuts-and-bolts course can turn them away.

Need for a Statistician

Many mathematicians realize that they cannot possibly teach this elementary, introductory statistics course. They may be able to teach themselves the actual statistical techniques. But they typically lack the statistician's experience and sense of data to teach such a course. Some mathematics departments have realized that the best solution to this problem is to hire statisticians with applied interests as members of their departments, just as they have algebraists, combinatorialists, topologists and people with other specialities. With statisticians on the staff, a mathematics department can handle introductory statistics at all levels of mathematical sophistication.

Having one or more statisticians as part of a mathematics department is not without its problems. The number of statisticians is usually small, which means that there is a certain amount of professional isolation. The other members of the department typically work in fields that have little overlap with statistics. It may well even be that the statisticians know more about mathematics than the mathematicians know about statistics. This does not make it simple for the statistician who wants to talk to somebody about a current research problem or a recent publication in the statistical literature. Formal evaluations for promotions and tenure decisions also get more difficult when the evaluations within the department are done by people who do not share the same professional interests.

But it also broadens the scope of a mathematics department to have statisticians on the staff. More than that, colleges and universities need to have statisticians on their faculty, and there is no more natural place for such people than in a mathematics department when a separate statistics department does not exist.

For a further discussion of who should teach statistics, see Moore [5] and the comments responding to his article.

Conclusions

There will always be a duality in statistics between the theory developed by statisticians and the practical needs of the users of statistics. This duality is rooted in the history and the very nature of statistics. Because of it, statisticians will never be the only people using statistics, and users other than statisticians will be developing new statistical theory and methods.

The duality between theory and practice has relevance for the teaching of statistics as well. Statistics will continue to be taught by practitioners in other fields and by mathematicians as well as by statisticians. In many ways statisticians have abdicated their responsibilities for the teaching of statistics, as seen by the many departments of statistics that have clung to mathematical statistical theory as their only area of interest. But even if they wanted to, statisticians will never be able to claim the field of statistics as belonging only to them, and they will therefore never be in full control of the teaching of statistics.

What statisticians can do instead is to exercise leadership and guidance in the teaching of statistics. Because of the close kinship between statistics and mathematics, a natural place to begin this leadership and guidance is with departments of mathematics in colleges and universities where there are no departments of statistics. It is in these departments that we find mathematicians teaching statistics as part of the curriculum in mathematics. The mathematicians who teach statistics there are the first to recognize their limitations in their teachings of statistics and are the first to realize that any help they can get will be most welcome.

Statistics departments should seek out and get to know neighboring departments of mathematics that teach statistics as part of their curricula. Most mathematics departments have a lecture series throughout the year, and volunteers that come and give talks about statistics will always be welcome. This would also give the statistics departments a direct contact with possible future graduate students in statistics. Such contacts can also be developed by inviting mathematicians to come and visit the statistics departments. There are possibilities for listening to talks, and it could even be a place to spend a sabbatical semester for a mathematician who wants to learn more about statistics.

The largest limitation for mathematicians who are asked to teach statistics is in their lack of experience dealing with data. After all, mathematics is a language of symbols, and most mathematicians hardly ever run across many numbers. One way to get around this limitation is for mathematicians to get some experience with data analysis. Short courses could be given by local statistics departments or by regional and national statistics organizations. One place for such courses is to have them in conjunction with regional or national meetings of the MAA or other mathematics associations.

Thus, the view from the mathematics department is that the department will continue to teach statistics for years to come. The ideal solution is to have statisticians on the staff to do this teaching. If that is not possible, then mathematicians will have to continue to teach the statistics courses. To do this well, we need help from the statistics community.

References

1. Albers, Donald, et al., *Undergraduate Programs in the Mathematical and Computer Sciences*, MAA notes #**7**, 1986.

2. Iversen, Gudmund. R., Statistics in liberal arts education, *The American Statistician*, 1985, **39**, pp. 17-19.

3. McKelvey, Robert, et al., An inquiry into the graduate training needs of two year college teachers of mathematics, *Rocky Mountain Math Consortium*, 1979.

4. Moore, David. S., *Statistics: Concepts and Controversies*, New York: W. H. Freeman and Co., 1979.

5. Moore, David. S., Should mathematicians teach statistics?, *The College Mathematics Journal*, **19**, pp. 3-7 (with discussion).

Gudmund Iversen is professor of statistics in the Department of Mathematics at Swarthmore College, where there is a very harmonious relationship between mathematics and statistics. His main interest is in the application of statistics to the social sciences, and some of his writings have been in the area of contextual analysis. He has become increasingly interested in issues surrounding the teaching of statistics and what can be done to improve it, both as it is taught by statisticians and by non-statisticians.

Remedial Statistics?:
The Implications for Colleges of the
Changing Secondary School Curriculum

Ann Watkins, California State University at Northridge
Gail Burrill, Whitnall High School
James M. Landwehr, AT&T Bell Laboratories
Richard L. Scheaffer, University of Florida

All Talk and No Action: The History of Statistics in Secondary Schools

Educators are convinced that statistics should be a prominent part of the secondary school curriculum. This enthusiasm is not new; almost every major report of the past thirty years has included a recommendation such as:

> "A student who has worked through the full thirteen years of mathematics in grades K to 12 should have a level of training comparable to ... one semester of ... probability theory." [Cambridge Conference on School Mathematics, 1963]

> "Elementary statistics and probability should now be considered fundamental for all high school students." [National Science Board Commission on Precollege Education in Mathematics, Science, and Technology, 1983]

> "Secondary school mathematics should introduce the entire spectrum of mathematical sciences: ... Data analysis, including measures of uncertainty, probability and sampling distributions, and inferential reasoning." [Mathematical Sciences Education Board, 1990]

> "Topics from statistics, probability, and discrete mathematics are elevated to a more central position in the curriculum for all students." [National Council of Teachers of Mathematics, 1989]

Why does every report continue to include such endorsements? Because very little statistics or probability has ever been taught in secondary schools. The situation that the National Advisory Committee on Mathematical Education reported in 1975 sounds dismayingly contemporary:

> "While probability instruction seems to have made some progress, statistics instruction has yet to get off the ground.... At the high school level, probability topics in Algebra I and II texts are commonly omitted. A one semester senior course in probability and statistics has gained only a small audience of the very best students. Furthermore, this course places a heavy emphasis on probability

theory, with statistics, if treated at all, viewed as merely an application of that theory. Though National Assessment gives reasonable attention to probability and statistics objectives, current commercial standardized tests do virtually nothing with these topics."

These earlier findings are mirrored in a 1987 study of Ohio schools which found that only about one-fifth of Ohio high schools offered a separate course in probability and statistics and 25% of those included no statistical inference. Although three-fourths of the schools reported teaching some probability or statistics in other mathematics courses, "typically this means that descriptive statistics is included in a general mathematics course and some elementary probability is taught in second-year algebra or precalculus. Less than one-fifth of the schools reported teaching any statistical inference topics, like confidence intervals, outside a separate probability-and-statistics course." [Kullman and Skillings, 1990]

The relatively small number of schools teaching a nontrivial amount of probability and statistics shouldn't be surprising. Until recently, the problems confronting the high school teacher who wanted to offer a statistics course were considerable. He or she may never have taken a statistics course, and if so, it was most likely a calculus-based course in mathematical statistics. Materials were scarce and most teachers were forced to adapt publications for college students. Until 1985 there was no up-to-date statistics textbook available for high schools. Further, there was little popular support for offering statistics and probability [*Priorities in School Mathematics*, 1981]. Statistics and probability were seen as frills:

"Primary school students in Westfield, N.J., are now spending all their time on 'basic' skills rather than probability, statistics and more esoteric topics. 'We tried to cut out all the extra junk,' explained David J. Rock, the district director of instruction." [*The New York Times*, June 9, 1986].

Statistics still is not required for high school graduation, for college admission, or for success on standardized tests. For example, a typical SAT contains at most one probability question and two or three questions concerning the average. Although "The College Board has long shared the concerns ... about the need for better and earlier instruction in statistics and probability and applauds the efforts of the National Council of Teachers of Mathematics to encourage such instruction, ... the SAT must be appropriate for all secondary school students, some of whom do not have the advantage of formal study of statistics" [Dietrich, 1991]. However, the College Board is considering adding an Advanced Placement course in introductory statistics.

The Push for Change

Although there is still too little probability and statistics in U.S. secondary schools, there has been an astounding increase since about 1985. This change has come about largely because of the influence of several major curriculum projects. For the first time, commercially available material that has been written with the help of secondary teachers and tested in secondary classrooms is available and selling well. Some of the more prominent of these efforts to inject statistics and probability into the secondary school curriculum are listed below.

Quantitative Literacy The Quantitative Literacy Project (QL) of the American Statistical Association and the National Council of Teachers of Mathematics has been active since 1981. With NSF funding for most of these years, QL gives workshops for secondary teachers based on its four booklets, *Exploring Data, Exploring Probability, The Art and Technique of Simulation*, and *Exploring Surveys and Information from Samples*. QL has been instrumental in convincing many teachers of the importance of data analysis in the secondary curriculum.

In 1991, NSF funded a three year continuation of QL, "A Data-driven Curriculum Strand for High School Mathematics," which is developing a series of modules on data analysis that can be injected easily into the traditional mathematics curriculum. The modules will use statistics to illustrate and motivate the math throughout the main components of high school algebra and geometry, as well as to teach some statistics. A second grant funded by NSF in 1991 is developing a teacher and administrator Quantitative Literacy workshop model for the elementary mathematics curriculum.

Used Numbers The Used Numbers Project wrote and tested a series of six booklets for elementary school, including *Statistics: The Shape of the Data, Statistics: Prediction and Sampling*, and *Statistics: Middles, Means, and In-Betweens*, that teach students to collect and interpret data.

North Carolina School of Science and Mathematics The Department of Mathematics and Computer Science at the North Carolina School of Science and Mathematics, a state-supported school for high ability students, has written a series of booklets with support from the Carnegie Corporation of New York. The booklets include *Data Analysis* and *Geometric Probability*.

Woodrow Wilson The Woodrow Wilson National Fellowship Foundation offers five-day summer workshops for high school teachers. The instructors are high school teachers trained at a month-long institute, "Focus on Statistics," held at Princeton University in 1984. A similar institute for middle school teachers is planned for the summer of 1992.

Illinois Institute The Illinois Institute for Statistics Education (IISE) is a summer workshop program funded by NSF for teams of teachers. Its textbook for high school students [Travers et al., 1985] develops statistical ideas by simulation.

American Statistical Association The Center for Statistical Education of the American Statistical Association organizes Quantitative Literacy workshops, distributes *Guidelines for the Teaching of Statistics K-12*, provides speakers and consultants, and sponsors three competitions yearly: the American Statistics Poster Competition for excellence in graphical display in grades K-3, 4-6, 7-9, and 10-12, the Quality and Productivity Scholastic Competition for grades 7-12, and the American Statistics Project Competition for team projects from grades 4-6, 7-9, and 10-12. The newsletter of the American Statistical Association-National Council of Teachers of Mathematics Joint Committee on the Curriculum in Statistics and Probability, *The Statistics Teacher Network Newsletter*, is available from the American Statistical Association.

NCTM In 1989, the 80,000-member National Council of Teachers of Mathematics published their *Standards for Curriculum and Evaluation in School Mathematics*. This document calls for strong and pervasive statistics and probability strands from kindergarten through grade 12.

ICOTS In August 1990, the Third International Conference on Teaching Statistics (ICOTS) was held in New Zealand. About 600 participants from over 30 countries discussed the teaching of statistics from kindergarten through graduate school and the importance of a statistically literate populace. A remarkable degree of unanimity exists among these leaders in statistics education that the teaching of statistics should be more practical than theoretical. They also generally agree on the central place of data analysis and on the importance of computer use. ICOTS 4 will be held in 1994.

UCSMP The University of Chicago School Mathematics Project (UCSMP) has written a series of textbooks for grades 7 - 12. An unusually strong emphasis on probability and statistics is maintained throughout.

Teaching Statistics *Teaching Statistics* is a British journal for teachers of students aged 9 to 19. Published since 1979, it includes book and software reviews, data banks, problems, ideas for lessons, and articles on statistics and on the psychology of learning statistics.

As such activities gain momentum in secondary schools, we in the colleges and universities will be faced with entering students already versed in statistical thinking. We will address the implications of this later in this article.

The Articles of Faith

The principles guiding the major statistics education projects of the past decade are remarkably similar. Perhaps this is because they are based on the conviction that statistics is not a branch of mathematics. (However, if statistics is to be taught in the secondary schools, the mathematics curriculum is the most promising entrée. There is no debate in secondary schools about whether mathematics teachers should teach statistics [see Moore, 1988]. They are currently both the best educated in statistics and the most willing.)

In the common emphases below, one can see both the belief that statistics should not be taught as a branch of mathematics and the influence of the reform movement in mathematics education:

1. **Data analysis should be central.** Data analysis, as the term is generally used in secondary schools, refers to defining a problem, collecting data, organizing and summarizing it, and interpreting it with the help of graphical analyses. The battle over whether the statistics curriculum should be taught from a data-analytic point of view has essentially been won. Almost every expert, panel, committee, and commission studying the problems in mathematics education agree that data analysis is central.

2. **Statistics is not probability.** The study of probability does not need to precede the study of statistics. Students should first learn elementary data analysis with probability being taught as the need for it arises.

Traditional topics taught in introductory statistics -- standard deviation, the normal distribution, least squares fit, t-test, chi-squared test, type I and type II errors, permutations and combinations, and Bayes' theorem -- should be taught after the more basic ideas of data analysis, informal probability, simulation, and an intuitive introduction to statistical inference.

3. **Resistant statistics should have a larger role.** Students more readily understand statistics based on the median and there are powerful statistical reasons for emphasizing resistant statistics (which aren't sensitive to outliers).

4. **There is often more than one way to approach problems in statistics and probability.** This means that discussion and evaluation of different approaches can take up a large part of class time. Students must be encouraged to attack problems from different angles and be prepared to support their conclusions.

5. **Real data that are interesting and important to the students should be used whenever possible.** Real data give the study of statistics both its legitimacy and its excitement. The data should drive the statistical methods rather than vice versa, and each set of data should itself be worthy of study. In addition, students, who are accustomed to the neatness of the numbers in the typical mathematics classroom, need experience in dealing with the messier numbers of the real world.

6. **The emphasis should be on good examples and on building intuition.** Showing how to lie with statistics and stressing probability paradoxes destroy a student's confidence.

7. **Students enjoy and profit from project work, experiments, and other activities designed to give them practical experience in statistics.**

8. **Students should write more and compute less.** The emphasis should be on interpreting plots and numerical summaries rather than on the technicalities of constructing plots and computing statistics.

9. **The statistics and probability taught in secondary schools should be important and useful to students in its own right and not serve just to prepare students for college courses.** The word "pre-college" should be purged from our vocabulary.

While there is a lot of agreement on the philosophy of teaching statistics and probability, the following questions are still being debated:

How can we assess student progress in learning statistics?

When and in what way should the normal distribution be introduced?

Should counting arguments be part of the probability curriculum? (The NCTM *Standards* includes finite probability in the discrete mathematics strand, not in the probability strand.)

Should there be an Advanced Placement statistics course?

Should topics such as standard deviation, binomial distribution, and hypothesis testing be in the curriculum for all students, or only for the college bound?

Concerns of Secondary Students and Teachers

Those of us involved in the Quantitative Literacy Project learned much about what secondary school teachers and students think about statistics. The concerns detailed below are serious ones; inservice workshops do not entirely eliminate them, but do help make teachers more comfortable and successful in teaching statistics.

Teachers do not know enough about what statistics is and what statisticians do. They need information about how statistics is different from mathematics, about how to do data analysis, about why the urn model of probability is limited, and about how statistics is used and why it is important. Because of lack of experience with statistics, teachers feel uncomfortable leading discussions and want complete "solutions" manuals with examples of "good" student answers.

In many cases neither students nor teachers see statistics as a part of the mathematics curriculum. American students are conditioned to believe that mathematics consists of contrived problems and exercises in computation. Consequently, students make comments such as these about statistics. (Student grammar and spelling is preserved.):

> *"I think this is more appropriate for a business class."*

> *"You shouldn't have to do this kind of stuff in a math class."*

Students and teachers know that learning statistics will not help on standardized tests or help get to calculus faster. Honors students are often so competitive about moving ahead in the precalculus curriculum that they resent taking time to learn some statistics:

> *"It was just busy work for us because they didn't want us to be ahead of the rest of our class."*

> *"...an excuse to keep me from advancing further in the math curriculum."*

On the other hand, because statistics is not perceived as part of the mathematics curriculum that they have been struggling with, some students, especially general mathematics students, love it:

> *"I learned a lot from this class and it made math fun and interesting too me."*

> *"I feel I will probably be able to actually use some of this knowledge in real life situations, where as most math that I am learning now seems to be fairly unnessarry for real life."*

Unless they have substantial training, teachers tend to focus on the mechanics of constructing plots rather than on the interpretation, and on calculation rather than on data. If teachers or students view statistics in this way, they think that it is trivial and that "it was just busy work for us."

Teachers often expect that a hierarchy of skills exists and must be learned: "What comes after box plots?" Teachers often miss the point that the same basic set of tools can be applied with different levels of sophistication to many different sets of data.

Students and teachers do not expect ambiguity in their mathematics classes. The shift from exact answers to approximation, and the emphasis on words such as "explore," "analyze," and "describe" are significant changes in the way most teachers and students deal with numbers. A simulation problem inevitably brings "What is the real answer?" The teachers and students have trouble adjusting to "mathematics" in which there is not necessarily a unique answer:

> *"Learned much but some questions were opinionated. Much esiar (sic) if only one possible answer."*

> *"The mean is the average. The median is not."*

The nontraditional classroom environment necessary when working with statistics and probability causes uneasiness. In a typical mathematics class, the teacher reads the answers to the homework at the beginning of each class period, lectures about the day's topic, makes the next assignment, and gives students time to start their homework. Students expect the teacher to tell them how to do each type of problem.

The structure of a statistics class is not as predictable. One set of homework exercises may take three days as students work their way through a problem or an experiment. The answers cannot be read, but must be discussed. Often new avenues are suggested by the students in the middle of a lesson. Unlike the solitary mathematics assignment, statistics is more appropriately done by students in groups where they can exchange ideas. The bell may ring before the teacher has time to pull ideas together and to make an appropriate homework assignment.

Another area of concern is assessment and grading. When the assignment is to analyze a set of data or to set up a simulation, student progress must be evaluated by new methods. Student work cannot be graded by counting the number of problems correct. Students are not used to writing sentences in their mathematics classes; mathematics teachers are not used to judging answers in written form. Quick quizzes for informal evaluations are not easy to make.

Teachers must learn how to use computers in instruction. Computer simulations fascinate most students and are an excellent vehicle for analyzing concepts such as variability or the effect of different sample sizes on a statistic. Computer graphics are an integral part of data analysis and make it more interesting for many students. However, if too much use is made of the computer too soon, students lose any understanding and sense of reality they originally brought to the process.

Use of software has had limited success except with those teachers who use software to teach everything. Teachers don't want to learn complicated packages and are uncertain how to use them to best effect in the classroom. As more computers become available both for laboratory and demonstration purposes, and as teachers are encouraged and assisted in using the software appropriately, the computer should become more visible. There remains, however, a shortage of good, very friendly, software.

Indications of Change

Since public education in the United States is controlled by the individual states, one way to measure the impact of the projects of the past decade is to observe recent changes in the probability and statistics strands of state guidelines for mathematics education. It is clear that the trend is toward more emphasis on statistics and probability.

Many new state guidelines are built directly on the Quantitative Literacy materials and training programs. The Wisconsin guidelines that mandate that all students graduating from high school must have instruction in statistics draws heavily from QL. The state's Mathematics Curriculum Task Force has recommended at least one semester of statistics for all students. Similar efforts to add statistics and probability to the mathematics curriculum can be seen in Connecticut, New York, California, Indiana, Illinois, Georgia, Texas, New Jersey, Delaware, Iowa, Minnesota, and Maryland.

The interest by teachers in statistics workshops and in funding agencies in sponsoring them continues to grow. For example, during the summer of 1991, week-long QL workshops were held in seven states with funding from outside sources.

The Impact on the College Curriculum: Remedial Statistics?

Much of the current introductory college statistics course is already considered high school material. Soon, and for the first time, many students who have had statistics and probability in high school will be entering college. They will have had experience in data analysis, solving probability problems by simulation, designing experiments and surveys, and should have an appreciation for the importance of statistics. They will also understand basic statistical measures and how to compute them.

Students should not have to repeat state-mandated high school material in college and so the college course will have to adapt. Colleges and universities should take a look at their state mathematics frameworks as a guide to what changes may be necessary. There is an obvious danger that many colleges and universities will be tempted to use the time gained to go into more esoteric statistical topics such as ANOVA or multiple regression.

Aside from modifying the curriculum to avoid repetition of high school material, college teachers will have to deal with students who have a slightly different view of statistics from that usually taught in the introductory course in college. For example, data analysis is the major component of the proposed high school curriculum. Students who enter college will expect the analysis of real, interesting data throughout the course. They will know that data analysis cannot be done on made-up data. They will also know that data analysis can't be confined to a section of the course called 'descriptive statistics' and then ignored when more complicated statistical methods are taught. The high school curriculum emphasizes treating each set of data as worthy of study in its own right. This implies that, for each example, enough background has to be given to show that the problem involves some interesting questions, that the data are relevant to the problem, and that the data were collected in such a way that they are worth analyzing.

In many cases, the standard graphical methods of data analysis will give clear and sufficient answers to the questions of interest, so students may protest that the more complicated method of statistical inference that the professor had planned on introducing is not needed. For example, secondary students may have learned to find the median (or Tukey) line to model a linear relationship. How will the professor now justify least squares regression? In fact, it is not easy to find examples where the conclusions are not clear unless more complicated methods for modeling or dealing with statistical significance are wielded.

Another area for change in the standard introductory course involves the use of the computer. It almost goes without saying that much of the data analysis should be done on the computer, as is often the case in courses now. In addition, students will be accustomed to using the computer to simulate situations they cannot analyze theoretically, and this facility should be taken advantage of. For example, consider the distribution of the chi-squared statistic from a

2×2 table. Assuming independence, the statistic could be simulated and studied by assuming some common population row proportion and two sample sizes. The following kinds of questions would be of interest. How does the distribution depend on the two sample sizes and the proportion? Is it close to a normal distribution? Is the distribution of some transformation of the chi-squared statistic close to a normal distribution? What is the distribution like if there is not independence, so that two different proportions and sample sizes are needed to generate the simulations? How can the distributions from such simulations be used to learn something about a particular observed 2×2 table? Some students will find these simulations more convincing than the mathematics, while others may gain a better appreciation for the role, elegance, and unifying value of mathematical theory.

A third area of change in the college course might involve an increased emphasis on student projects, experiments, and data that the students collect themselves. The high school materials emphasize the value of statistical methods for helping students to learn about questions and issues that are important to them; moreover, doing this requires writing conclusions and summary reports in English. It seems natural and desirable to carry this emphasis over to an introductory course in college.

Finally, there will arise the problem of entering students who, for one reason or another, have not learned high school-level statistics and probability. Should colleges do as they do with calculus and enforce prerequisites for the introductory statistics course? Will there be placement tests, and, eventually, remedial courses in high school statistics and probability?

A Challenge

The projects of the past ten years have been successful, by almost any measure, in raising interest in the teaching of statistics. Much work must still be done in teacher training and curriculum development before statistics will be widely incorporated into the mainstream of secondary school mathematics. What can you do? Volunteer to speak about statistics before a class or the math club at your local high school. Work to make statistics a required part of the training of teachers at your college or university. Suggest that your local science fair give a special prize for the best use of statistics; volunteer to judge and get the word to teachers that good statistical analysis is expected. Arrange for secondary school students and teachers to visit statisticians in industry. Organize a workshop for secondary teachers (contact the American Statistical Association for help).

However, perhaps the strongest action is to get the message to local high school teachers that students are coming to your college or university unprepared to take either the introductory statistics course or courses in other departments on statistical methods. It is likely that more students coming to your campus take a statistics course than a calculus course. Show high school teachers the numbers; impress them with the fact that the precalculus curriculum doesn't help their students with statistics; detail the background in probability and statistics that the students should have. High school teachers are generally very concerned with how well their students do in college and will do everything they can to prepare them for your classes.

The twenty years of successful projects by the American Statistical Association-National Council of Teachers of Mathematics Joint Committee on the Curriculum in Statistics and Probability (*Statistics: A Guide to the Unknown*, the Quantitative Literacy workshops and publications, and *Statistics by Example*) can be duplicated on a local level if statisticians and secondary teachers work *together* to design workshops and write materials for the secondary curriculum in statistics and probability.

References

Cambridge Conference on School Mathematics. *Goals for School Mathematics*. Boston: Houghton-Mifflin, 1963.

Department of Mathematics and Computer Science, North Carolina School of Science and Mathematics. *Data Analysis* and *Geometric Probability*. Reston, VA: National Council of Teachers of Mathematics, 1988.

Dietrich, Frederick, vice president for guidance, access and assessment services at the College Board in New York, Letter to the Editor, *Los Angeles Times*, March 31, 1991.

Gnanadesikan, Mrudulla, Richard L. Scheaffer, and Jim Swift. *The Art and Techniques of Simulation*. Palo Alto, CA: Dale Seymour, 1987.

Kullman, David and John Skillings, "Current Practices in Teaching Statistics," Reader Reflections, *The Mathematics Teacher*, **83** (1990) 7-8

Landwehr, James M. and Ann E. Watkins. *Exploring Data*. Palo Alto, CA: Dale Seymour, 1986.

Landwehr, James M., Jim Swift, and Ann E. Watkins. *Exploring Surveys and Information from Samples*. Palo Alto, CA: Dale Seymour, 1987.

Mathematical Sciences Education Board (MSEB). *Reshaping School Mathematics: A Philosophy and Framework for Curriculum*. Washington, D.C.: National Academy Press, 1990.

Moore, David S., "Should Mathematicians Teach Statistics?," *The College Mathematics Journal*, **19** (1988) 3-34.

Mosteller, Frederick et al., Eds. *Statistics by Example*. Reading, MA, Addison-Wesley, 1979.

National Advisory Committee on Mathematical Education (NACOME). Overview and Analysis of School Mathematics, K-12. Washington, D.C.: Conference Board of the Mathematical Sciences, 1975.

National Council of Teachers of Mathematics (NCTM). *Curriculum and Evaluation Standards for School Mathematics*. Reston, Va: NCTM, 1989.

National Science Board Commission on Precollege Education in Mathematics, Science, and Technology. *Educating Americans for the 21st Century*. Washington, D.C.: National Science Foundation, 1983.

Newman, Claire M., Thomas E. Obremski, and Richard L. Scheaffer. *Exploring Probability*. Palo Alto, CA: Dale Seymour, 1987.

Priorities in School Mathematics: Executive Summary of the PRISM Project. Reston, VA: National Council of Teachers of Mathematics, 1981.

Statistics Teacher Network Newsletter. Center for Statistical Education, American Statistical Association, 1429 Duke Street, Alexandria, VA 22314-3402.

Teaching Statistics. Longman Group UK Limited, Subscriptions Department, Fourth Avenue, Harlow, Essex CM19 5AA.

Tanur, Judith M. and Frederick Mosteller, Eds. *Statistics: A Guide to the Unknown.* Third Ed, Belmont, CA, Wadsworth, 1989.

Travers, Kenneth J., et al. *Using Statistics.* Menlo Park, CA: Addison-Wesley, 1985.

Used Numbers Project. *Statistics: The Shape of the Data, Statistics: Prediction and Sampling, Statistics: Middles, Means, and In-Betweens.* Palo Alto, Dale Seymour, 1989.

Ann Watkins has been interested in statistics ever since a professor at UCLA told her that it would make her fortune, as it had his. While waiting, she works as a professor of mathematics at California State University, Northridge, co-edits the *College Mathematics Journal*, and gives workshops for the ASA-NCTM Quantitative Literacy (QL) Project. She is a past Second Vice-President of the MAA. She is also co-author of a series of books on statistics and data analysis and co-editor of the books *New Directions in Two-Year College Mathematics* and *A Century of Calculus*.

Gail Burrill teaches at Whitnall High School in Greenfield, Wisconsin. She is a recipient of a Presidential Award for Excellence in Teaching Mathematics. She is past chair of the NCTM-ASA Joint Committee on Statistics and Probability and served as principal investigator of the ASA-NCTM QL Project. She is currently directing the NSF-funded project, *A Data Driven Curriculum*, to develop materials for the high school mathematics curriculum which links statistics and mathematics.

James M. Landwehr is a member of the Statistics Research Department at AT&T Bell Laboratories and has worked on a variety of statistical research and application problems. He has also been active in the ASA-NCTM QL Project to encourage statistical education in grades K-12. He is a Fellow of both ASA and AAAS.

Richard Scheaffer is professor of statistics at the University of Florida. As a former chair of the ASA-NCTM Joint Committee on Statistics and Probability, he was active in the development of the QL Project which provided much of the background and impetus for the emphasis on data analysis in the NCTM Standards. He was also a developer of the Center for Statistical Education at the ASA which oversees curriculum development and teacher enhancement projects for the K-12 curriculum.

Diagnostic Testing for Introductory Statistics Courses

Elizabeth M. Eltinge
Texas A&M University

Many students experience frustration when taking an introductory level statistics course. This frustration can lead to poor performance in the class and a negative attitude toward statistics. These reactions, while not shared by all students, are widespread enough to cause concern to those who teach statistics.

There are two primary reasons for these reactions:

1) inadequate preparation of the students in the mathematical and logical concepts which form the basis of statistical reasoning;

2) the method by which most introductory statistics courses are taught.

The second topic is addressed elsewhere in this volume. This paper addresses the first topic by reviewing several issues in the assessment of students' preparation. First, a rationale for the assessment of students' entering skills is developed. This is followed by a discussion of the skills that should be included in the assessment. The final section discusses ways to use the results of these assessments.

Why Entering Skills Should be Assessed

To perform well in a statistics course, students must have a minimum level of competence in basic algebra and logical reasoning. Possession of these skills does not guarantee success in a statistics course, but lack of these skills certainly leads to difficulty. Students who lack certain algebraic skills often experience frustration in learning statistics because they have difficulty in mastering the basic steps involved in computing means, standard deviations and confidence intervals. Students who lack certain logical reasoning skills have difficulty assigning meaning to computed means and standard deviations or interpreting relational phrases, such as "at least" or "at most." In addition, students with low levels of algebraic and reasoning skills have difficulty understanding the reasoning behind confidence interval construction and hypothesis testing. It is unreasonable to expect such students to grasp the implications of fairly abstract concepts such as the variability of a sample statistic.

It would be useful to both students and instructors to be able to identify quickly students who, due to their limited skills in algebra and reasoning, are unlikely to do well in statistics without an exceptional expenditure of time and effort. This could be done with a pre-test covering basic algebraic and reasoning skills given to students at the beginning of the semester. Such a diagnostic test would be useful to the students because it would help them to identify weaknesses in their algebraic skills and to budget their time accordingly. A good diagnostic test would be useful to instructors because it would help them to identify extremely weak students who could then be counseled to postpone the course until they have improved their algebraic and reasoning skills. A diagnostic test should be administered very early in the semester to allow those students who perform very poorly to make necessary changes in their course schedules.

There are also several administrative advantages of a diagnostic test. At Texas A&M University, the introductory, algebra-based statistics courses carry a prerequisite of college algebra. In some departments, prerequisites are not considered binding and thus are often ignored. Prerequisites are printed in the course catalog, but many students do not refer to the catalog when planning their schedules. Consequently, many students enroll in courses without first completing the necessary prerequisite courses. Some students who enroll in our calculus-based introductory statistics course for engineers (which carries a prerequisite of two semesters of calculus) do not know what "that little squiggle sign" means when asked to compute an integral of a function. At a much more elementary level, we see liberal arts, business and education students enrolled in our algebra-based introductory statistics courses who claim not to know that the product of two negative numbers yields a positive number. The system of stating prerequisites in catalogs, on syllabi and during lecture on the first day of class is not delivering the necessary message to all of the students.

A more effective way to communicate the information about prerequisite courses is to administer a short quiz during the first week of class. This quiz should cover some of the material that students are expected to have mastered before enrolling in the course. This places the students in a more active role of verifying that they have indeed satisfied the course prerequisites. It also allows for testing of prerequisite knowledge, regardless of courses taken. We have also found this approach useful in addressing the concerns of other departments across campus. The departments which require their students to take statistics are generally very interested in the performance of their students on the pre-tests which we administer. These other departments also recognize that by removing those students who are underprepared, we are in a better position to teach a stronger course in statistics.

There are also research advantages of giving a diagnostic pre-test. Identification of particular concepts which are essential for success in statistics can be useful in elucidating how students learn statistics. This could be used to improve the delivery of statistics courses. Difficulties with mathematical concepts often lead to problems in learning statistical concepts. For example, students who have difficulty working with decimals sometimes cannot determine if 0.01 is larger or smaller than 0.006. This leads to confusion in comparing p-values with various levels of alpha in the discussion of hypothesis testing.

Students sometimes develop methods of coping with their mathematical weaknesses which can lead to further difficulties. For example, students may become very dependent on calculators for computing "exact" values and not be able to make reasonable estimates. Students with high dependence on calculators may also have difficulty in assessing whether or not a final answer, such as a negative value for a variance, is reasonable [17]. Dependence on calculators also leads to lack of familiarity with operations involving fractions, which can lead to difficulties computing values for certain F-statistics or working with simple probabilities. If one can identify these common mathematical weaknesses and predict the effect of these weaknesses on the learning (or impediment to learning) of statistical concepts, then management and delivery of statistics courses can be adjusted accordingly.

Pre-Test Content

I started using a pre-test in an introductory statistics course in the Spring of 1989. The first pre-tests we used were administered to about 60 students, had an open-ended format and were hand scored. We used the student responses from the open-ended format to develop a multiple-choice format for the pre-test which could be computer scored. We have since administered pre-tests to approximately 4,000 students enrolled in algebra-based introductory statistics courses at Texas A&M. We currently use a pre-test which combines some of the items in the quiz presented by Pisani [13] with some items I have gathered based on observation of students. Sample questions from this pre-test are presented in the Appendix. To encourage the students to give the pre-test serious consideration, we weigh the pre-test as one homework assignment in their final grade assessment. The average scores for the pre-tests have consistently been around 70% correct. Correlations of pre-tests scores with final grades have ranged from slightly above 0.2 to slightly below 0.5.

A good diagnostic test should contain items which test accurately for knowledge critical to success in statistics. The scores on such a test should also be useful in predicting success in introductory statistics courses. These are two separate issues to consider during test development. A pre-test which tests accurately for knowledge of critical material may or may not be a good predictor of success in the course itself. This is because a single instrument cannot measure important extraneous variables such as effort expended by students or effectiveness of an individual instructor. Hence, a pre-test may measure critical prerequisite skills but the score on the pre-test alone may not be a good predictor of success in a course. Similarly, an item or variable which is a good predictor of success may or may not indicate mastery of critical prerequisite material. A good predictor of success in a statistics course is a high overall grade point average in previous courses. A high grade point average should indicate that a student has a well rounded knowledge base but, in itself, does not guarantee knowledge of essential skills for a statistics course. Thus, both content and predictive validity need to be addressed.

One approach I have used to develop pre-tests with content validity is to note particular mathematical problems as they occur during the semester. While meeting with students, grading examinations, and discussing problems other instructors are having with weak students, I pay particular attention to those problems which appear to be based on mathematical weaknesses. I keep a file of these problems as I encounter them. For example, on one examination I noticed a few students had difficulty solving a problem because they could not properly convert units. Unit conversion can be viewed as a problem of setting up and solving equalities involving proportions and ratios. Difficulties in solving this type of problem could stem from a lack of competency in basic algebra and in working with fractions, which in turn could lead to difficulties in solving basic problems in probability. Questions 1, 3, 5, 6, 7, 8, 10, 14 and 16 in the Appendix all require students to manipulate ratios and fractions. Question 5 specifically requires students to perform unit conversion, and has consistently been the most missed problem on the pre-test.

Some beginning students of statistics have difficulty assigning abstract meaning to the numbers they compute. These students also have difficulty understanding relationships between numbers. Consider, for example, the relationship in hypothesis testing that as the absolute value of a z-statistic increases, its associated p-value decreases. Some students do not associate the test statistic with the value of a random variable; the test statistic is simply a number they compute. Nor do they relate the p-value associated with the test statistic to a probability statement. Probability is viewed by some only as the topic at the beginning of the textbook involving dice, cards, and balls in urns. Consequently, p-values do not fit into their conception of probability. When discussing hypothesis testing, we often present two possible decision rules for testing the null hypothesis. One decision rule is based on the critical value, while the other is based on the p-value. With the first approach, a large absolute value leads to rejection of the null hypothesis; with the second approach, a small number leads to rejection of the null hypothesis. Students who have difficulty attaching abstract meaning to numbers become very confused with the relationship between these two decision rules. Because they have difficulty thinking about numbers (and sometimes even determining which number is larger), the decision rules of rejecting the null hypothesis with large or small numbers seem arbitrary and confusing, no matter how many normal curves with shaded tails we draw for them. They view the entire process of hypothesis testing as performing various arithmetic operations without attaching any meaning to the final answer.

Some statisticians argue that the major emphasis in teaching beginning statistics should be placed on data analysis and not on hypothesis teaching [7, 11, 14, 19]. Hypothesis testing is an abstract process that is difficult to teach to beginning students. Developing the foundation for hypothesis testing can be tedious and boring to students. Presenting the formulas without any foundation is meaningless computation of numbers. However, even if we never teach hypothesis testing in any introductory statistics course, students still need to be able to associate' meaning to numbers.

"Story" or "word" problems provide one way to test for the ability to associate meaning with numbers. Students who have difficulty identifying meaning with numbers often cannot assess the relationships among numbers presented in a word problem. This leads to difficulty in determining how to write appropriate equations to solve a problem. Once the equations are written students can often complete the mechanical operations to compute the correct answer. Thus, the principal difficulty is in determining symbolic representations of the relationships among numbers, rather than in the arithmetic operations. Questions 5, 14 and 16 in the Appendix require students to attach abstract meaning to numbers.

Ability to read information from graphs and tables provides another indication of quantitative reasoning ability. I have seen students presented with information in a table who are unable to identify which numbers they need to answer a question. Other students have difficulty understanding a graphical representation of the relationship between two variables. I have not yet incorporated examples of such questions into the Texas A&M pre-tests. However, Pisani [13, pp. ix.-xi.] presents good examples of questions from a set of graphs.

The pre-test has been very useful to us in that it requires students to take cognizance of course prerequisites, as outlined above. However, the pre-tests I have used do not exhibit strong predictive validity. One reason for the low predictive power of the pre-test could be the limited range of the scores on the pre-test. The pre-tests have been very short (ten to fifteen items), and there appears to be a "ceiling" effect in that the distribution of the scores is skewed to the left. In addition, a few students with very low pre-test scores drop the course. The deletion of these weaker students further restricts the range of the pre-test scores. The predictive power of the pre-tests may be enhanced by increasing the length of the pre-test and thus increasing the range of possible scores. The current pre-test, however, already requires approximately twenty minutes to administer. The advantage of greater predictive accuracy would have to be weighed against the disadvantage of using more classroom time.

A more likely explanation for the low predictive power of the current pre-tests could be that pre-test scores cannot measure subsequent effort or general study skills of the students. Some students with very low scores on the pre-test may exert extra effort in the course and thus earn high grades. Alternatively (but with the same effect of reducing predictive validity), students with high scores on the pre-test may conclude the course is very easy, and exert little effort, resulting in a lower grade than they are capable of earning. I have found that in certain introductory statistics courses, one of the best predictors of success on examinations is students' average homework scores. If one considers average homework score to be an indicator of effort exerted in a course, then increased effort does result in higher grades. This is not a surprising result, but nonetheless is not fully internalized by all students.

Further predictive models are being developed which incorporate more variables about the students. Important variables for consideration in the model include cumulative grade point averages, success in previous mathematics and statistics courses and number and types of courses transferred to the University. A good predictive model could be used to alert students to potential problems they might have in the course.

Recommendations to Students Based on Test Results

Once a valid and reliable method of predicting success, or lack of success, in a statistics course is developed, consideration needs to be given to those students who appear to be at risk. Related work has been done by Jannarone with graduate students in psychology [8]. The graduate students in the study are expected to enter graduate school with a certain amount of knowledge about statistics before they enroll in a graduate level course in statistics. Jannarone points out that it is difficult to teach statistics effectively to students lacking basic skills in the same classroom with students who are well prepared. He recommends offering a diagnostic examination during the first week of classes, and giving those with clear deficits the choice of taking an undergraduate course in statistics before they take the graduate course. The diagnostic test used is actually a final examination for an undergraduate course in statistics. Remediation of skills is approached with sensitivity to the feelings of the students involved and efforts are made to associate remediation with long-term benefits.

Variability in mathematical abilities in beginning statistics students is also addressed by Bashaw [2]. He recommends directing most students toward self-study of the prerequisite material. Depending on the extent of the lack of prerequisite skills, self-study or remedial coursework may be recommended. For any form of remediation to be successful, students must be actively aware of the benefits associated with their efforts.

Students with severe weaknesses in mathematical ability should probably be removed from the statistics classroom until they can remediate their skills. Trying to teach regression analysis, for example, to students who have a great deal of difficulty graphing a straight line on a Cartesian coordinate system reduces the level at which one is able to teach students who do possess the prerequisite knowledge. It is a disservice to the many capable students to sacrifice academic rigor for the sake of some students who are severely hindered by their poor mathematical skills.

Furthermore, an introductory statistics course is not the proper setting in which to remediate basic mathematical skills. The job of remediation is better handled by someone who is prepared for the unique problems faced in that situation. I recommend that students with severe mathematical deficiencies take a remedial algebra course before they enroll in a statistics course.

There is evidence that certain remedial programs are successful in preparing students for college courses [4, 9, 10, 12]. There is a need for more complete data about subsequent college performance of students enrolled in remedial programs to allow for more detailed evaluation of the effectiveness of remedial programs [20]. Data collected for particular programs indicate that students who enroll in and complete remedial courses with passing grades go on to have higher college grade point averages than those students who enroll in and do not complete or who do not receive a passing grade in a remedial course [4, 10, 12]. One complication in evaluation of remedial programs is that not all students who are advised to take remedial coursework actually enroll in the recommended courses [16].

Some researchers in remediation recommend a diagnostic/prescriptive model of remedial teaching [1]. With this model, concepts and procedures to be retaught are identified, then appropriate instructional strategies are prescribed. Further, remediation is more successful if the remedial course is designed to prepare students for a specific course [9]. Thus, in order to increase the likelihood of successful remediation for a statistics course, we should identify those algebraic and reasoning concepts and procedures which are essential to success in statistics, accurately diagnose the entering students' level regarding those specific skills and prescribe specific remedial work to improve those areas which are weak.

During remediation, it should be made clear that statistics is not the same as mathematics. This can help to prepare the students for what to expect in a statistics course. Many students do not have a clear concept of what statistics is. Some students never even hear of statistics as a formal discipline until they enroll at a college or university campus and learn that they must complete a statistics course in order to graduate. They do not know how to distinguish statistics from mathematics, and hence may transfer their mathematics anxiety to their statistics course. One hopes that this lack of pre-college exposure to statistics is changing, thanks to efforts such as those by the Quantitative Literacy Project of the American Statistical Association [15], the PBS television series *Against All Odds: Inside Statistics* [6] and the 1989 NCTM Standards [3], which place a greater emphasis on statistics and probability.

Discussion

A common reaction of many students to their first statistics course is one of frustration and lack of understanding of basic statistical concepts. Part of the reason for this negative reaction is student inadequacies in algebraic and logical reasoning skills. Steen [18] identifies one of the marks of mathematical maturity as the ability to abstract or to glean the essential structure from a complex situation. Gaining mathematical maturity requires students to expand beyond strict numerical computations to interpretation of the abstract meanings implied in numerical values. A good course in statistics can thus enhance mathematical maturity, provided students possess the prerequisite skills for the course. Students lacking these skills should be identified and referred to remedial work to prevent them from becoming frustrated and confused in a statistics course. One way to identify these students is through the use of a pre-test covering basic skills. Effective use of an accurate pre-test can thus improve the learning environment in a statistics classroom by helping to ensure that the students possess the essential skills for learning statistics.

Effective use of a pre-test can also contribute to research on how students learn statistics. Garfield and Ahlgren [5] cite the need for collaborative, cross-disciplinary research on how students think about probability and statistics. One element of research in this area is the identification of algebraic and reasoning skills which are essential to statistical thinking. Identification of these skills can be accomplished in part through the use of pre-tests.

We generally can do little to change the level of mathematical preparation of the students enrolling in our introductory statistics courses. It is, however, inappropriate to ignore this lack of preparation. We need to develop ways to assess the entering knowledge, and make recommendations based on the outcomes of those assessments. Students with severe mathematical illiteracy should be encouraged to seek remediation of their deficit skills. Students who do possess the prerequisite skills should be encouraged to learn to think about statistics and to reason about decision making in the face of uncertainty. We must work within the current academic systems to produce the greatest improvements we can in statistical literacy.

Acknowledgements

The author thanks James A. Calvin, David S. Moore, H. Joseph Newton, and the Editors for helpful comments made on an earlier version of the manuscript.

References

1.   Abbott, Gypsy Anne and Elizabeth McEntire (1985). *Effective remediation strategies in mathematics* Research Triangle Park, N.C.: Southeastern Regional Council for Educational Improvement. ERIC document ED 262 459.

2.   Bashaw, W. L. (1969) *Mathematics for statistics*. New York: John Wiley and Sons.

3.   Curriculum and Evaluation Standards for School Mathematics, (1989). Reston, Virginia: National Council of Teachers of Mathematics.

4.   Design Team of Butte College (1985). Evaluation of remedial programs: Pilot study final report. Oroville, CA.: Butte College. ERIC document ED 263 963.

5.   Garfield, Joan and Andrew Ahlgren (1988). Difficulties in learning basic concepts in probability and statistics: Implications for research. *Journal for Research in Mathematics Education*, **19**, No. 1.

6.   Guthrie, Donald (1990). Review of *Introduction to the practice of statistics* by David S. Moore and George P, McCabe and *Against all odds: Inside statistics* David S. Moore, content developer. *Journal of the American Statistical Association*, **85**, No. 409.

7.   Hogg, Robert V. et al. (1985). Statistical education for engineers: An initial task force report. *The American Statistician*, **39**, No. 3.

8.   Jannarone, Robert J. (1988). Preparing incoming graduate students for statistics. Ware, Mark I. and Charles L. Brewer, Eds., *Handbook for Teaching Statistics and Research Methods*. Hillsdale, New Jersey: Lawrence Erlbaum Associates.

9.   Kolzow, Lee C. (1986). Study of academic progress by students at Harper after enrolling in developmental courses, Vol. **14**, No. 9. Palatine IL: Office of Planning and Research, William Rainey Harper College. ERIC document ED 265 914.

10. Kraska, Marie F., Martin H. Nadelman, Arnold H. Maner, and Reenie McCormick (1990). A comparative analysis of developmental and nondevelopmental community college students. *Community/Junior College Quarterly of Research and Practice*, **14**, 1.

11. Moore, Thomas L, and Rosemary A. Roberts (1989). Statistics at Liberal Arts Colleges. *The American Statistician*, **43**, No. 2.

12. New Jersey State Department of Higher Education (1986). Effectiveness of remedial programs in New Jersey public colleges and universities. Fall 1983-Spring 1985. Report to the Board of Higher Education. Trenton. NJ: New Jersey Basic Skills Council. ERIC document ED 304 999.

13. Pisani, R. (1985). *Statistics: A tutorial workbook.* New York: W. W. Norton.

14. Scheaffer, Richard L. (1990a). Toward a more quantitatively literate citizenry. *The American Statistician*, **44**. No. 1.

15. Scheaffer, Richard L. (1990b). Popularity of quantitative literacy programs in schools on the upswing. *Amstat News*, No. 164.

16. Skinner, Elizabeth Fisk and Stephen Carter (1987). A second chance for Texans: Remedial education in two-year colleges. Tempe, AZ: National Center for Postsecondary Governance and Finance. ERIC document ED 297 783.

17. Smith, Philip T. (1987). Levels of understanding and psychology students' acquisition of statistics. in Sloboda, John A. and Don Rogers, (Eds), *Cognitive processes in mathematics.* New York: Oxford University Press.

18. Steen, Lynn Arthur (1983). Developing mathematical maturity. in A. Ralston and G. S. Young, (Eds), *The future of college mathematics: Proceedings of a conference/workshop on the first two years of college mathematics.* New York: Springer-Verlag.

19. Tanner, Martin A. (1985). The use of investigations in the introductory statistics course. *The American Statistician*, **39**, No. 4.

20. Wright, Douglas A. and Margaret W. Cahalan (1985). *Remedial/developmental studies in institutions of higher education policies and practices.* Rockville, MD.: Westat Research Inc. ERIC document ED 263 828.

Elizabeth Eltinge is an assistant professor in the Department of Statistics at Texas A&M University. Her research interests include student misconceptions of statistics and methods of improving statistics instruction. She currently serves as the Director of Undergraduate Academic Affairs within the Department of Statistics.

Appendix

Sample Questions from the Pre-Test

Items marked with an asterisk are adopted from Pisani (1985) pp. ix.-xi.

1*. Of the following, the pair that is not a set of equivalents is:

    a) 0.021%, 0.00021   b) 1.5%, 3/200   c) 110%, 1.10
    d) ¼%, 0.0025   e) 225%, 0.225

2. Solve the following equation for x: $y = (x-1)^2$.

    a) $(y-1)/(x-2)$ b) $\sqrt{(y+1)}$ c) $1+\sqrt{y}$
             d) 1   e) $y-1$

3*. 5/16 is closest to:

    a) 1/2 b) 1/3 c) 14   d) 1/5   e) 1/6

4. Solve the following equation for x: $x = 3!$

    a) 3   b) 27 c) 9   d) 6   e) 81

5. If a rectangle has the dimensions 3 ft. by 5 ft. 4 inches, what is the area of the rectangle?

    a) 16.0 sq.ft.   b) 16 ft. 8 in.   c) 16.2 sq.ft.
        d) 192 sq.ft.   e) 200 sq.in.

6*. 13/38 is approximately:

    a) 15%   b) 35%   c) 45%   d) 55%   e) 75%

7*. The relationship between 0.01% and 0.1 is:

    a) 1 to 1   b) 1 to 10   c) 1 to 100
        d) 1 to 1,000   e) 1 to 10,000

8. Solve the following equation for x: $\dfrac{a}{b} = \dfrac{c}{x}$

    a) $\dfrac{a-c}{b}$ b) $\dfrac{a}{bc}$ c) $\dfrac{cb}{a}$ d) $\dfrac{b}{ac}$ e) $\dfrac{a}{b}$

9. Given: $x_1 = x_2 = x_3 = x_4 = x_5 = 5$, compute $\sum\limits_{i=1}^{5} x_i$.

    a) 20   b) 25   c) 5   d) 4   e) 5^5

10*. Of the following sequences of fractions, what set is arranged in increasing order?

a) $\dfrac{7}{12}$, $\dfrac{5}{6}$, $\dfrac{2}{3}$, $\dfrac{3}{4}$

b) $\dfrac{7}{12}$, $\dfrac{7}{11}$, $\dfrac{8}{11}$, $\dfrac{4}{5}$

c) $\dfrac{7}{12}$, $\dfrac{8}{11}$, $\dfrac{8}{12}$, $\dfrac{9}{12}$

d) $\dfrac{7}{12}$, $\dfrac{1}{2}$, $\dfrac{7}{15}$, $\dfrac{1}{3}$

e) $\dfrac{7}{12}$, $\dfrac{3}{4}$, $\dfrac{1}{3}$, $\dfrac{9}{12}$

11*. The square root of 100,000 is about:

a) 30   b) 300   c) 1,000   d) 3,000   e) 10,000

12. Solve the following equation for x: $x^2 - x = 6$.

a) 3   b) 3, -2   c) 6   d) $\sqrt{(6+x)}$ e) 3x

13*. 72.376% of 416.9327 is approximately:

a) 100 b) 200   c) 300 d) 400   e) 500

14*. Vodka is 40% alcohol. If one quart of vodka is mixed with 0.6 quarts of orange juice, what percentage of the mixture is alcohol?

a) 15%   b) 20%   c) 25%   d) 34%
e) cannot be determined from the data given.

15. Solve the following equation for x: $x = \sqrt{100}$.

a) 100 b) ± 10 c) ± 25 d) 4   e) $\pm 10^2$

16*. A is older than B. With the passage of time, the ratio of the age of A to the age of B:

a) remains the same
b) increases
c) decreases

Part II

Innovative Curricula

for

Statistical Education

Data, Discernment and Decisions:
An Empirical Approach to Introductory Statistics

Richard L. Scheaffer
University of Florida

Introduction

Introductory statistics becomes interesting, even exciting, to students when they perceive that it can help them understand and improve their world, however broad or narrow that world may be. Since much of anyone's world is described quantitatively, this understanding can come about through enhanced skill at collecting and interpreting data. Improvement comes when understanding leads to wise action.

An introductory statistics course, then, should emphasize data collection and interpretation in the context of solving real problems. When the problems involve chance outcomes, simulation can be employed to generate data. Simulation can lead to an intuitive development of some procedures for statistical inference, but the emphasis should be on what statistical inference is and what it is not, rather than on any specific set of methods.

Purpose and Overview

The growing interest in statistics on all academic levels -- from kindergarten through graduate school, from mathematics and science to business and health -- and the increasing use of statistical techniques in industry have caused a surge of activity related to the content of statistics courses and the methods of teaching this content. Many, if not most, recommendations on content suggest more "data analysis". This article presents one point of view on what data analysis concepts should be taught, how to teach those concepts and why this emphasis is important. The points made here were honed by years of experience teaching statistics at a large university, years of effort on the American Statistical Association's Quantitative Literacy Project to introduce statistics into school mathematics, and conversations with many statistics instructors from institutions large and small.

Most scientific investigations begin with a question and follow with an exploration of data to focus the question. Then, an attempt is made to list all factors that affect the question and collect data on as many of them as possible. After the possible causal factors are studied, a carefully planned experiment may be set up to determine precise answers to the question under consideration. The "answer" is then checked with more data to see if it holds under new conditions. Any "answer" always leads to new questions, and the investigation continues.

This seven-step model consists of:
1. General question
2. Specific problem

3. Analysis of causes
4. Root cause and possible solutions
5. Results
6. Implementation to prevent recurrence
7. Future work on related problems

It is a general problem-solving model employed in various types of quantitative investigations, but is best known as a model for industrial process improvement. Versions of it can be found in the work of Kume [4] and in the quality assurance manuals of various industries.

As an example, suppose I have a general question about why I have no time for reading novels. A quick check of my daily schedule tells me I spend much (perhaps too much) time in the car. I analyze the causes by collecting data on where and when I drive. The basic cause seems to be too much time chauffeuring children. Suggested solution -- carpooling with the neighbors. This solution seems to provide a savings of 1.5 hours a week, and the savings appear to hold up for a series of weeks. Good! Now, perhaps I should investigate if by changing the hour I drive home from work, I can reduce time in traffic jams!

What is the point of introducing this model? First, real data analysis in a statistics course must be *real* to the students. It must reflect a problem solving atmosphere that would be found in their "real" world. Just throwing some actual data into a textbook is not enough. Second, most of real data analysis is simply *exploratory* in that the data is looked at carefully by using a variety of techniques to see what patterns may emerge. At some point, a carefully designed experiment or sample survey may be run and *confirmatory* analyses conducted to estimate a parameter, predict a future value, or fit a mathematical model, but this level of detail generally requires background far beyond the introductory statistics course. At the introductory level, students should be taught how to intelligently explore data, and what types of questions might require a deeper analysis. They do not need an encyclopedic knowledge of statistical methods.

In summary, then, the introductory course should prepare students to be logical problem solvers, guide them through a variety of exploratory techniques, and provide them with a basic understanding of how confirmatory techniques work so they have some perspective on what statistics can do and what it cannot do. The remaining sections of this article address specific exploratory techniques that should be part of any introductory statistics course, show how randomness is essential to statistical practice, and tie data analysis and probability together in a simulation approach to statistical inference.

Exploring Data

Each stage of the problem-solving model given in the previous section may involve data. In fact, most important decisions in a student's life will involve data. Thus, it is initially important that students be taught to gather data intelligently and to look carefully and critically at all data presented to them. Whenever confronted with new data, the first questions to ask are:

Where did the data come from?
What do the numbers mean?

Several secondary questions should also be asked:

Was the data collected by a reputable source?
Are they precise measurements or fuzzy guesses?
Are the figures in millions or billions, or are they rates, proportions or averages?

Once these two sets of questions are answered satisfactorily, a systematic look for pattern and departures from pattern may begin. For univariate data sets, the most informative graphical displays are one or more of the following: dotplot, stemplot, boxplot, histogram.

Country	Total Tax Revenues (in billions)	Per Capita Tax Revenues (in dollars)
United States	$1,134.1	$4,740
Australia	50.6	3,213
Austria	28.1	3,714
Belgium	38.0	3,854
Canada	17.3	4,621
Denmark	28.5	5,573
Finland	20.2	4,116
France	232.2	4,216
Greece	11.7	1,178
Italy	146.6	2,565
Japan	375.2	3,107
Netherlands	56.2	3,879
New Zealand	7.5	2,304
Norway	27.7	6,668
Portugal	6.5	676
Spain	47.3	1,226
Sweden	50.6	6,064
Switzerland	29.7	4,554
United Kingdom	171.3	3,025
West Germany	236.0	3,869

Table 1. Tax Revenues Around the World

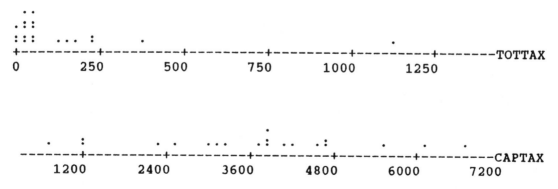

Figure 1: Dotplots of Tax Revenues

The data in Table 1 are total tax revenues and per capita tax revenues for 20 free market economies for 1985. Dotplots of these data (Figure 1) clearly demonstrate the skewness of the total revenue figures and the symmetry, but wide spread, of the per capita tax figures. The

```
Stem-and-leaf of TOTTAX    N = 20      Stem-and-leaf of CAPTAX    N = 20
Leaf Unit = 10                         Leaf Unit = 100


  (13)   0  0012222234555                 1    0  6
    7    1  147                           3    1  12
    4    2  33                            3    1
    2    3  7                             4    2  3
    1    4                                5    2  5
    1    5                                8    3  012
    1    6                               (4)   3  7888
    1    7                                8    4  12
    1    8                                6    4  567
    1    9                                3    5
    1   10                                3    5  5
    1   11  3                             2    6  0
                                          1    6  6
```

Figure 2: Stemplots of Tax Revenues

stemplots (Figure 2) show similar patterns but have the advantage of preserving all of the data values. (Also, notice the truncation in the stemplots. This helps teach the importance of significant digits in a practical context. Is anything lost in the truncation?)

	N	MEAN	MEDIAN	STDEV
TOTTAX	20	140.8	48.9	253.5
CAPTAX	20	3658	3862	1572

	MIN	MAX	Q1	Q3
TOTTAX	6.5	1134.1	27.8	165.1
CAPTAX	676	6668	2680	4604

Table 2: Numerical Summaries of Tax Revenue Data

Boxplots (Figure 3) show the quartiles as the ends of the boxes, the median within the box, and outliers, or extreme observations. Notice how two countries (the U.S. and Japan) clearly become outliers on total tax revenue as compared to the other countries. (Why are no outliers identified in the per capita tax revenue data?) We will return to further discussion of boxplots a little later.

Histograms are most useful for large data sets and, consequently, will not be shown for the data of Table 1. The importance of histograms can be seen from a classic quality assurance example of W. E. Deming, as shown in Figure 4. Machined steel rods had a lower specification limit (LSL) of 1.0 cm. Any rod larger than that could be reworked and used, but rods under 1.0 cm had to be scrapped. Inspectors looked at 500 rods, and their quality check measurements were combined and plotted. A problem was immediately recognized. What do you think it was?

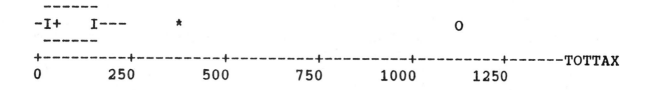

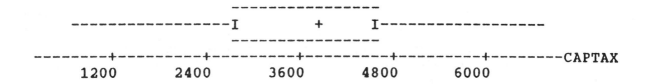

Figure 3: Boxplots of Tax Revenues

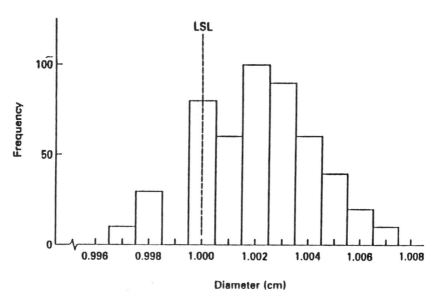

Source: Deming, W. E. (1978). "Making Things Right," Statistics: A Guide to the Unknown, 2nd ed., Holden-Day, San Francisco.

Figure 4: Diameters of Steel Rods

From graphical displays, we proceed to numerical summaries of data. The numerical summaries do not provide as much information as a graph, but they are useful as a brief short-hand sketch of key information that is easy to carry around, talk about, and print. Also, numerical summaries are the basis for the confirmatory techniques of classical statistics, which do serve a useful purpose.

Traditionally, numerical summaries attempt to measure centrality and variability. Table 2 shows the common summaries for the data on tax revenues of Table 1. The mean and median measure centrality and the standard deviation and interquartile interval (Q_1, Q_3), or interquartile range of Q_3 - Q_1, measure variability. Notice that the mean for the total revenue data (140.8) is quite different from the median (48.9). When both are plotted on Figure 1, the median falls within the large cluster of values on the left, but the mean is to the right of most of the data points. For the per capita revenue data, the mean and median are close together and both fall close to the middle of the distribution displayed in Figure 1.

With regard to variability, the standard deviation supposedly measures the "typical" deviation between a data point and the mean of all the data in the set under study. The mean deviations for total tax revenues and per capita tax revenue (Table 1) are plotted in Figure 5. For total revenue, the standard deviation is 253.5, but all of the mean deviations except one are *smaller* than this. Thus, the standard deviation does not do a good job of measuring typical deviation from the mean when the data is skewed. On the other hand, the interquartile interval (27.8, 165.1) does tell us that half of the observations are contained in an interval of length IQR = Q_3 - Q_1 = 137.3. Thus, a typical deviation from the median is closer to 100 than to 253.5. It appears that the median and IQR form a better summary of centrality and variability than does the mean and standard deviation.

For the per capita revenue, the standard deviation is 1572, and Figure 5 shows a number of deviations around this value (some larger, some smaller) on both the positive and negative sides. The IQR = Q_3 - Q_1 = 4604 - 2680 = 1924, so most deviations from the median are around 1000 or so. In this case, either the mean and standard deviation or the median and IQR form reasonable summaries of centrality and variability.

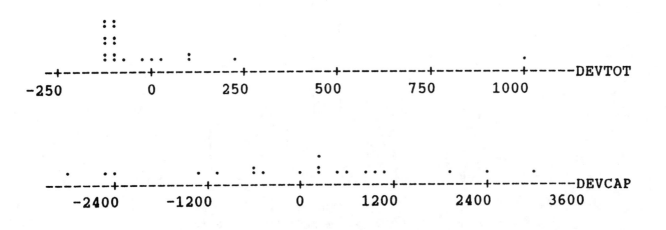

Figure 5: Dotplots of Deviations from the Mean for Tax Revenue Data

In comparing centrality and variability among two or more sets of data, parallel box plots (focusing on medians and interquartile intervals and ranges) provide an elegant and powerful graphical technique. Figure 6 shows how boys and girls compare on exam scores (percentage scale) for one class. Figure 7 shows median purchase prices (in thousands of dollars) for existing one-family houses in U.S. metropolitan areas across some recent years. Discussions of patterns and changes in patterns for such data sets can lead to lively classroom discussion or project work.

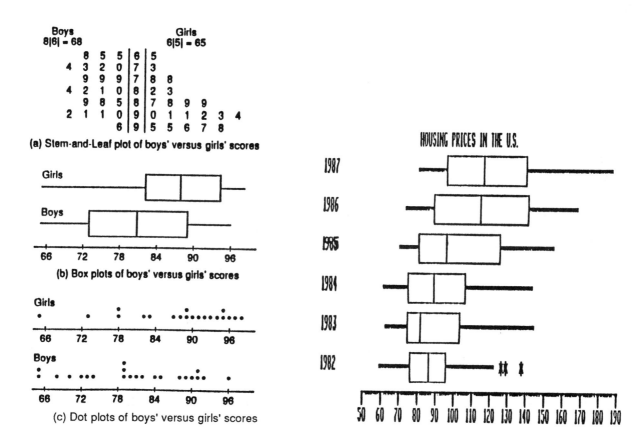

(a) Stem-and-Leaf plot of boys' versus girls' scores

(b) Box plots of boys' versus girls' scores

(c) Dot plots of boys' versus girls' scores

Figure 6: Test Scores

Figure 7: Housing Prices in the U.S.

Association between two variables is investigated most efficiently by beginning with a scatterplot and looking for a linear trend. In the general question of how to describe alligator populations in Florida, one specific problem involved a study of association between weight and length. (Length is fairly easy to approximate when viewing a "gator" in the wild; weight is not easy to guess.) The data for 25 alligators is shown in Table 3. A quick and easy exploratory tool for summarizing a linear trend is the median fit line, M, as described in [5], which is plotted in Figure 8. Note that this line captures the trend in most of the data but misses the two right-hand points rather dramatically. Still, an increase of 2.84 pounds per inch of length is not a bad approximation for small to medium sized alligators. The median fit line simply captures the trend in median y-values for convenient groupings of data according to the x-values.

Classical least squares can be used as an exploratory tool, even before the statistical properties of regression estimators are studied. The least squares line, L, (also shown on Figure 8) shows a growth of 5.9 pounds per inch, but this obviously looks too high. The regression line does not capture the trend too closely, as is amplified in the plot of residuals (differences between y-values and points on the regression line for each x-value) shown in Figure 9. The curvature in the data plot and in the residual plot suggests that a transformation might improve the fit. A plot of ln (weight) versus ln (length) is given in Figure 10, and this plot shows a nice linear trend that could be fit well by either a median fit or regression line. (The depth of exploration to be done here depends on where the students are in their mathematical sophistication.

Alligator	l	w
1	94	130
2	74	51
3	147	640
4	58	28
5	86	80
6	94	110
7	63	33
8	86	90
9	69	36
10	72	38
11	128	366
12	85	84
13	82	80
14	86	83
15	88	70
16	72	61
17	74	54
18	61	44
19	90	106
20	89	84
21	68	39
22	76	42
23	114	197
24	90	102
25	78	57

(Weight is in pounds; length in inches.)

Table 3: Lengths and Weights of Alligators

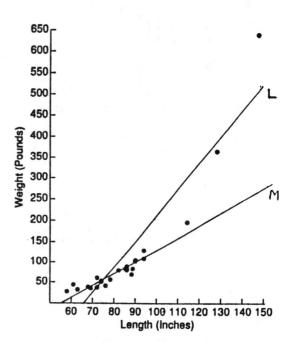

Figure 8: Median Fit and Least Squares Lines:
M: W = -158 + 2.84L, L: W = -393 + 5.90L

Something can be said without logarithms, but more can be said with logarithms.)

Basic concepts of designed experiments can be introduced within the exploratory framework, since much can be learned by merely taking a careful look at the data from such an experiment. A full exposition of these ideas will not be attempted here, but one brief illustration may help make the point. Two-level comparative experiments are the building blocks for much research activity in industry, agriculture, and other areas. Graphical investigation of main effects and possible interactions from such experiments can produce great insight into possible improvement of many problem situations. Suppose a student wants to select the route between home and school that takes the least time to traverse. He can narrow choices down to two for an experiment that forces him to try each choice on the way to school for a number of days, and each choice on the way home from school for a number of days. Dotplots or stemplots of data from the four path-direction combinations can help him decide if one path is always better (no interaction) or if one path is better when going to school and the other is better when coming home (interaction). More paths can be added to the experiment, but pairwise comparisons are the clearest way to begin the discussion of the usefulness of designed experiments coupled with data exploration. No analysis of variance is needed, in many cases, to see a pattern developing in the data, and such a pattern can be a valuable aid in decision making.

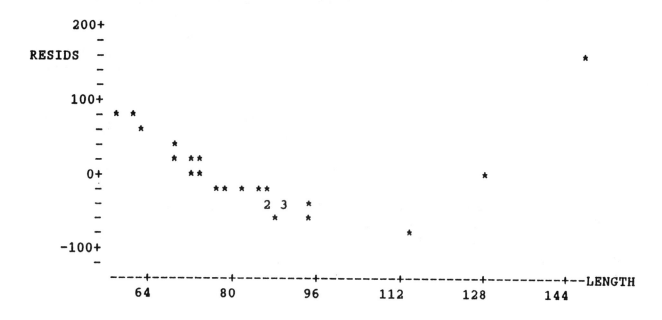

Figure 9: Residuals from the Least Squares Fit

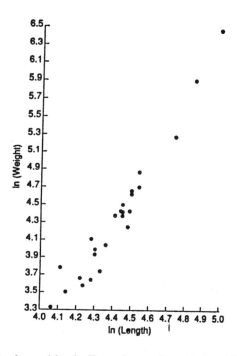

Figure 10: Logarithmic Transformation of the Alligator Data

From Exploring to Confirming: The Role of Randomness

Once solid skills in identifying pattern and departures from pattern in data are mastered, the notion of probability and randomness can be introduced. Still in the spirit of data analysis, these concepts should be introduced by empirical processes that generate data.

In answer to the question, "What is the chance that a balanced coin will come up heads when flipped?", almost everyone will say "one half". It surprises many, however, that the actual deviations between the sample fraction of heads and ½ can be rather large for small sample sizes. To see how the pattern for "proportion of heads" develops and changes with the sample size, the students must actually flip sets of 2, 5, 10, and 20 coins and graph the results. Then, they must do the same experiment again to see that essentially the same patterns are generated.

To move away from the probability of ½, a similar experiment should be conducted with dice, spinners, or another manipulative. Then, an experiment should be conducted for which the probability of interest is not known in advance. One possibility is to toss paper cups, which may land on the closed end, open end, or side, but not necessarily with probability ⅓ for each. Here is the perfect opportunity to demonstrate that "randomness" has two components; each cup tossed must have the same opportunity to land each of three ways and each toss is a new beginning (no carryover effect is allowed). If a cup becomes rounded on the bottom, thereby making that landing less likely, the pattern of results will change (or there may be no recurring pattern). Randomness will produce a regular pattern (which can be anticipated after some experience) if each trial is conducted in the same way with unchanging chances for each possible outcome. (In other words, the trials must be identical and independent.)

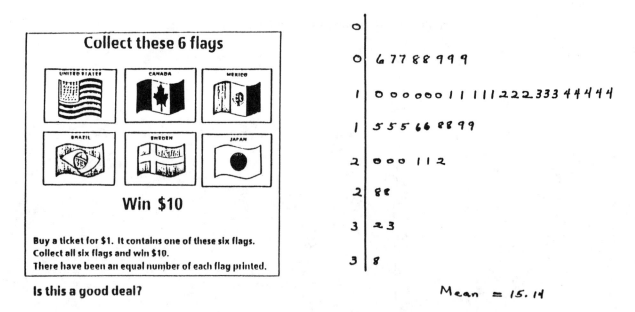

Figure 11: A Simulated Solution in Probability

The game shown on the left side of Figure 11 requires an assessment of the expected number of purchases required to obtain a full set of six different tickets. This situation can be *simulated* quite easily by a die model (or a model using random selections of six digits). The pattern presented by one set of 50 trials is shown on the right side of Figure 11. The statistic of interest is the mean number of tickets purchased to obtain the full set, and this mean will turn out to be around 15 time after time. In fact, the consistency of this distribution (pattern) from experiment to experiment amazes students. Again, through simulation it can be shown that randomness generates pattern and that such patterns are useful in assessing "practical" situations. (Not all examples need be *that* practical.)

At this point, *bias* can be brought into the discussion as a principle cause of pattern going awry. To illustrate, one student can be assigned the task of "sampling" students from the school by any method that comes to mind. (Call this the "subjective" sample.) A second student is given careful instructions on how to select a random sample and then is assigned the task of selecting one. Repeated performances will show the subjective samples to have too many (or too few) participants with high incomes, high interest in sports, or high academic achievement, while the random samples produce a more balanced view as compared to the known facts about the school. From a question on how much time is spent watching sports on TV, the subjective samples might produce averages consistently higher (or lower) than the true school averages, whereas the random samples will produce averages that are more evenly spread around the true average. Subjective samples tend to produce biased estimates because of the opinions of the sampler, whereas random (probability) samples are completely objective and tend to produce unbiased estimates.

To summarize, data exploration looks for pattern and departures from pattern. Randomness produces data with pattern that can be assessed by exploratory techniques and used to make decisions in areas involving chance outcomes. In the next section, we discuss how to make use of these ideas to build models for purposes of estimation and prediction.

Confirming Models: The Basis of Statistical Inference

Suppose a retail store was concerned that so many customers seemed unhappy with the service. After a detailed analysis of causes, their solution was to hire more employees. Now, the store wants to estimate the proportion of customers that like the service. To do this, management takes a random sample (not necessarily an easy task) of customers and asks each to respond "yes" or "no" to a simple question: Are you satisfied with the in-store service? They receive a sample proportion of "yes" answers, but what does that say about the proportion of all customers who might say "yes"?

In the spirit of the preceding section, Figure 12 (from reference [6]) shows the pattern produced by repeatedly taking samples of size 20 from situations with fixed probabilities of a "yes" response. The open boxes show the middle 90% of the sample proportions for "yeses". For example, if 40% of all customers would say "yes", then the middle 90% of the sample proportions of "yeses" in repeated samples of size 20 ranges from 20% to 60%. Using Figure 12 in reverse, if 11 out of 20 samples customers say "yes", then this sample proportion of 55% is within the "likely" range of sample outcomes for any population of customers with "yes" probabilities between .35 and .70. In other words, the sample proportion is consistent with the true proportion of satisfied customers being between 35% and 70%.

In using this approach, the store is assuming that the true proportion of satisfied customers remains fairly constant (after the new policy was instituted) so that the one sample provides good information. This is a simple model and the estimation of one parameter tells us about all we need to know.

A similar approach to the one identified by Figure 12 for estimating proportions can be used for estimating means, medians, and other single parameters. All is accomplished by simulation and a study of patterns generated by randomness. In fact, a study of hypothesis testing statistics such as the chi-square statistic can proceed in a similar way. Building the interval estimates from knowledge of simulated distributions is certainly slower than throwing out a formula,

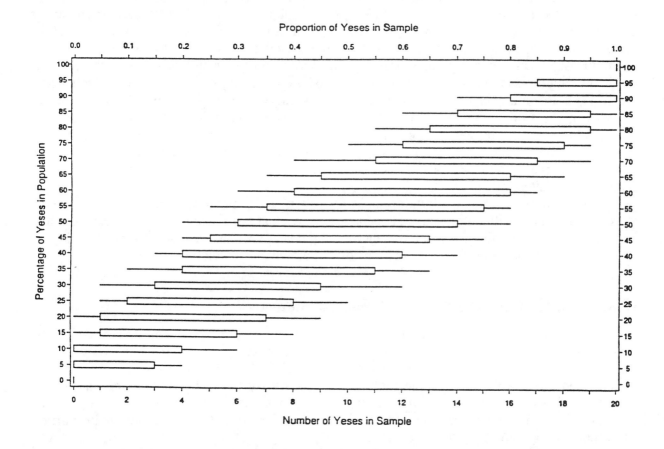

Figure 12: 90% Box Plots from Samples of Size 20

but the extra effort is worthwhile if the goal is to have students learn statistical principles rather than memorize statistical methods.

After going through the simulation process, students understand how statistical decisions are made and appreciate that statistics helps to guide decision making but cannot provide concrete answers in many cases. Perhaps most important is the fact that students will then see that classical statistical procedures *require* an underpinning of randomness. For those who only memorize methods, estimating the difference between mean scores for boys and girls in Figure 6 or between median housing prices for 1986 and 1987 in Figure 7 presents no difficulty. They substitute numbers into a formula. For those who have come through a simulation approach, these problems pose great difficulty because it is not clear if anything is random. The data in Figure 7 does not appear to be; no inference statement is necessary or appropriate. The data in Figure 6 could be randomly selected, but from what population? More information is needed before classical inference procedures would have any meaning.

The alligator weights and lengths of Table 3 could be thought of as a random sample from a population of alligators in north Florida and used as the basis for building a linear model to predict weight from length by regression techniques. But, the data must be transformed to show linearity. After the logarithmic transformation (Figure 10), the variation in ln(weight) seems to change with ln(length), so a weighted regression could be used. The point is that use of the classical regression technique requires randomness, assumptions, and adjustments that are far beyond the level of introductory statistics. The well-trained student will recognize this (to some degree), be willing to settle for an exploratory explanation of the main features of the data, and call in an expert if more detailed analyses are needed.

Conclusion

Why use an exploratory approach to introducing statistical concepts to students? Not only is the approach easy, entertaining, and wide open to creativity, but it is also the way scientific investigation should begin. "I have a question. Let's collect some data and see if we can explore some possible answers." Most of what a student should learn first about data collection, summarization, and interpretation can be taught through this approach, unencumbered by a plethora of mysterious assumptions. Thinking skills, as well as basic operational skills, can be taught along the way. Teaching only classical statistics as an introduction to the subject not only gets the cart before the horse in the scientific method, but also supplies students with a set of fancy tools designed for skilled craftspeople and specific jobs before they've had any experience with the simple tools that work on many jobs.

References

1.   Cleveland, William S., *The Elements of Graphing Data*, Wadsworth Advanced Books, Monterey, CA, 1985.

2.   Deming, W. E., "Making Things Right," *Statistics: A Guide to the Unknown*, 2nd ed., Holden Day, San Francisco, 1978.

3.   Granadesikan, M., R. L. Scheaffer, and J. Swift, *The Art and Techniques of Simulation*, Dale Seymour Publications, Palo Alto, CA, 1987.

4.   Kume, Hitoshi, *Statistical Methods for Quality Improvement*, The Association for Overseas Technical Scholarship, 1985.

5.   Landwehr, J. M. and A. E. Watkins, *Exploring Data*, Dale Seymour Publications, Palo Alto, CA 1986.

6.   Landwehr, J. M., J. Swift, and A. E. Watkins, *Exploring Surveys and Information from Samples*, Dale Seymour Publications, Palo Alto, CA, 1987.

7.   Newman, C. E., T. E. Obremski, and R. L. Scheaffer, *Exploring Probability*, Dale Seymour Publications, Palo Alto, CA, 1987.

8.   Scheaffer, R. L., "Why Data Analysis?" *The Mathematics Teacher*, **83**, No. 2, February 1990.

Richard L. Scheaffer is professor of statistics and former department chair at the University of Florida. His research interests lie primarily in sampling theory and applications, but he has spent much of the past ten years actively engaged in statistical education projects. As a former chair of the ASA-NCTM Joint Committee on the Curriculum in Statistics and Probability, he was instrumental in the development of the QL Project, which provided much of the background and impetus for the emphasis on data analysis in the NCTM Standards for School Mathematics. He was also one of the developers of the ASA's Center for Statistical Education which continues to oversee curriculum development and teacher enhancement projects for the K-12 curriculum.

Providing a Statistical "Model":
Teaching Applied Statistics using Real-World Data

John B. Willett and Judith D. Singer[1]
Harvard University Graduate School of Education

Introduction

Service courses in applied statistics abound in college and graduate school curricula around the world [34]. In a recent survey of United States graduate programs in health professions, for example, Cockerill and Fried [12] found that *all* programs required their students to take at least one statistics and research methods course. In a similar survey of psychology departments, Aiken, West, Sechrest and Reno [1] found that the same was true for nine out of every ten doctoral programs.

The clear consensus, among students and professors alike, is that too many of these applied statistics courses are far from successful [3, 9, 25, 33]. In Dallal's [13] words, "[t]he field of statistics is littered with students who are frustrated by their courses, finish with no useful skills, and are turned off to the subject for life" (p. 266). Joiner [27] gave service courses in statistics "a grade of F" for being unmitigated failures (p. 53). These courses frequently receive the worst evaluations in a school. And is it any surprise? Most of them are abstract, mechanical and boring, with antiquated and formulaic pedagogy that is little more than a presentation of the "statistic of the day."

Many students enrolled in applied statistics courses are hamstrung further by false perceptions of their own inadequacy. They regard "stats" courses as a necessary evil, an unavoidable rite of passage, and they view data analysis with fear and trepidation. For those who last saw a logarithm or summation sign a decade earlier, such anxiety is understandable. But when it so cripples our students that they cannot learn, we need better methods for reaching them and teaching them as well.

How can learning applied statistics be made more interesting, more palatable, more successful? We believe that one approach is to capitalize on students' fascination, not for statistics itself, but for the *substantive problems that statistics can address.* College students are interested in discovering whether there is race or gender bias in achievement testing, whether children's pre-natal experiences influence their later inhibition, whether better economic incentives lengthen how long teachers stay in teaching. They take applied statistics courses not to become statisticians, but to learn how to address questions such as these or to learn how to

[1] The order of the authors has been determined by randomization. Earlier versions of parts of this paper were presented at the annual meetings of the American Educational Research Association, New Orleans, Louisiana (April, 1988) and Chicago, Illinois (April, 1991). An earlier version of the third and fourth sections was published in *The American Statistician*, 1990, **44**(3), 223-230.

read others' research on such topics. They are less interested in the algebraic ins-and-outs of mathematical statistics than they are in learning how to *use* statistics. Our job, as educators, is to provide courses that meet their needs.

Put yourself in your students' shoes. It is Friday afternoon at 2:00 pm, and you have survived another week of classes. By next Monday, you must complete your statistics homework. You open your textbook to find the following problem:

> *Here are a set of X and a set of Y scores ...*
> X: 2 2 1 1 3 4 5 5 7 6 4 3 6 6 8 9 10 9 4 4
> Y: 2 1 1 1 5 4 7 6 7 8 3 3 6 6 10 9 6 6 9 10
> *Calculate:*
> (a)   *The means, sums of squares and cross-products, standard deviations, and*
>   *the correlation of X and Y.*
> (b)   *The regression of Y on X.*
> (c)   *Regression and residual sums of squares.*
> (d)   *The F ratio for the test of significance of the regression of Y on X, ...*
>
>   [37, p. 43]

As a student of psychology, business, the health sciences, education, or for that matter, mathematics or statistics, would you be motivated by this assignment or would you be turned off by the jargon and lack of context? Would completing your homework help you understand how regression analysis can answer interesting questions about relationships among substantively important variables? Would you remember any of this two years from now, when you have to analyze your thesis data or tackle a problem on the job?

Now suppose you opened your textbook and found a different sort of problem:

> *The cost of a college education has been rising rapidly during the past decade;*
> *at many private schools in the northeast, annual tuition now exceeds $10,000.*
> *David Breneman, president of Kalamazoo College, has suggested that some*
> *colleges are charging high tuition not just to raise revenues, but to create an aura*
> *of high prestige [39]. So with tuition at an all-time high, the question arises as*
> *to what the money actually buys. Better trained faculty? Better student/faculty*
> *ratios? Better students? Table 1 presents tuition rates and selected characteris-*
> *tics of faculty and students for a random sample of 34 private colleges in the*
> *northeast. Use multiple regression analysis to examine the relationship between*
> *tuition and two potential predictors:*
>
> >   MEANSAT: *Mean total SAT score for matriculating freshmen.*
> >   PCTDOC: *Percent of faculty holding a doctorate or the highest degree*
> >   *in their field.*
>
> *Report your findings using non-technical language in a letter to the editor of our school's*
> *Alumni Gazette. Organize the data-analytic evidence supporting your conclusions into*
> *a statistical appendix to be submitted with your letter.*

These data are real! You might actually learn something interesting by completing the assignment. Which schools are overpriced? Which schools are bargains? You would begin to see links between research questions and statistical models. You would begin to learn how to translate statistical findings into readable prose. You might even begin to think about how to use regression analysis to examine the data you have been collecting all semester.

We believe that artificial data sets do little to help our students become competent data analysts. All they do is perpetuate the myth that statistics is dry and dull. After "analyzing" the data, students have not experienced the thrill of doing research, nor have they been challenged to express their results in non-technical terms. We believe such methods for teaching applied statistics should be purged from the pedagogic repertoire.

In their place, we recommend that instructors and textbook authors use real-world data, so that students can learn skills in a realistic and relevant context. In addition to being more interesting, real data sets provide a practical arena in which students can learn how to link research questions to statistical models. Real data sets illustrate how statistical methods can inform the current research debate. By using real data, we can teach not only *how* we analyze data, but also *why* we do so [4, 10, 11, 24, 35, 49, 50, 51, 53].

Use of real data sets has another advantage -- we can teach applied statistics in the way that statistics are applied. Real data allow instructors to "model" good data-analytic practice, making statistics more palatable to students and empowering them as well. In this paper, we describe how we use real data in the classroom and we identify characteristics of data sets that make them particularly good for teaching. We also identify advantages and disadvantages of this approach, and offer suggestions for overcoming the obstacles. In a separate section of this volume, we provide an annotated bibliography that lists several hundred primary and secondary data sources that teachers may use in their own courses (see also [15, 22]).

How Can We Use Real-World Data to Teach Applied Statistics?

Before high-speed computing and statistical packages were widely available, computational burden assumed instructional priority in applied statistics classes. After all, analytic "success" hinged upon the analyst's ability to execute the requisite calculations. Because computation was time-consuming and tedious, many instructors and textbook authors reduced the burden by using arithmetically simple artificial data sets. They used observations that were integers, often chosen so that summary statistics were also integers. Articles describing methods for constructing such data sets were published periodically (e.g., [6, 14, 19, 40, 41, 45]), and they were common fare in statistics textbooks (e.g., [23, 32, 59]).

Although artificial data reduced the time spent manipulating numbers, the drudgery of hand computation remained. Calculations were easier, but they still had to be performed. In the hope of keeping student attention focused on statistical concepts, not arithmetic details, many textbook authors and classroom teachers provided step-by-step formulae that decreased the computational burden. This inevitably led to an emphasis on confirmatory analyses that could be explicated as a rigid sequence of concrete steps, to be followed as one might follow a recipe. Applied statistics courses often became "cooking" classes in which students memorized the computations instead of learning the concepts.

Although use of artificial data sets and cookbook strategies stemmed from a desire to improve instructional quality, the result usually fell far short of that goal. This approach confirmed students' expectations that statistics was boring, unrelated to their substantive interests. Data analytic "cookbooks" seduced them into believing that analyses could be reduced to a set of predefined steps, conducted blindly by a robot. Concern for numerical accuracy took precedence over the acquisition of conceptual insight [11, 53]. In the end, researchers trained this way would often use methodology, rather than substance, to design their research. They would ask, for example, how to design a project so that they could use analysis of variance rather than asking how best to address a specific research question.

In today's computer age, we can, and we should, change the way we teach applied statistics [15, 18, 54]. Computers eliminate the need for simplified arithmetic. Tedious calculations can be relegated to the machine. Students need not memorize formulae whose sole purpose is computational. Exploratory and descriptive methods, once avoided because they were messy and time-consuming, can be incorporated into the students' analytic repertoire. Data analysis can become a partnership of exploration and confirmation, of induction and deduction.

Relieving students of the computational burden helps us focus their energies on the tasks that only human beings can perform: stating research questions, selecting appropriate statistical models, interpreting parameter estimates, writing up results, contemplating the implications for policy and practice. Just as computers have revolutionized the way in which we *analyze* data, so, too, should they revolutionize the way in which we *teach* how to analyze data. We should not be training researchers who see themselves solely as technicians; we should be training researchers who can use their technical skills to address real-world problems [36].

But decades of educational research tells us that teachers tend to teach as they themselves were taught [47]. Teachers use as their models the teachers who impressed them the most when they were students. But the teachers who most impressed mathematicians and statisticians when they were students may be inappropriate models when teaching non-mathematicians and non-statisticians [27]. For statisticians in training, the methods are paramount; for applied researchers in training, the *application* of the methods should reign supreme.

How were most of us -- the mathematician and statistician teachers -- taught? Most of our teachers used the *theorem*, *proof*, and *worked example* method. Class began with a statement of a theorem or formula. The instructor presented a mathematical proof, or if time was short, the proof was assigned for homework. Then came a worked arithmetic or algebraic example that illustrated a specific property of the theorem or formula. There was little, if any, contact with real data. As Moore and Roberts [35] noted, this approach remains popular today because it is "seductively easy to teach" (p. 81).

The "theorem-proof-example" strategy may have worked fine for us, but it is a poor model for applied statistics instruction. Our students care little about methodology for its own sake, nor do they care for theorems, formulae and proofs. They are bored by variables called X and Y. Their interests are substantive: they want to know what X and Y represent in the real world, why X and Y might be related, and what the implications are of any relationship that might exist between X and Y. Our job, as teachers, is to convince them that statistics will help them address these substantive questions and to persuade them that without knowing how to address these questions, they are doomed to naively accept everything they read.

One particularly effective way to achieve this goal is to have the pedagogy of applied statistics courses simulate the practice of applied research. What do applied researchers do when they conduct their own research? Beginning with a substantive question or hypothesis, they obtain relevant data, either through primary data collection or by accessing an existing database. They then posit a statistical model whose parameters represent critical phenomena, or relationships, reflected in the research question or hypothesis. After selecting relevant statistical analyses, they implement them on the computer, fitting models, estimating parameters, examining diagnostics, as appropriate. In the end, they write up their findings for the research community, describing the entire enterprise from research questions and statistical results to implications for future research and they submit papers to journals or colleagues for peer review.

This process -- or any other conceptualization of the research process -- provides a wonderful framework for teaching applied statistics. We have yet to find a statistical concept or technique that cannot be taught in this way, in the context of a real research question that can be addressed with real data. By adopting this approach, the teaching process itself becomes a model of how research ought to be conducted.

Figure 1 illustrates how we use this paradigm to teach log-linear modeling. We offer it as an example, not the "last word" in pedagogy. We use a similar strategy, with the same data, to introduce classical contingency table analysis in introductory classes; indeed, we use this paradigm with all types of statistical content and all levels of student.

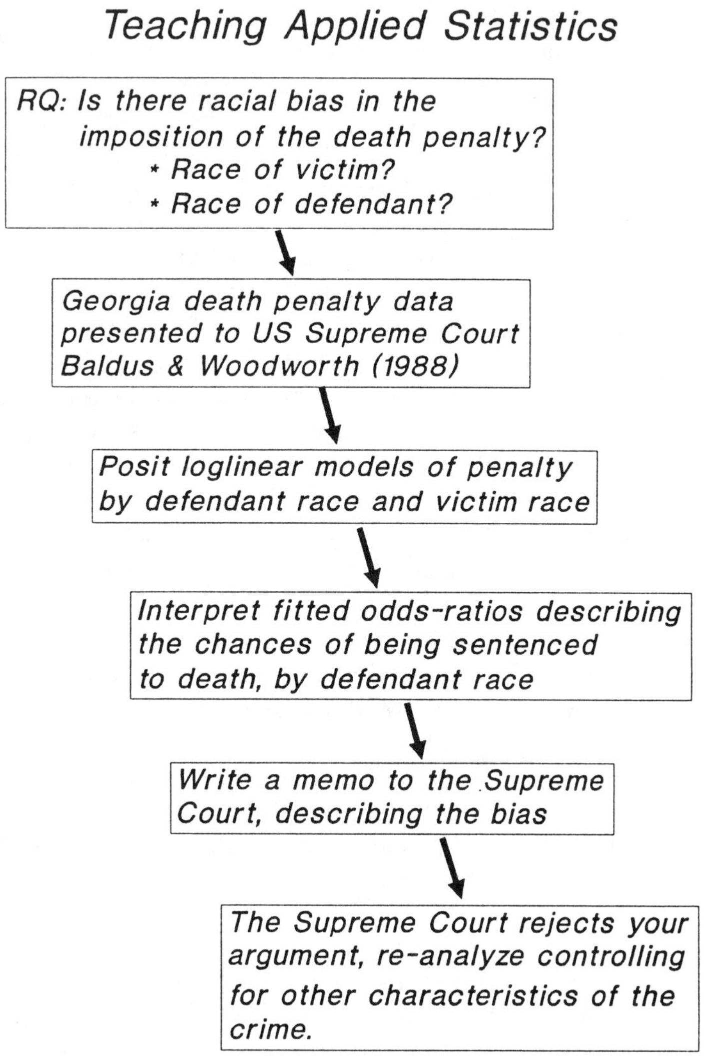

Teaching Applied Statistics

RQ: Is there racial bias in the imposition of the death penalty?
* Race of victim?
* Race of defendant?

↓

Georgia death penalty data presented to US Supreme Court Baldus & Woodworth (1988)

↓

Posit loglinear models of penalty by defendant race and victim race

↓

Interpret fitted odds-ratios describing the chances of being sentenced to death, by defendant race

↓

Write a memo to the Supreme Court, describing the bias

↓

The Supreme Court rejects your argument, re-analyze controlling for other characteristics of the crime.

Figure 1

We always begin by introducing the *substantive* topic, not the statistical one. In this case, a black man named McClesky was sentenced to death in 1978 for killing a white policeman in Georgia. Defense lawyers appealed the case all the way to the U. S. Supreme Court, arguing that McClesky received the death penalty for reasons of race. We ask the class whether there might have been racial bias in the imposition of the death penalty in Georgia during and prior to the McClesky verdict and how they might go about detecting it. This topic never fails to engage students, and it often leads to heated and memorable argument -- particularly after the data analyses are conducted.

Contextual materials spark interest further. We provide students with newspaper articles, research reports, papers from scholarly journals, and in this case, transcripts of the Supreme Court testimony and opinions. We never reveal the findings in advance; they evolve over time as they do in real data analysis. Because lawyers for McClesky argued that there was a disparity in death-penalty sentencing in Georgia based on the race of the victim and, to a lesser extent, the race of the defendant, the question the class addresses is whether the data support this claim [2]. The data we used were published by *Chance* magazine in 1988.

Prior to class, we conduct a sequence of data analyses and prepare handouts that display the output (sometimes with annotations). During class, we distribute the handouts to students and display them on an overhead projector at pedagogically appropriate moments. The particular analyses vary, of course, but they typically include: (1) *exploratory analyses*, including graphical displays and descriptive statistics; (2) more focused *confirmatory analyses* intended to answer the research question; (3) *diagnostic analyses* that reveal model deficiencies and failure of assumptions; and (4) *follow-up analyses* rectifying any problems that arise.

Class presentations simulate the research environment as closely as possible. We spend most of the time discussing why we conducted the analyses we did and interpreting the analytic results. Student participation is encouraged, and we solicit suggestions for alternative analyses. When we have been prescient, we have supplementary handouts ready; when students offer a new suggestion, we prepare handouts (or overheads) for the next class. Underlying assumptions are studied seriously, especially when we analyze data sets selected because of their severe violations. We outline appropriate sensitivity analyses and, when possible, present fix-up strategies. With a helpful media laboratory, instructors can even conduct analyses interactively in front of the class.

The Georgia death-penalty data handouts begin with exploratory analyses, including univariate summaries of each variable and graphical displays of the relationships between penalty awarded (the conceptual "outcome") and the race of the victim and of the defendant (the two "predictors"). They then contain the results of fitting a taxonomy of competing hierarchical log-linear models each potentially capable of answering the research question. We always follow up with details on selected models -- residual plots, diagnostics, parameter estimates, standard errors, etc. -- so that assumptions can be checked and, if there is one, a final fitted model can be interpreted.

Class discussion focuses on the reasons why specific models were included in the taxonomy and which model, if any, could be considered "best-fitting." Discussion and debate among students reveals the competing constraints of parsimony and fit, substance and statistics. The pedagogic process can sometimes be circuitous, but is always profitable because students learn much more than the technique of "log-linear modeling" -- they learn how to use log-linear models to do research.

In the Georgia death-penalty data, the major finding -- summarized as odds-ratios obtained from parameters estimated in the "best-fitting" model -- is that a defendant was about nine times more likely to be sentenced to death if s/he killed a white person rather than a black person. The effect was statistically significant at conventional levels and formed the statistical basis of the McClesky Appeal. The evidence of bias revealed in analysis leads to energetic discussion and permits us to ask the class to consider how they might report the findings authoritatively to a non-statistical audience (such as the Supreme Court). As an assignment, students might write a brief to the Supreme Court with statistical evidence attached. By submitting their work for grading and feedback (on the writing and on the content), students get exposure to the "peer review" process. They then read the Supreme Court decision in which Justice Powell wrote that "this study does not demonstrate a constitutionally significant risk of racial bias affecting the Georgia capital sentencing process" and in which Justice Brennan disagreed, writing about the role that statistical evidence plays in the courts [7].

This research-paradigm pedagogic approach allows the student to assume the role of researcher, exploring data that address a real research question. Class examples and homework exercises become "trial" runs in which students grapple with problems and anomalies that they will inevitably encounter in their own work. Real data sets and the research-paradigm pedagogy bring students close to an actual research experience -- warts and all.

We use this research-paradigm approach as a vehicle for teaching all types of statistical techniques. In introductory classes, for example, we acquaint students with univariate descriptive statistics by examining the distribution of governor's salaries across the fifty states (take a look at it; the variation will astound you!). We introduce simple correlation by looking at Cyril Burt's "data" on the IQs of identical twins reared apart. And we develop classical contingency table analysis through simpler views of the Georgia death penalty data.

The research-paradigm approach has at least four advantages over traditional pedagogic strategies. First, it reduces anxiety and empowers students because their initial energies focus on substance, not statistics. This allows easier access to the material, and once it is accessed, the students become engaged and motivated. Second, the realistic learning context shows students that statistical methods are relevant to their own work. The data sets themselves are memorable, and often they become the mnemonics for recalling techniques. Third, differing interpretations and misinterpretations allow the instructor to address a broad range of methodological issues dealing with research design, measurement and analysis. The student begins to realize not that one can "lie with statistics" but that it is much easier to lie *without* them. Fourth, the instructor can focus on the *why* of data analysis, not just the *how*. Extensive use of the computer frees up class time for understanding and interpretation -- it is not enough to learn how to do the computations, it is what the numbers mean that is important. Students begin to realize that the computer's output is only as the good as the instructions it has been given and that research findings can only make a difference if they can be communicated to others.

Which Real Data are the Best, Pedagogically-Speaking?

Not all real data sets are equally effective vehicles for teaching applied statistics. In this section, we discuss eight attributes that we believe enhance a data set's instructional suitability. We have found that the best data sets come in raw form, are authentic, include background information, have case-identifying information, are intrinsically interesting or relevant, are topical or controversial, offer substantive learning, and lend themselves to a variety of statistical analyses.

The importance of using raw data

Of these eight criteria, the most important is that the data be in raw form, not summarized using sufficient statistics. Rich information is lost when raw data are replaced by means and covariance matrices. Students further from data are less able to practice "real-life" data-management skills.

When raw data are available, students can adopt the data-analytic approach preferred by many practicing statisticians, be it the exploratory approach advocated by Tukey [55] or the initial data examination approach advocated by Chatfield [8]. This allows students to look for high-leverage cases, non-linearity, heteroscedasticity, and other problems that all too often arise in real data. For example, when missing values lead to data loss, instructors can introduce notions of data-imputation and sensitivity analysis [31]. Analyzing summary statistics may seduce students into believing that such problems do not exist, or if they do, that they are of little consequence.

Authenticity

A "real" data set must be authentic; it must consist of real measurements taken on a real sample of cases. Attaching life-like variable names to artificial data will not do. Consider the following exercise from Hays [23]:

> *An experimenter was interested in the possible linear relationship between the measure of finger dexterity X, and another measure representing general muscular coordination Y. A random sample of 25 persons showed the following scores: ... Compute the correlation coefficient, and test its significance. (p. 490).*

Why should a student believe that these data are real? How were dexterity and muscular coordination measured? From what population was the sample drawn? Is the sample homogeneous with respect to age, physical development being a factor that might influence general muscular coordination and therefore the relationship between coordination and finger dexterity? Why is the investigator only interested in a linear relationship? Students easily see through the artifice of "life-like" data. As a result, they may not ask questions like those raised above, because it is clear that the data were not actually "collected." Yet these are precisely the questions we want thoughtful students to raise when reviewing other peoples' research and when conducting their own.

Background information

Each data set should be accompanied by sufficient background information on the purpose and design of the research, the source of the data, measurement techniques, variable definitions and so on. This information enables the student to assume the role of researcher. If the data come from a published paper or published tabulations, students should be given access to the original document. As Cobb [11] wrote when assessing the data examples used in 16 introductory textbooks: "a data set is no longer alive if it is uprooted from its context like a pulled tooth. (What would you think of a dental school whose students only practiced drilling individual teeth that their instructor had already extracted?) To make a data set feel alive, the author must tell enough about what the numbers mean so that analysis is a search for meaning, not just an exercise in arithmetic" (pp. 331-332).

Case identifiers

When available, one particularly helpful piece of background information is the case identifier. Case identifiers allow students to use their own knowledge about the specific cases in their data analyses, thereby enriching the exercise. The data sets we use, for example, often include state, school district and school identifiers, all of which have meaning for students. Case identifiers are particularly helpful when detecting outliers and high-leverage observations. Students analyzing data on the citation frequencies of prominent psychologists, for example, can use their background knowledge to understand why Sigmund Freud might be an outlier [21].

Interest and Relevance

Many statistics texts are brimming with real data, but on topics of no interest to students in a wide range of disciplines. Snedecor and Cochran [52] present data on the calcium concentration in turnip greens (p. 239) and the average daily weight gain of swine (p. 303). Draper and Smith [17] provide data on the viscosity of filled and plasticized elastomer compounds (p. 228) and on the effects of temperature on the growth rates of ice crystals (p. 66). "Classic" data sets, such as Fisher's iris data [20] and Brownlee's stack loss data [5] also fail to inspire most students today.

Intrinsic interest is obviously in the eye of the beholder, but we find it helpful to use data from the students' disciplines. For example, the annual salary survey conducted by the American Association of University Professors (published annually in *Academe*), includes data of interest to most of the students we teach: the average salaries of faculty members by institution and academic rank. The survey of school districts conducted by Education Resources Corporation is another useful source; it provides information on teacher's and administrator's salaries by district for a nationwide stratified random sample of districts. (Full citations for these sources are given in the annotated bibliography at the end of this volume.)

Topicality and controversy

Topicality and controversy can help motivate students. The Georgia death-penalty data always rouses interest in our students even though it has nothing to do with education, the subject they are studying. We have also found that Powell and Steelman's [38] analysis of the relationship between state SAT scores and the percent of students taking the test sparks interest, particularly when we provide newspaper accounts of Department of Education "wall charts" that rank states according to these scores and critiques of state comparisons of SAT scores [42, 56, 57, 58].

Older data sets *can* motivate students, especially if the topic is controversial enough. Burt's data on the IQs of identical twins provide a wonderful pedagogic vehicle [26], when analyzed in the context of Burt's views on the nature/nurture debate and Dorfman's [16] and Kamin's [28] evidence that Burt falsified data to support the nature argument. Analyzing controversial data sets show students not just the statistical techniques, but also how the techniques can support or undermine a hypothesis.

And very old data sets can sometimes be as interesting as up-to-the-minute ones. The early volumes of journals such as *Child Development*, *Journal of Educational Psychology*, and *Journal of Genetic Psychology*, are useful sources. Although they do not always address fascinating research questions, students are interested in seeing how researchers used to analyze data. This provides an opportunity to compare findings using "modern methods" to those obtained under the older and simpler methods in the original sources.

Substantive learning

Empirical researchers analyze data because they want to learn something about the way the world works, not because they want to conduct statistical analyses for their own sake. When students analyze a real-world data set, they often "accidentally" learn something of substance, discovering just how useful statistics can be. The substantive learning involved does not have to be on a grand scale, but it must be real. One of our most popular data sets, for example, comes from a local magazine. Every few years, *Boston* magazine surveys local school districts and publishes district by district data on per-pupil expenditures, teacher salaries, student demographics and so on. *The Boston Globe* regularly publishes similar data sets. Students analyze these data and learn how their home town compares to others in the area and how district characteristics are related to each other. They gain new insight into the on-going political debate as to why some school districts are perceived to be "better" than others.

Possibility of multiple analyses

Empirical researchers often use more than one type of analysis to address their research questions; so, too, should instructors working with real-world data. When a data set is used in multiple analyses, students learn that research questions can be answered in many ways. Repeat analyses often come from showing that a key assumption may have been violated in an earlier presentation; revisiting the data facilitates the discussion of fix-ups and the notion of sensitivity analysis. This allows the teacher to stress that, in the real world, not all analyses of the same data will agree. The investigator must consider the nature of the question, the structure of the data and the suitability of the analytic method used. These are critical lessons for budding researchers.

Subsequent analyses of the same data allows a hierarchy of more complex questions to be answered -- perhaps questions suggested by earlier analyses. We might introduce simple linear regression, for example, by examining the relationship between state SAT scores and the percentage of students taking the test. Noting that the percentage-of-students-taking-the-test is nonlinearly related to the outcome introduces the notion of a polynomial regression model. Several weeks later, we might return to the data and use influence statistics to identify Alaska as an outlying and high leverage observation. The analytic sequence can be spread over many weeks, with additional findings being revealed over time.

No experience reinforces the importance of multiple analyses as much as the discovery of previously unknown findings. For example, in a course on categorical data analysis, we discuss a paper by Scarcella [43], who used classical techniques to examine the relationship between language background, language proficiency and an individual's choice of writing device (repetition, paraphrase, explanation). When students re-analyze the data using log-linear modeling, they discover a previously unnoticed effect -- that it is language proficiency, not language background, that predicts writing device.

What Are the Drawbacks to Using Real-World Data?

Using real data and the research-process paradigm to teach applied statistics is not without shortcomings. Although we have not found that this approach uses more *class* time than traditional lecture methods, it does take more *preparation* time. Data sets tend to be small and lack statistical power. Aggregate data sets and data collected on self-selected samples are often the best we can do. In-class testing is more difficult. Vagaries of using the computer can overwhelm all other considerations. Below we discuss each of these problems and we offer some remedies for overcoming them.

The workload of finding real data sets

A major motivation for using artificial data is that an instructor can readily create data sets with the requisite characteristics. Dayton [14], for example, showed how to construct data sets with suppressor variables. Searle and Firey [45] suggested that instructors could reduce student plagiarism by generating dozens of data sets and giving each student one to "analyze." Producing a variable that is normally distributed, but with an outlier or two, is a simple programming problem; identifying a real data set with the same features can take hours.

We have no doubt that using real data sets increases the amount of time required to prepare classes, homework and exams. To identify a single data set which permits illustration of a specific statistical technique, an instructor must spend hours analyzing different data sets, some of which do not support interesting findings, others of which present analytic problems out of line with the curriculum. This is especially true when developing materials for elementary courses, in which students are still learning basic skills, not how to cope with non-standard problems.

For these reasons we provide, in the annotated bibliography at the end of this volume, an extensive list of references to hundreds of data sets. Although instructors must still examine the data sets to determine which are best suited for teaching a specific concept with the particular types of students in the class, we hope that this annotated bibliography will facilitate the planning process.

Small data sets and statistical power

In introductory and intermediate courses, we prefer small data sets with sample sizes in the 35-75 range. Small data sets allow students to become intimately acquainted with each case, fostering a deeper understanding of the relationship between data and analysis. Once students have developed these skills, we introduce larger data sets.

The problem is that small data sets falsely represent the effect sizes typically found in the real world. Because null findings tend to be dull, we use data sets with large enough effect sizes that yield "statistically significant" results despite the sample size. Although we, as instructors, know that such effect sizes are rare in practice [30, Ch. 8], the students do not see much evidence of this in their class problems or their homework. Thus, when they read journal articles that report R^2 statistics with magnitudes of 20%, many students conclude that such effect sizes are small and rare -- and they are, relative to *their* in-class experience.

This problem is not unique to real data sets; most artificial data sets used by textbook authors and college instructors are also small. But real data sets *appear* representative of the larger class of statistical problems arising in the real world. Because we see little means of eliminating this problem, we specifically focus our students' attention on it by discussing the concepts of statistical power, effect size, and the distinction between statistical significance and practical significance.

Aggregate data and self-selected samples

Aggregate data or data on self-selected samples, such as the SAT state data set, are among the easiest data sets to access. While some variables in these data sets are measured at the aggregate level -- college tuition, student/faculty ratio, number of students -- most are aggregates of lower levels of data, creating a host of problems.

The question is whether the gains are worth the drawbacks, and in most instances, we believe they are. Aggregate data sets are among the most readily available, intrinsically interesting data sets we use. The observations contained in such data sets often have meaningful identifiers -- names of towns, cities, counties, school districts, states or countries -- enabling students to become more intimately acquainted with each data point. And in more advanced classes, we return to these data sets and illustrate the problems involved in analyzing aggregate summaries or data on self-selected samples.

In-class testing

It is difficult, although not impossible, to test students in-class using real data unless computer terminals or personal computers are available for each student in the classroom. We use multiple homework assignments and take-home exams in place of in-class testing. In both cases, data sets are made available to students on the computer for analysis and students must write an account of their work in the form of a journal article or research paper.

Teachers who prefer in-class examinations might conduct a series of analyses on the computer in advance of the examination and distribute computer output to the students for interpretation. Contextual material, research questions and so forth could be provided during (or in advance of) the exam. In doing so, though, note that the students are not choosing the analyses to be conducted -- they are simply interpreting your output -- and thus such an exam may not be testing all the skills you have taught during the semester.

The use and abuse of the computer

Our pedagogic approach relies heavily on the computer. This has advantages and disadvantages. Using the computer to shoulder the computational burden frees up time for thoughtful class activity. When students conduct their own analyses, however, computers can produce the reverse effect unless activities are carefully monitored. We have found that some students become so engrossed in "hacking" that their conceptual thinking suffers. Their attention and creative energies become devoted almost entirely to writing code, debugging and executing programs. The mindless, mechanical production of endless computer output becomes their sole objective [29, 44]. Instructors can avoid these problems, by crafting carefully-worded assignments and examinations that emphasize the importance of non-programming activities, including framing research questions carefully, choosing appropriate statistical models and methods of estimation, interpreting parameter estimates, and communicating findings.

Postscript

Real-world data and an empirical research paradigm can be an applied statistics instructor's strongest ally in motivating students to learn how to analyze data. Although the use of real data sets is not without problems, the strengths far outweigh the weaknesses. Perhaps the only way of discovering the advantages of authentic data sets is to try one in your classes. We think you will see the difference.

References

1. Aiken, L. S., West, S. G., Sechrest, L., & Reno, R. R. (1990). Graduate training in statistics, methodology, and measurement in Psychology. *American Psychologist*, **45**, 721-734.
2. Baldus, D., Pulaski, C., and Woodworth, G. (1983) Comparative review of death sentences: An empirical study of the Georgia experience, *Journal of Criminal Law and Criminology*, **74**, 661-753.
3. Brightman, H., & Broida, M. (1975). On problem solving, motivation and statistics. *The American Statistician*, **29**, 164-166.
4. Brogan, D. R. (1980). A program of teaching and consultation in research methods and statistics for graduate students in nursing. *The American Statistician*, **34**(1), 26-33.
5. Brownlee, K. A. (1965). *Statistical Theory and Methodology in Science and Engineering*. New York: John Wiley.
6. Carmer, S. G., & Cady, F. B. (1969). Computerized data generation for teaching statistics. *The American Statistician*, **23**, 33-35.
7. Chance (1988). Supreme Court Ruling on Death Penalty, *Chance: New Directions for Statistics and Computing*, **1**, 7-8.
8. Chatfield, C. (1985). The initial examination of data. *Journal of the Royal Statistical Society, Series A*, **148**, 214-253.
9. Chervany, N. L., Collier, R. O. Jr., Fienberg, S. E., Johnson, P. E. & Neter, J. (1977). A framework for the development of measurement instruments for evaluating the introductory statistics course. *The American Statistician*, **31**, 17-33.
10. Chottiner, S. (1991). Using real (intimate) data to teach applied statistics. *American Statistician*, **45**, 169.
11. Cobb, G. W. (1987). Introductory textbooks: A framework for evaluation. *Journal of the American Statistical Association*, **82**, 321-339.
12. Cockerill, R., & Fried, B. (1991). Increasing public awareness of statistics as a science and a preofession -- Reinforcing the message in universities. *The American Statistician*, **45**, 174-178.
13. Dallal, G. E. (1990). Statistical computing packages: Dare we abandon their teaching to others? *The American Statistician*, **44**, 265-269.
14. Dayton, C. M. (1972). A method for constructing data which illustrate a suppressor variable. *The American Statistician*, **26**, 36.

15. Dayton, C. M. (1988). Integrating analyses of data bases into statistical instruction. Paper presented at the annual meeting of the *American Educational Research Association*, New Orleans, April.

16. Dorfman, D. D. (1978). The Cyril Burt Question: New Findings. *Science*, **201**, 1177-1186.

17. Draper, N. R., & Smith, H. (1981). *Applied Regression Analysis*, 2nd edition. New York: John Wiley.

18. DuMouchel, W. H. (1979). Comment on Thisted. *The American Statistician*, **33**, 30-31.

19. Edwards, B. (1959). Constructing simple correlation problems with predetermined answers. *The American Statistician*, **12**, 25-27.

20. Fisher, R. A. (1936). The use of multiple measurements in taxonomic problems. *Annals of Eugenics*, **7**, 179-188.

21. Gordon, N. J., Nucci, L. P., West, C. K., Hoerr, W. A., Vguroglu, M., Vukosavich, P., & Tsai, S. L. (1984). Productivity and citations of educational research: Using educational psychology as the data base. *Educational Researcher*, **13**, 14-20.

22. Halperin, S. (1988). Real and contrived examples in statistics instruction. Paper presented at the annual meeting of the *American Educational Research Association*, New Orleans, April.

23. Hays, W. L. (1981). *Statistics*, 3rd edition. New York: Holt, Rinehart and Winston.

24. Herzberg, P. A. (1991). Comment on Singer & Willett. *American Statistician*, **45**(2), 169.

25. Hogg, R. V. (1972). On statistical education. *The American Statistician*, **26**, 8-11.

26. Jensen, A. R. (1974). Kinship correlations reported by Sir Cyril Burt. *Behavioral Genetics*, **4**, 1-28.

27. Joiner, B. L. (1988). Let's change how we teach statistics. *Chance: New Directions for Statistics and Computing*, **1**(1), 53-54.

28. Kamin, L. J. (1974). The Science and Politics of IQ. Potomac, MD: Erlbaum.

29. Levin, J. R. (1991). Teaching statistics conceptually: The case *against* technology. In J. P. Stevens (Chairperson), *On the teaching of applied statistics*. Symposium conducted at the annual meeting of the *American Educational Research Association*, April, Chicago.

30. Light, R. J., Singer, J. D., & Willett, J. B. (1990). *By Design: Planning Better Research in Higher Education*. Cambridge, MA: Harvard University Press.

31. Little, R. J.A., and Rubin, D.B. (1987). *Statistical Analysis with Missing Data*. New York: Wiley.

32. McCall, R. B. (1975). *Fundamental Statistics for Psychology*, 2nd edition. New York: Harcourt Brace Jovanovich.

33. Minton, P. D. (1983). The visibility of statistics as a discipline. *The American Statistician*, **37**, 284-289.

34. Minton, P. D. & Freund, R. J. (1977). Organization for the conduct of statistical activities in colleges and universities. *The American Statistician*, **31**, 113-117.

35. Moore, T. L. & Roberts, R. (1989). Statistics at liberal arts colleges. *The American Statistician*, **43**, 80-85.

36. Mosteller, F. M. (1988). Broadening the scope of statistics and statistics education. *The American Statistician*, **42**, 93-99.

37. Pedhazur, E. J. (1981). *Multiple Regression in Behavioral Research*, 2nd edition. New York: Holt, Rinehart and Winston.

38. Powell, B. & Steelman, L. C. (1984). Variations in state SAT performance: Meaningful or misleading? *Harvard Educational Review*, **54**, 389-412.

39. President says 100 private colleges follow crowd: the higher their prices, the more students apply. *The Chronicle of Higher Education*, 2 March 1988, p. A29.

40. Read, K. L. Q. (1985). ANOVA problems with simple numbers. *The American Statistician*, **39**, 107-111.

41. Read, K. L. Q., & Riley, I. S. (1983). Statistics problems with simple numbers. *The American Statistician*, **37**, 229-231.

42. Rosenbaum, P. R. & Rubin, D. B. (1985). Discussion of "On State Education Statistics": A difficulty with regression analyses of regional test score averages. *Journal of Educational Statistics*, **10**, 326-333.

43. Scarcella, R. C. (1984). How writers orient their readers in expository essays: A comparative study of native and non-native english writers. *TESOL Quarterly*, 671-688.

44. Searle, S. R. (1989). Statistical computing packages: Some words of caution. *The American Statistician*, **43**, 189-190.

45. Searle, S. R., & Firey, P. A. (1980). Computer generation of data sets for homework exercises in simple regression. *The American Statistician*, **34**, 51-54.

46. Sechrest, L. (1987). Data quality: The state of our journals. In L. S. Aiken & S. G. West (Chairpersons), *Adequacy of methodological and quantitative training: Perspectives of the disciplines*. Symposium conducted at the annual conference of the *American Psychological Association*, New York, April.

47. Shulman, L. S. (1987). Knowledge and teaching: Foundations of the new reform, *Harvard Educational Review*, **57**, 1-22.

48. Singer, J. D., & Willett, J. B. (1988). Opening up the black box of recipe statistics: Putting the data back into data analysis. Paper presented at the annual meeting of the *American Educational Research Association*, New Orleans, April.

49. Singer, J. D., & Willett, J. B. (1990). Improving the teaching of applied statistics: Putting the data back into data analysis. *The American Statistician*, **44**(3), 223-230.

50. Singer, J. D., & Willett, J. B. (1991a). Providing a statistical model: Teaching applied statistics the way that "statistics" is applied. In J. P. Stevens (Chairperson), *On the teaching of applied statistics*. Symposium conducted at the annual meeting of the *American Educational Research Association*, April, Chicago.

51. Singer, J. D., & Willett, J. B. (1991b). Reply to Herzberg and Chottiner. *The American Statistician*, **45**(2), 170.

52. Snedecor, G. W., & Cochran, W. G. (1980). *Statistical Methods*, 6th edition. Ames, Iowa: Iowa State Press.

53. Stevens, J. P. (1990). On the teaching of applied statistics and applied statistics textbooks. Paper presented at the annual meeting of the *American Educational Research Association*, Boston, April.

54. Thisted, R. A. (1979). Teaching statistical computing using computer packages. *The American Statistician*, **33**(1), 27-30.

55. Tukey, J. W. (1977). *Exploratory Data Analysis*. Reading, MA: Addison-Wesley.

56. Wainer, H. (1986a). Five pitfalls encountered while trying to compare states on their SAT scores. *Journal of Educational Measurement*, **23**, 69-81.

57. Wainer, H. ed. (1986b). *Drawing Inferences from Self-Selected Samples*. New York: Springer-Verlag.

58. Wainer, H., Holland, P. W., Swinton, S., & Wang, M. H. (1985). On "State Education Statistics". *Journal of Educational Statistics*, **10**, 293-325.
59. Winer, B. J. (1971). *Statistical Principles in Experimental Design*, 2nd edition. New York: McGraw Hill.

John B. Willett and Judith D. Singer are Associate Professors at the Harvard University Graduate School of Education specializing in quantitative methods. Collaborators since 1985, they have written and presented dozens of talks, workshops, and papers on the application of statistical methods in education and the social sciences. Together with other colleagues, they have written two books, *By Design?* and *Who Will Teach?*, both published by Harvard University Press. Their current research interests center on applications of survival analysis in the social sciences. They teach a five-semester graduate sequence in quantitative methods using the research process approach described in this paper. They have recently received the Raymond B. Cattell Award and the Palmer O. Johnson Award from the American Educational Research Association and an NSF Visiting Fellowship from the ASA.

Low-tech Ideas for Teaching Statistics

Robin H. Lock   and   **Thomas L. Moore**
St. Lawrence University   Grinnell College

"The job is not to cover the material but to uncover it."
George Miller

That pithy quotation is a good place to begin an article about teaching statistics. It has a dual meaning. First is the obvious meaning so strikingly presented: the age-old tension between covering the syllabus and imparting real understanding to a preponderance of our students. Most of us, at one time or another, in almost any course have felt this tyranny of the syllabus. But here, within this context of ideas for mathematicians teaching statistics, we can interpret the quotation in another way.

We assume that the reader of this article is a mathematician who, by choice or duress, is also a teacher of statistics. A mathematician coming to the teaching of statistics needs to uncover the fundamental differences between mathematics and statistics. The tendency to treat statistics as "just another mathematics course" fails to recognize these differences and too often results in the common complaint that statistics is dull and boring. Many authors have written about the differences between statistics and mathematics and about the role of statistics within a department of mathematics. See, for example, [7], [8], [12], and [13].

One of the most basic distinctions is the vital role of data in statistics. While the framework in mathematics is often theorem/proof/example, the emphasis in statistics is on the dialog between models and data. In fact, as George Box [1] explains, this dialog between models and data is not only fundamental to scientific advance, but also to the advance of statistical theory. In large part, to teach statistics is to expose students to the essential role of applications and data to statistics. It is our conviction that a first course in statistics (for either mathematics majors or non-majors) should include many real examples that illustrate the full process of statistical problem solving: the initial question, the design of the study, the production or collection of the data, the analysis of the data, the communication of the results, and the evaluation of what questions or work constitute the next step. An acronym (due to Walter Shewhart) that summarizes this process more succinctly is PDCA: Plan, Do, Check, Act. With this in mind, examples become much more than merely vehicles for illustrating the latest statistical recipe: "Apply the two-sample *t*-test to these data to see if...." Rather, examples become vehicles for reinforcing this PDCA paradigm.

If this sounds more time-consuming, you are right. A first course in applied statistics should no longer be merely a probability course with a few hypothesis tests tacked on at the end. Furthermore, we are convinced that properly teaching this paradigm, at the expense of a headlong rush through a long list of statistical recipes, is a more effective way to introduce students to statistical thinking. The concrete suggestions we offer here are the beginning of a collection of ideas we have been gathering. Some we can attribute accurately to their originators, but not all. This list is not intended to be exhaustive. Our goal is to suggest some ways to enliven the statistics class, encourage student interest, and impart understanding of statistical concepts. We hope that our suggestions will be a catalyst for your own invention of low-tech ideas. Although we strongly support the use of computers in teaching statistics, we have deliberately segregated ourselves from computers in this discussion of low-tech ideas, since that subject is adequately covered elsewhere in this volume.

Low-Tech Simulations of Statistical Concepts

Computers are great simulators of randomness, but can easily provide overkill and intimidation to the beginning student. Sometimes a simple class activity can make the point more forcefully and convincingly.

When introducing the basic concept of probability as long-range relative frequency, it is a good idea to have the students play with real coins, dice, and cards. Besides simple problems, you might also include some real simulations of non-trivial problems. For example, the famous "game show problem" of Marilyn vos Savant [17] has quickly become a favorite with students. Recall the problem. A certain game show asks the contestant which of 3 doors hides a fabulous prize; the other 2 doors hide goats. The contestant must guess the correct door to get the prize. Suppose the contestant guesses door number 1. Then, before revealing the contents of door 1, the game show host reveals either door 2 or door 3 to be hiding a goat and gives the contestant the opportunity to change his or her guess. The question is whether it is advantageous for the contestant to switch to a new door. This problem has generated much discussion in the press and should be controversial with your students as well.

It is a problem that can easily be simulated in class. Have students work in groups and generate several instances of using both a switch and a no-switch strategy. By pooling the class's results, the sample should be large enough to convince the students that the switch strategy is better. Jeff Witmer reports success with this classroom activity. [20]

Another non-intuitive example is provided by David Moore [12, p. 344]. Spin a normal penny several times to estimate the probability that the penny will land "heads" after coming to rest. The penny should be spun by flicking it with the index finger of one hand while holding it on edge with the other index finger. Most pennies have a definite tendency to land "tails" more often than "heads" in this experiment. In this example, the a priori analysis of probability is impossible and so the exercise in long-range relative frequency is better motivated. Laurie Snell [18] recently showed us an apparent variant of this example. Stand 10 pennies on edge on a hard table. (This is not an example for the shaky of hand.) Jar the table so that the pennies fall. Note that almost all of the pennies land "heads". (Laurie likes to bet on 8 or more.)

A favorite of ours, suggested by Hahn [4], effectively illustrates the central limit effect. Have each of your n students write down the day of the month (1, 2, ..., 31) of his or her birthdate and that of his or her 2 closest relatives. At the blackboard, construct two histograms (or dotplots): first a histogram of the $3n$ individual birthdates and second a histogram of the n family averages of birthdates. The first histogram will exhibit the underlying uniform distribution on the numbers 1,2, ..., 31 while the second histogram will exhibit the clearly normal-looking distribution of the sample averages.

Dice are also useful for illustrating random phenomena. To obtain a larger sample size demonstration of the central limit effect, one might use a collection of perhaps ten dice (or repeated throws with a single die). Each student can roll all the dice, recording the frequencies of individual numbers and calculating the mean. The frequencies can be pooled to approximate the expected uniform distribution for dice rolls, while a histogram of the means exhibits the decreased variability and beginnings of a normal distribution. Further calculation can give some empirical evidence to support the formula for calculating the standard error of the mean. Having some familiarity with dice and generating the data themselves can make the demonstration more convincing for students who might not be so trusting of "random" phenomena being generated and displayed by a computer.

You can also challenge their faith in dice with the following experiment. Have students roll two distinguishable six-sided dice, say one red die and one white, a large number of times. After each roll, the die with the larger value showing is declared the winner (ignoring ties). Assuming the dice are fair, the hypothesis that the proportion of red "wins" is 50% should rarely be rejected. Imagine the students' surprise when the data convincingly show a preference in favor of the white die. What could be the reason? Students offer many conjectures but, despite having rolled the dice themselves, they never notice the three extra spots which were added to the white die, turning its "two" into a "five." Fortunately, the statistics are rarely fooled -- provided you use a large enough sample size (over 300 rolls) to minimize the chances of an embarrassing Type II error. Here is a simplistic but effective illustration of the regression effect told to us by Colin Sacks [16].

Suppose there are 25 students in class. Number index cards 1, 2, ..., 25 and hand out one per student. Call this their "anxiety score" (25 = highest anxiety, etc.). Collect cards and shuffle. Proclaim that you have now given the students a program for "anxiety reduction" (you might embellish the story by saying that, when a lecture is interrupted for a class activity, anxiety goes down.) Redistribute the cards. You now note that most high anxieties went down. Well, yes, most low anxieties went up, but perhaps that is an acceptable "price to pay" for lowering those at the top end. Although this only shows the regression effect in the overly simple situation of zero correlation, it seems to help students begin to see what is going on. In fact, with slight positive correlation, the effect will be almost as strong; as the correlation becomes closer to 1, this effect will become less pronounced, but will not go away entirely.

Another card-based demonstration was suggested to us by Florence and Sheldon Gordon [3]. Do some students have a degree of extra-sensory perception (ESP)? Create a deck of 25 cards consisting of five cards each for five different symbols (or perhaps use the suits of standard playing cards). One student, the "transmitter", goes through the deck, concentrating on each card in succession, while the rest of the students serve as "receivers" trying to guess each card's symbol from among the known set. When the answers are revealed, students can figure their own scores, calculate the probability of getting that many correct (using a binomial table or normal approximation), or test whether their proportion correct is significantly different from what one would expect under a "no ESP" hypothesis. A class of 30 students will often contain one or two members who get an unusually high score. Have we discovered those who truly have some ESP or merely fallen victim to the multiplicity of the experiment?

Ann Watkins [19] uses M&M's to motivate statistical lessons other than by giving them to students with high exam scores. Students love to use M&M's to construct sampling distributions. M&M's are also good for the lesson that demonstrates that we expect 95% of confidence intervals to contain the population percentage. Students can each be given a small bag of M&M's (the smallest size is available at Halloween) and construct their own confidence intervals for, say, the percentage of red.

The theoretical percentages of each color for M&M's (as given by the M&M's company) are:

	Red	Yellow	Green	Orange	Tan	Brown
Plain	20	20	10	10	10	30
Peanut	20	20	20	10		30

The M&M's company says,

> *"We can assure you that the colors are blended mechanically according to the formula. Our present means of color dispensing should place a fairly uniform blend of colors in every package; however, occasionally an unusual assortment of colors may occur."*

(These percentages do change from time-to-time.)

Another food-centered activity illustrates a method for estimating the size of a population of (for example) fish in a lake. Biologists often use the so-called capture-recapture method which works as follows. A sample of n fish is initially captured and tagged in some way. The sample is released back into the lake and allowed time to become mixed back into the population. Then a second sample of m fish is re-captured from the lake. Suppose t fish in the second sample bear tags. If we let N denote the unknown population size, then a reasonable estimate of N (in fact, a maximum likelihood estimate, assuming a hypergeometric model) is that $N = nm/t$, which is the solution to the proportionality $n/N = t/m$.

Jeff Witmer [20] uses the little goldfish crackers made by Pepperidge Farm for the fish and a large bowl for the lake. Since the crackers come in different flavors and hence colors, one can replace the first sample of crackers by a like number of different colored crackers to represent the tagged fish before replacing them into the bowl. As in the M&M's example, the students can end the exercise with a well-deserved snack.

This capture-recapture method is also at the heart of proposals to adjust for undercount in the U.S. Census in which capture is the original census and recapture is the post-enumerative survey to estimate the number of people missed by the census. Since census adjustment is controversial, this aspect of capture-recapture should be an area for lively class discussion.

In-Class Data Collection

Collecting data from students in the classroom gives them a microcosmic overview of statistics and they will enjoy using the data. A particularly interesting article on using in-class data collection is given by Loyer [10]. He suggests using an opening day questionnaire to provide data for examples throughout the semester. His idea of having students generate random 3 digit sequences is particularly entertaining and helps bring home the point that true randomness is not easily emulated by the human brain. Ask students simple questions that they can relate to, like whether they prefer Wendy's, McDonald's, or Burger King.

Mark Johnson [5] of Georgia Tech likes to ask his students two different versions of the same question. For example, group A is asked "I expect this course to be a) excellent, b) good, c) fair, d) poor" while group B is asked "I expect this course to be a) totally awesome, b) neato, c) so-so, d) yuk." Ask how results would compare if math/science majors got version A and others got version B. Johnson also likes to ask the students to guess the instructor's age and also give an interval that they would bet even money on the age being within.

Of course, if you are going to collect the data, you had better analyze it soon thereafter or else the students are getting a message we don't want to send them. So, for example, we look at their 3 digit sequences and note the dearth of zeros or the shortage of sequences with repeat digits (Loyer) or we observe the proportion of intervals actually containing the instructor's age (Johnson). Doing "live" calculations is one area where a computer, or at least a calculator, is useful. One might also collect the data one day and present the results at the next class.

Another means of collecting data in class is to run small experiments. These get the students to be active, reinforce statistical concepts, and often have interesting outcomes. For example, the Minitab Handbook [15] gives the case of an in-class calisthenic experiment in which pulse rates are measured before and after a one-minute interval in which half the class exerts themselves and half stays at rest. The groups are determined by a coin flip and interestingly, in the Minitab data, the number who end up at rest is surprisingly large.

Sample size may be an impediment to some in-class data collection schemes. For instructors who are "blessed" with a large lecture class, sufficiently large samples are no problem, although the mechanics of collecting the data can be tricky. For smaller classes, one might pool results over several semesters, allowing each class to contribute new data. We have done this for several years with the dice central limit demonstration mentioned earlier to provide enough data to more convincingly illustrate the tendency towards normality of the distribution of the means. Another example is a data set we collect as an initial example for a chi-square test. Students are hypothetically offered a choice among four pitchers of beer: regular, lite, dark, or root beer. Are the proportions chosen about the same or is there a preference for or against any of the beer types? It might be difficult to generate a sufficiently large sample to test this in a single class, but relatively easy to compile the results from several classes.

The work on biases in human judgment and decision making by Kahneman and Tversky [6] provides very interesting classroom examples. For instance, we have commonly provided the class two versions of a question like:

Version A: Imagine that you have decided to see a play where admission is $20 per ticket. As you enter the theater, you discover that you have lost a $20 bill. Would you still pay $20 for a ticket?

Version B: Imagine that you have decided to see a play where admission is $20 per ticket. As you enter the theater, you discover that you have lost the ticket. Would you pay $20 for another ticket?

Kahneman and Tversky found that, for version A, 88% answer "yes" while for version B, only 46% answer "yes." Class results are usually similar. The chance to reinforce the notion of random allocation is obvious. By using random allocation of the two forms of the question, the students see concretely that there is little chance that confounding factors are causing the differences in responses -- it must be the questions themselves. A bonus is to colorfully show the students the subtle importance of wording in asking a question (on a survey).

Local campus issues can supply interesting opportunities for data collection. For example, Grinnell College recently changed the standard annual teaching load for faculty from 6 courses to 5. Two years prior to this, a student project had examined the general issue of class size. Their goals were to compare class sizes between academic divisions and between the course levels. In particular, for the semester they studied, the mean class size was 17 students. Last fall we asked each of the students in the class to compute the average enrollment for the classes they were taking. The campus was experiencing contentious student claims that the reduced teaching loads had led to increased class size. Here was a chance to find out. We plotted the data on the board and indeed the average was over 24. Were the student claims verified? Students then came up with various threats to validity: (1) confounding factors (other things have changed between the original study and our study); (2) clearly the sample (our class) may be biased; (3) the "before" and "after" averages are based upon different sampling units (the first average is an average over courses, the second is over students; clearly the latter will be inflated).

Another example of in-class data collection involves pulse rates and comes from P. F. Lock [9]. Sometime during a lecture, take a minute out of class to have students determine their pulse rates. At this point, you might compare the rates of those who drink coffee to those who don't or the exercisers to the non-exercisers. Then collect another sample, say right before a quiz. You can use these data to show the difference between paired and independent samples and the effect of pairing in reducing variability.

Our final idea for an in-class survey is to illustrate a randomized response technique for asking a sensitive question. (Fox and Tracy [2] describe this technique in much detail.) Most students are quite intrigued by how one might obtain an accurate estimate on some issue such as drug use or sexual habits for which the subjects are likely to be less than truthful. For example, you might be interested in the proportion of students who have used a fake I.D. card. Have each student flip two coins. All those with "heads" on the first flip should answer the question truthfully; the others should respond "yes" or "no" depending on the result of the second flip. A quick calculation to estimate the desired proportion can yield some interesting results.

Items from the Media

Keep your eye out for interesting data in newspapers, magazines, and newscasts. Polls on issues of current interest, reports on medical studies, and examples of good or bad graphics are relatively easy to find. Local media or a campus newspaper can often provide examples which are particularly relevant to students from your area. As with projects, you might have students find their own examples and obtain a resource to draw on for future classes. Also look for discussable statistical quotes like "Two winters out of three the snowpack is below average." (from Ann Watkins [19]). Or Ralph Nader was once quoted as being incensed that "... only 75% of all nuclear reactors are operating at above average levels of safety."[3] We once found a local headline which claimed that a change in the legal drinking age resulted in a 33% decrease in local traffic fatalities. Upon reading the article, it turns out that indeed the number of fatalities had dropped from three in the six months before the change to only two in the same period afterwards. Is the headline justified?

Projects

We have our students collect their own data. They are asked to find a question, design a study, and carry it through: PDCA. The idea did not originate with us; it has been used successfully at many levels. (See, for example, Harry Roberts' article in this volume.) Often our students say the projects were the highlight of the course and helped them assimilate statistical concepts better than any other activity. Students can get very excited about data that is close to their lives.

At first glance, the impediments to letting students choose their own projects seem insurmountable -- the burden of grading all of those papers, evaluating vastly different projects, and dealing with the individual problems that students encounter. One easy way to decrease the work load is to have the students work in teams. This has the added advantage that the students often can work through difficult parts of the project on their own and your consulting time is reduced. Not only does a project teach quite vividly the statistical paradigm, but your teaching life is enlivened as well. Rather than grading 30 nearly identical homework assignments of some routine sort, you have the entertaining task of reading 15 (or fewer) unique student papers on subjects that have, for the most part, gotten the students involved in something they care about. Invariably, each semester, we get many solid end-products and a small number of real gems.

Here's one example: The use of library books. Several years ago, a student asked the simple question: How often are books checked out of the college's library? She worked part time in the library, so qualified as an expert of sorts. She collected a sample of roughly 50 books and measured the time in years since the book had last been checked out (her operational definition of "how often"). What could be simpler?

Yet this project illustrates that even simple projects can teach statistical lessons. (In fact, a teacher's role in assigning projects is to help the student identify a project that is at the right level -- with simplicity usually being a virtue.) The student chose to obtain a simple random sample of books, using the card catalog (this being before the days of on-line cataloging) to define the frame which brought her up against questions like multiple cards per book. After obtaining the sample, she encountered other problems: some books were neither on the shelf nor traceable (lost books), some books had never been checked out, etc. And so even before the analysis, she had encountered the types of problems that befall almost every statistical investigation, but which are traditionally ignored in teaching introductory statistics.

Students are often interested in their fellow students. Comparisons based upon gender are popular. Do female students tend to eat fewer meals in the dining hall, spend more time in the library, have higher long-distance phone bills, wash their clothes more often, wake up earlier, or take longer showers than male students? Other comparisons can be made based upon variables like year in school, major, or location of campus abode (e.g., on-campus vs. off-campus).

On the other hand, students need not restrict their attention to studies of their classmates as we have received projects on the randomness of bridge hands (computer dealt vs. hand dealt), a comparison of summer and winter trading volumes for stock in a well-known beer company, and the strategy for playing "The Price is Right".

In all cases, it is important to have the students submit to you and have approved a written proposal of their project so that it may be screened for both feasability and appropriateness. We also recommend the project be assigned early in the semester to allow ample time for its completion and so that topics covered throughout the semester have the motivational backdrop of the student's project. You may contact the authors of this paper for copies of assignment sheets for projects.

Perhaps the best advantage in assigning student projects is that you, as a teacher, are supplied with an ever-increasing store of real examples for future generations of introductory statistics courses. Consider again the project on the check-out intervals of library books. Like the "gift that keeps on giving", this project has become a "project that keeps on teaching." We have since used this project many times in the introductory course as a lively classroom example. The typical approach goes like this. The instructor brings to class a large armful of books and asks the question posed by our student: how often are books checked out of the college's library? After agreeing on a measure of "how often" and agreeing that we need to get a sample of books from the library, the instructor exposes his sample. It takes the students little time to expose the sample as a fraud. (For example, the sample is suspiciously heavy in math/stat books, and, of course, all of these books have been checked out.)

The discussion progresses from there to having the students in the class design the sampling plan and, after doing this, we describe the aforementioned project. Despite a rather bland subject matter, we are continually amazed at the level of enthusiasm the students bring to the discussion. Happily, the conclusion to the project is also interesting to the students. The distribution of times since last being checked out is a right-skewed distribution with a median of 6 years.

The One Minute Drill

Our final low-tech idea is not about teaching statistics per se, but just about teaching. Mosteller [14] describes the following simple but effective way of getting frequent student feedback about the course. The final minute of each class period is reserved for the students to evaluate that day's lesson. They write short answers to 3 questions: (1) What was the most important point of the day? (2) What was the muddiest point in the lecture? And (3) What would you like to learn more about? The students hand in their answers anonymously. The teacher summarizes results before preparing the next class and deals with the responses appropriately. We have used this in our teaching recently and feel that it is a tremendous quality control device for the course. Consult Mosteller [14] for details.

Conclusion

We hope that these examples illustrate some simple ways to bring real data into the classroom and let them teach us what they can about statistics. The goal is to provide our students with more interesting and stimulating experiences rather than simply more fodder for our latest statistical recipe. Some lecture time will necessarily give way to class discussion and experimentation, but as James Landwehr [8] has said: "Perhaps such use of class time is not appropriate for teaching pure mathematics, and thus a statistician teaching in this way within a math department might face some problems. However, considering issues about questions behind the data, the data collection process, and appropriate ways to analyze the data **is** statistics; ... we should not feel apologetic about taking class time to deal with these issues."

Bibliography

1. Box, G. E. P. "The Importance of Practice in the Development of Statistics", *Technometrics*, **26** (1984), 1-8.

2. Fox, James Alan and Paul E. Tracy, *Randomized Response: A Method for Sensitive Surveys*, 1986, SAGE Publications, Beverly Hills.

3. Gordon, Florence and Sheldon Gordon, Personal correspondence, New York Institute of Technology, Suffolk Community College.

4. Hahn, G.J. "Improving Our Most Important Product," *Proceedings of the Second Conference on the Teaching of Statistics* (SUNY at Oneonta, 1987), 7-13.

5. Johnson, Mark E., Personal correspondence, Georgia Institute of Technology.

6. Kahneman, Daniel and Amos Tversky, "The Framing of Decisions and the Psychology of Choice", *Science*, **211** (1981), 453-458.

7. Kempthorne, O., "The Teaching of Statistics: Content Versus Form", *The American Statistician*, **34** (1980), 17-21.

8. Landwehr, J. M., "Discussion on the Role of Statistics at 4-year Undergraduate Institutions", *Proceedings of the Section on Statistical Education of the American Statistical Association*, (1990), to appear.

9. Lock, P. F., "In-class Data Collection Experiments to Use in Teaching Statistics", *Proceedings of the Second Conference on the Teaching of Statistics* (SUNY at Oneonta, 1987), 83-86.

10. Loyer, M. W., "Using Classroom Data to Illustrate Statistical Concepts", *Proceedings of the Section on Statistical Education of the American Statistical Association.* (1984), 57-62.

11. Moore, D. S., *Statistics: Concepts and Controversies*, 3rd edition, 1991, W.H. Freeman and Company, New York.

12. Moore, D. S., "Should Mathematicians Teach Statistics?" (with discussion), *The College Mathematics Journal*, **19** (1988), 3-35.

13. Moore, T. L. and J. Witmer, "Statistics Within Departments of Mathematics at Liberal Arts Colleges", *The American Mathematical Monthly*, **98** (1991), 431-436.

14. Mosteller, F., "Broadening the Scope of Statistics and Statistical Education", *The American Statistician*, **42** (1988), 93-99.

15. Ryan, B. F, B. L. Joiner, and T. A. Ryan, *Minitab Handbook*, 2nd ed., 1985, PWS-Kent, Boston.

16. Sacks, Colin, Personal correspondence, Grinnell College.

17. vos Savant, Marilyn, "Ask Marilyn", *Parade Magazine*, September 9, 1990, 16.

18. Snell, J. Laurie, Personal conversation, Dartmouth College, Hanover, NH.

19. Watkins, Ann E., Personal correspondence, California State University, Northridge.

20. Witmer, Jeff, Personal correspondence, Oberlin College, Oberlin, OH.

Robin Lock is an associate professor of mathematics at St. Lawrence University. His research interests are in applied statistics, particularly methods for approximating permutation tests, and in statistical education. Despite the "low-tech" orientation of this paper, he is especially interested in ways to use computers in teaching statistics. He is also an avid collector of data sets and always welcomes new contributions.

Thomas Moore teaches statistics in the mathematics department at Grinnell College. He is co-founder of the Statistics in the Liberal Arts Workshop (SLAW) and has been active in the statistics education committees of MAA and ASA. He has also spent a year's leave as an applied statistician at the Eastman Kodak Company.

Student-Conducted Projects in Introductory Statistics Courses

Harry V. Roberts
Graduate School of Business
University of Chicago

Use of Projects in Teaching Statistics

I have done most of my statistical teaching in an MBA program at a major business school. However, I have frequently taught an introductory course for which the objectives are similar to those of undergraduate statistics courses and students are similar in quantitative preparation.

Introductory statistics courses have a reputation, often deserved, for being dull. In part, this reputation comes from the common dislike of anything smacking of mathematics, but this is only part of the problem. For statistics is widely taught with major emphasis on theory and only token attention to applications to real problems. Hence the fascination of statistical thinking -- which lies in its usefulness in thinking about and acting on real problems -- does not get communicated to most students.

Many statisticians are drawn to statistics, in part at least, because statistical theory is fascinating to them. Most students are drawn to statistics courses because they are required to take them. It is not surprising that students do not automatically share the instructor's interest. But instructors and students can share a common interest: the application of statistics to important problems.

The distinguished statistician, George Box, sums it up in one sentence: "*In my view statistics has no reason for existence except as the catalyst for investigation and discovery.*" Box likens the preoccupation with theory in statistical teaching to the teaching of swimming by theoretical training alone. Unfortunately, many statistics teachers don't even let students get their feet wet with nontrivial applications, let alone require that they actually try to swim. This is understandable: in many academic subjects besides statistics, students aren't required to swim. Moreover, statistical applications are not easy for instructors to direct, manage, and provide feedback and evaluation, especially when classes are large.

In my early years of teaching the introductory course, I began to assign projects in which students worked on real statistical problems of their own choosing, using the tools developed in the course as these proved appropriate. The results were mixed. Students liked and learned from the projects, but there were three major obstacles:

1. Computation was tedious and very time consuming for students.
2. Supervision of projects was very time consuming for me.
3. The statistical coverage of the courses I taught gave little guidance for the analysis of time series data.

Over many years, I have gradually learned how to substantially overcome these obstacles. Improvements in statistical computing have reduced and, by now, almost eliminated the computational tedium. Gradually I learned better to manage projects so that effective supervision became less time consuming. Finally, I introduced time series concepts into the introductory course, thus permitting the students to tackle a much wider range of projects.

I now use projects not only in the introductory course but also in more advanced courses. Moreover, for about the last ten years, the project has been the only basis for grading; I have eliminated all quizzes and examinations and made exercises optional. (Frequent progress reports on the project are strongly encouraged.) Thus, I evaluate students by what they are able to *do* using statistics. The result, I believe, is that both the students and I learn more, and that we both have fun in doing it. (For me, each project is different and presents its individual interest and challenge; contrast this with reading examination papers!)

Projects also help to meet the needs of students coming to statistics courses from different academic disciplines; the closer the application to the student's own field of study, the more easily the statistical ideas can be grasped. (This is a conclusion that I have come to reluctantly, since, for me, the universality of statistical ideas and their wide range of applicability has always been a key to the charm of statistics.)

The kind of narrow memorization required for "programmed problem solving" -- examinations -- discourages the frame of mind needed for the "unprogrammed problem solving" entailed in the effective application of statistics. Moreover, the project approach requires cooperation between teacher and student, whereas examinations have a strong adversarial component. It is a pleasure to have gone through the last decade without once hearing the question, "Are we responsible for this point on the examination?"

In statistics teaching, projects are the best way I know to achieve learning through doing. In their book, *American Business: A Two Minute Warning*, C. Jackson Grayson and Carla O'Dell [1] make the following comments about "Training and Continuous Learning" in the context of business competitiveness. Their principles apply to universities as well as to companies.

It is not only *more* training that is needed, but also a *different kind* of training.

▸ The principal ingredient missing from most training programs is relevancy.
▸ Some fundamental principles of training and learning will have to be recaptured:
  1. People learn best when directly involved in real problems and issues.
  2. One-shot classroom training doesn't change behavior. Practice and feedback work far better.
  3. The best learning is active, not passive. In an organization, active problem-solving works extremely well in cross-functional groups and in unfamiliar situations.
  4. Learning by example -- seeing it done -- is more effective than hearing about it.

Student performance on projects provides excellent feedback to the teacher on statistical points that are not understood by students; often this feedback reveals flaws in teaching. When students stop making a particular type of mistake on their projects, then -- and only then -- can I be reassured that my teaching of that point is not seriously flawed.

I have also learned from the projects that many students have somehow acquired basic misunderstandings about statistics before they come to introductory courses. These misunderstandings seem to be a part of contemporary intellectual culture. For example, many students believe that approximate normality of distribution of data *implies* randomness of time sequence. As a result, they often waste time on histograms and normal probability plots of data that are severely nonrandom in sequence. They also think that summary numerical measures (for example, levels of significance or *R*-square statistics) are more important than inspection of data plots in making sense out of data. Projects provide the opportunity to set things straight.

But the most important benefit is that projects provide the best way I know to make students aware of the relevance of statistics to the real world. Without that awareness, any statistics course must be judged a failure, however well, say, the students learn to construct and interpret a 95 percent confidence interval.

From the instructor's perspective, projects are a continuing source of interest and challenge. They also provide a rich source of examples for use in lecturing and discussion.

What Types of Projects Are Feasible?

The following classification is helpful to an understanding of what kinds of things students can do with projects.

Sample Surveys

Surveys are potentially very useful, but implementation of a questionnairebased survey not only requires a team but also a good deal of instruction in survey procedures, including essentials of probability sampling, design of questionnaires, use of follow-ups, editing and coding of responses, etc., that is not included in most statistics courses. Writing a good questionnaire, for example, is not an easy task. If I were teaching a sampling course, in which there would be time to develop the necessary background in survey procedures, a survey would be an obligatory project.

In most introductory courses, surveys are less feasible, because the instruction in survey procedures could preempt most of the time. However, if the surveys are focused on groups for which lists are available (e.g., students or faculty at a college) and on simple questions (e.g., evaluation of courses or administrative procedures at a college), useful projects can readily be formulated. Simple random sampling or systematic sampling from lists can be used. Followup of nonrespondents should be stressed.

Predictive Time Series Projects

Here I am writing from my business school experience. Because understanding of processes that generate data through time is vital for management decision and for quality improvement, and because I approach statistics from a time series perspective, I suggest that students consider working on time series data from an important process, such as company monthly sales for the last five years. The general mission is to provide a statistical *explanation* of the past and/or *prediction* of the future based on the available data. Usually some important business lessons are learned as a byproduct.

It is easy to find time series data on topics of interest to non-business students: for example, weather data (temperature, rainfall), demographic data (population, mortality), accidents, or sports performances. Almanacs or the *Statistical Abstract of the United States* are excellent sources.

Process Improvement Projects

A slightly different type of project, also based typically on time series data, is aimed at process *improvement*, in the spirit of modern quality and productivity improvement [2]. In a statistics course for business students, the obvious processes to be improved are those within business organizations. But the idea of process improvement applies also to non-business organizations, such as universities, and also to personal processes, such as typing, studying, control of weight or hypertension, or free throw shooting. Personal improvement projects require close guidance in the design stage to get the student on the right track for data collection. The students must be reminded that they are trying to improve something, not just to collect data to be run through a computer.

Process improvement projects often entail statistical comparison of means. Comparisons of means also arise in contexts where a full improvement study is not feasible but access to existing data or collection of new data is possible. Students may find it interesting, for example, to compare SAT scores in a high school for males versus females; gain in weight-lifting capability at a health club for smokers versus non-smokers; tips to a waitress in a marina restaurant left by sailboaters versus motorboaters; student performance in a statistics course by those with strong mathematical backgrounds versus the others; and so forth. And the same idea can be extended to studies using regression analysis if regression is included in the scope of the course. Sports data are often interesting and useful from this perspective: for example, how well do expert forecasts predict the actual point spreads in professional football games?

There are other types of projects. For example, simulation studies can give excellent insight into the behavior of data and the application of statistical analysis. For students in laboratory sciences, the study of measurement error can be helpful. But for maximum student motivation, I think that process improvement projects in general and personal improvement projects in particular are likely to be most useful in introductory undergraduate courses.

Logistical Details

Students should be responsible for finding or collecting their own data, and I urge that the data be brought up to the current date whenever possible: the projects should take place in real time. It would not be hard to supply data sets from past projects, but much of the excitement and realism would be lost. There is more to a statistical application than the analysis of a canned data set, even a good canned data set. But a good canned data set is much better than no application at all.

The instructor has to instill awareness of the problems arising from messy, incomplete, inaccurate data. The importance of operational definitions has to be stressed. The student must learn early that carelessness in data collection will compromise the usefulness of the results.

The student's analysis should be accomplished in a series of installments summarized in progress reports. For example, there could be three progress reports with due dates at any time before the end of, say, the fifth, eighth, and eleventh weeks of a term. The first installment should include a proposal and, if possible, some preliminary data analysis. The second installment should present substantial data analysis. The final installment should include not only a summary of data analysis and report of findings but also an executive summary, in completely nontechnical language, emphasizing what has been learned about the problem studied (*not* what has been learned about statistics or statistical computing). Some guidance on good report writing, especially the executive summary, is helpful for students.

The requirement that the project be completed in installments is essential. It greatly reduces the temptation to procrastinate, and it provides the opportunity for the teacher to give timely feedback to students. One of the hazards of projects is to be trapped by statistical pitfalls, and the instructor must help the student to avoid or to recover from these pitfalls. Thus, I give detailed comments and suggestions for each installment as quickly as possible -- always within a week, and often within hours. I return each paper immediately after reading, rather than waiting to finish all papers and then returning them as a batch. I also encourage students to get papers to me between regular class meetings in order to even out my reading load and reduce my response time.

When an installment would merit a low grade, I give the grade "REDO", which means that changes or extensions, mentioned in my comments, are necessary. Resubmission carries no penalty, but it must be prompt. The grade REDO is especially common on the first assignment, which includes the proposal. The REDO device usually leads in the end to a successful project, and the fraction of failing grades is usually very low.

Even with the requirement of doing the project in installments, students are sometimes tempted to procrastinate on the project, for example because of an impending midterm examination in another course. Since procrastination is very costly on projects, I have sometimes applied a modest penalty for late submissions. It is seldom necessary to carry out the penalty: almost all students meet the deadlines, and the few who don't almost always have genuine emergencies that permit me to be flexible. But the instructor must make the penalty credible. (However, I am increasingly reluctant to rely on any form of penalization; the aim should be to "sell" statistics so effectively that students will have the intrinsic motivation to do what is required.)

Maximum opportunity for contact with the instructor throughout the term is essential, and I provide many options for reaching me, including phone calls to me at the office or at home, e-mail, periods before and after classes, office hours, FAX messages, etc. A short conversation early in the term can often put a project on a much better track. The few project failures seem to occur when students try to do everything on their own, without seeking guidance.

Statistical Computing on Projects

Good interactive data analysis, with emphasis on model identification, fitting, and diagnostic checking, is essential. The statistical package Minitab has served my needs well, and seems to be widely used for student computation in beginning statistics courses. Although my students have not used it, the package Statistix also seems suitable. Other statistical packages can of course be used, but the instructor has to be sure that they can be quickly learned and that they have good interactive capabilities.

Although Minitab is relatively easy to use, providing computing help is more challenging with projects than with standard assignments because each student is working on a different data set. Students can make procedural errors without being aware of them, and the instructor must be alert to this possibility.

Frustration with computing can blunt the advantages of projects, so I try to provide as much backup as possible on procedural questions about computing. To do a really good job, I try to provide concise "buffer documentation" that tells the students exactly what they have to know, and no more.

Project Management

Projects place a substantial, almost intimidating, responsibility on the teacher, but no teacher, however inexperienced in applications, should hesitate to try. Even if there is some initial embarrassment, the teacher's skills are likely quickly to improve with practice. Even experienced statisticians learn a lot from their mistakes. And if the teacher initially has little experience in applications, projects provide accelerated learning in skills of project design and data analysis.

Given a choice, most of my students have opted to do individual rather than team projects. With the continual improvement that comes from experience, I now can supervise 60-85 concurrent individual projects with relative ease. I do not use teaching assistants. An instructor can learn on the job, but few teaching assistants can learn fast enough to be helpful to students on projects.

In the very large classes sometimes encountered in undergraduate courses at state universities, the only realistic approach is to require team projects. Using teams, Douglas Zahn at Florida State has incorporated projects into an introductory undergraduate class with enrollments in the hundreds.

I estimate that, on average, the direct instructor time -- evaluation and making suggestions about the proposal, reading and providing feedback on installments and the final report, and answering various questions raised by students -- is between 1 and 2 hours per project. For me, the mean is now closer to 1 than 2 because of improved instruction that has come from experience: the projects show clearly what concepts are not well understood by the students, improved classroom instruction results, and the teacher spends much less time in correcting student mistakes.

There are offsets to the added time demands that projects impose on teachers. Since I do not give examinations, I do not have to write and grade them. Moreover, students rarely argue about the grading of projects, while they are very prone to argue about the grading of examinations. Further, working with projects is interesting, while reading examinations is the dullest and most discouraging part of teaching. In my experience, students live up to their potential on projects but fall far short of that potential on examinations.

How Inexperienced Teachers Can Get Started on Projects

The answer is simple: Just jump in and do them, and learn as you go. The rewards will be immediate, and they will continue indefinitely. Never-ending improvement is possible. Don't worry about making mistakes, but learn from them.

However, some practical realities must be faced:

1. Before projects can be used, there must be flexibility to modify existing course content and modalities of teaching. The teacher must be free to supplement examinations by projects or even to replace examinations entirely, as I have done.

2. Class time must be made available for discussion of possible project topics and general problems arising as students develop their projects.

3. The statistical coverage of the course must provide at least some tools that can be readily and effectively applied to student projects. Some introductory courses fail to do this.

4. Finally, teachers must have enough time to make the necessary modifications of existing courses.

It is easy to use these practical realities to rationalize the avoidance of projects altogether. But one must keep in mind that projects offer a way -- by far the most effective way known to me -- to bring statistics courses to life and so convert students from stoic captives to interested coworkers.

Team Projects

Team projects emphasize the essential role of teamwork, which extends far beyond statistics courses. Depending on the size of student project teams, the number of distinct projects to be supervised can be controlled, with consequent economy of the instructor's time. However, many instructors don't know how to manage teams. Poor management of the teams can spoil the whole project idea and confirm the common feeling that team projects necessarily entail the frictions and frustrations associated with committees, and that the only way to get anything done is to do it yourself. Good management of teams can confer great benefits that go far beyond statistics. The students will get an invaluable lesson in how teamwork can expand the capabilities of individual team members.

Although I am not an expert on management of teams, I have had a little experience and can offer some useful advice.

Under favorable conditions, groups can accomplish much more than the same number of individuals working on their own. Under unfavorable conditions, groups can accomplish little or nothing and frustrate themselves in the process. The word "team" in place of "group" creates the right frame of mind.

Some suggestions for creating favorable conditions for team functioning are:

▶ Each team member must make a commitment, without reservation, to the team process. "I may be smarter than anyone else on the team, but the team is smarter than I am."

▶ Team members must be good listeners. Understanding points of view must precede discussion of their merits.

▶ The first task is to define the nature and scope of the project, including responsibilities of individual members.

▶ Brainstorming techniques -- expression of ideas without discussion or criticism -- are often useful, especially in the early stages when the project is being defined.

▶ Brainstorming -- and other team discussion -- can be facilitated by designating one member (possibly rotating) as scribe. The scribe's duty is to summarize all discussion, as it proceeds, by summary statements on a flip chart. As flip chart sheets are filled, they are taped, in sequence, to the wall or blackboard. This permits viewing, and reviewing, salient features of the discussion as it proceeds; it also greatly facilitates writing up a summary of the discussion afterwards, or presentation of the team's thinking to others.

▶ The scribe should be as nondirective as possible, contributing to the discussion only when he or she thinks that a salient point is being overlooked by the team.

▶ It is possible to combine the role of scribe and team leader. (It keeps the leader from trying to dominate the team). The leader/scribe, however, must formulate in advance the questions to be discussed.

▶ All discussion should be aimed at clarification and suggestions before attempting persuasion.

▶ Disagreements should be resolved by appeal to data rather than by debates and appeals to authority.

▶ If the data aren't available, the team should arrange to get some. Don't try to jump from problem to solution without studying causation; study of causation is facilitated by data!

► Teams should have *short* meetings at regular intervals, with assignments to team members for tasks to be accomplished before the next meeting.

► The leader of the team should be responsible for a written agenda in advance of each meeting.

► One member of the team should be designated as secretary, with the task of preparing compact and timely minutes of meetings.

► At each meeting, one member should be designated as "time keeper" with the responsibility of declaring that the discussion has become too repetitious or contentious and that the subject should be switched to what the team should do next. For example, the team might summarize agreements reached and tasks assigned, and then adjourn; or it might switch to the next item on the agenda.

My experience with team projects has been mainly based on MBA students who are relatively mature and well motivated. It may be harder to organize and implement projects for freshman/sophomore level students. However, I mentioned earlier the encouraging experiments with undergraduate students by Douglas Zahn at Florida State.

Modifications of Course Coverage Entailed by Projects

In contemplating the addition of projects, the instructor is likely to find that some extension of course coverage will be needed to provide students the necessary tools. Simple ideas of time series analysis are one example; simple time series tools can greatly expand the scope for projects.

Concepts of modern quality and productivity improvement are another. The basic idea is that *all processes, including personal processes, can be improved, and that data -- and statistical ideas -- can contribute to the improvement.*

The usual reaction of teachers to proposed addition of new materials is to ask what must then be removed from the course. My belief is that teaching is simply another process that can be improved. Just as businesses learn to produce better products for less, so teachers can teach more in less time. They can adapt ideas from quality and productivity improvement to teaching.

Process Improvement Using Statistics: Personal Projects

To bring students in an introductory course to the point where they can do something useful about process improvement, some material on quality and productivity improvement must be added to the course.

I have found the book by Mary Walton, *The Deming Management Method* [2], an easy way to bring in important ideas of quality and productivity improvement. I often require Walton's book as a supplementary text, to be read at the outset of the course. Even though the book is oriented towards business application, it presents a philosophy -- that of W. Edwards Deming -- that should be of great interest to anyone, not just to business students. Walton's first 19 chapters present Deming's philosophy vividly; students find them easy, enjoyable, and rewarding reading. The key methodological chapter, Chapter 20, is called, "Doing It With Data". Here a set of simple tools is given, tools that the students can learn to use simply by reading and then trying them out on their own projects; little, if any, class time is required.

These tools heavily emphasize simple graphics. Some of the tools are (or should be) covered in any introductory course: time series plot, histogram, scatter plot. Others are prestatistical: flow chart, cause-and-effect diagram, Pareto chart. Another, the control chart, is an extension of the time series plot that can be introduced effectively in introductory courses. The graphical tools need to be supplemented by such nongraphical tools as checklists and simple data recording forms.

(The remaining chapters of Walton's book give business applications, mostly to manufacturing; these are well done but would not be assigned if the book were used as a supplementary text in an undergraduate introductory course.)

For a reasonably full set of statistical tools for process improvement, a substantial commitment to tools of statistical process control (including time series analysis), intervention analysis, and design of experiments would be needed to extend "Doing It With Data". This would go beyond the scope of most introductory undergraduate courses. But a small subset of these tools, including several topics already taught in undergraduate introductory courses, can provide a foundation for useful student projects in personal improvement.

In particular, one-sample and two-sample procedures, combined with the idea of randomization, can be applied to simple experiments in which a process modification is tried out with the aim of improvement. Suppose, for example, that a student interested in golf wanted to try out two grips for putting, the standard reverse overlap -- A -- (current technique) and the baseball grip -- B -- (suggested by Lee Trevino). After perhaps some practice trials with B to gain reasonable proficiency, the student might launch into a series of trials, each comprising ten successive putts from a fixed distance, with the grip used to be determined randomly (the instructor would include, in class lectures or readings, procedures for achieving randomization). This leads to the type of two-sample comparisons taught in introductory courses: point estimates, confidence intervals, and tests for the difference between the mean numbers of successful putts using the two grips.

An alternative design would be randomized pairs, in which trials of 10 putts are paired, with A and B randomized between the first and second trial of each pair. This leads to one-sample procedures based on the differences between results within each pair: point estimates, confidence intervals, tests.

There should also be some "diagnostic checking" to see if the data conform to the statistical assumptions underlying the procedures. These checks include checks for normality and also examination of possible departures from independence through time. In an introductory undergraduate course, independence checks might be restricted to visual examination of simple time series plots, supplemented by a runs count. (If not already included in the course, the runs check is a nice addition because it serves as a very useful illustration of the nonparametric approach to inference.)

Confining attention to these two simple designs limits the range of possible statistical tools, but it still permits many exciting projects. To see this, imagine that you are a golf enthusiast and would dearly love to cut a few strokes off your score. Also, this type of project makes students think carefully what really promising improvements they might try out. A student recently did an interesting golf project in which change in stance -- placing of the feet relative to the ball -- resulted in a nearly 10 percent improvement in the proportion of putts made. Later he studied the effect of the baseball grip and found that, even without a great deal of practice to get used to it, it was no worse that the standard grip.

It is not very hard to go from study of a single process modification to *simultaneous* study of two modifications, such as both grip and stance in the golf example. This leads to the idea of factorial experiments. The analysis of these experiments can be accommodated by only a slight extension of the topics usually taken up in introductory courses.

Another direction of extension entails some kind of intervention in a process that is in a state of statistical control, in the hopes of improving the process. Explicit randomization is not used. Rather there is reliance on the assumption that the process is in a state of statistical control to begin with: technically, that the results behave as if the data were "independent and identically distributed" or "IID". Again, this would require extension of course coverage: simple concepts of statistical process control would have to be introduced. I can offer two personal examples, innovations tried in order to improve my own performance in distance running.

1. Reduced daily training mileage from 9 miles to 6 miles, all run at a relatively constant speed within the workout and from day to day. But the reduced distance was run about 30 seconds per mile faster, with no long training runs, even before marathons. Racing performance improved.

2. No drinking of fluids during races, including marathons, even in hot weather; modification is to become supersaturated by high fluid intake shortly before the start of the race, with cooling during the race by dousing oneself with lots of water. Racing performance improved.

Both these interventions were contrary to commonly accepted doctrine among runners:

1. *For training*, accepted doctrine says that at least a few long training runs (say 12-20 miles) before marathons are essential, yet I did not run farther than 6 miles a day. My interpretation is that the *average* quantity and quality of training is important, not what one does in a few individual workouts.

2. *For cooling during a race*, the doctrine says that failure to drink during long hot weather races is not only unwise but dangerous. My interpretation is that the goal is to keep cool, and that drinking during the race is only one way to do this, not necessarily the best.

Some Types of Personal Projects Done by Students

1. Fitness
2. Sports skills
3. Weight loss
4. Health (e.g., control of borderline hypertension)
5. Skills (e.g., learning foreign language)
6. Efficiency in daily tasks, such as getting started in the morning or planning good work routines
7. Study habits
8. Raising children or training a dog
9. "How to do my job better". (A useful approach is the personal quality checklist, in which defects are assigned to inefficiencies, such as being late to meetings, failing to respond quickly to phone messages and letters, permitting one's desk to be cluttered with old paper, forgetting an important task, or misplacing some essential working materials. The number of defects by days is then recorded and analyzed to see if improvement can be attained.)

The student may start with observation of an ongoing process, but must design a new process to attain improvement. In designing such a process, students must keep in mind both their personal circumstances and their psychology. With fitness projects in particular, I urge them to try to work out a process that can be applied on a life-long basis, even under adverse circumstances, like traveling.

Examples:
1. Improving free-throw percentage in basketball.
2. Improvement of fitness, strength, or flexibility.

Some Useful Steps

Suggestions to students:
1. Find something you really want to improve. If it is a complicated process, start by flow-charting and/or cause and effect diagrams.
2. Decide on a few simple operational definitions of performance by which you can judge improvement.
3. Be sure that you can implement these definitions by accurate measurements. Even such simple measurements as pulse and weight must be carried out carefully and consistently.
4. If necessary, design a new process.
5. Devise a sampling plan by which the process will be studied during a base period, during which you will make no special effort at improvement, but pretest the basic performance measures that you have decided on. Include provision for auxiliary measurements that may facilitate interpretation of the performance measures.
6. For the base period, analyze the performance measures carefully to find out if the process is close to being in a state of statistical control; if so, find and remove any special causes.
7. Based on what you have learned during the base period, including the discovery of possible special causes, aim to introduce at least one intervention designed to improve performance. Your success will depend in large measure on your ability to think of good interventions.
8. Sometimes it will be possible to try several interventions at once, using principles of factorial designs. (In some applications, such as modifying one's swimming stroke, a single inter vention may be all that you can handle at one time.)
9. Analyze the results of your intervention. (Two-sample comparisons of performance after intervention versus performance before.)
10. Decide whether the intervention or interventions have succeeded and what might be done for further improvement.

Example: Improvement of Cardio-Vascular Fitness

Suggestions to my students:
Possible basic measurements: resting pulse, exercise pulse, and post exercise pulse. (Lower is better.)

Design an exercise process aimed at improving fitness, for example, the "step test": step up and down from, say, a 16 inch surface for a fixed duration (say two minutes) to a fixed cadence (say 96 steps up or down per minute). Check your pulse immediately after finishing, counting the number of beats for 15 seconds, then again after one minute and after two minutes. (15 second periods would be much too short for good readings of resting pulse, which should be done for at least one minute, but are needed since the pulse rate will drop rapidly after strenuous exercise. Pulse monitors are expensive, but they can greatly increase what you learn, since you can monitor heartrate during actual exercise.)

You might repeat this several times during an exercise session, keeping track of all measures for each repetition.

If you find that you are tempted to speed up the cadence as fitness improves, you can do so, but then you must measure not only pulse but cadence. The statistical analysis would become more challenging. You would now have to establish whether, for any given cadence, your pulse tends to be lower as the study progresses.

You could increase both the duration and cadence as fitness improves. The statistical analysis would become more challenging still.

Set an exercise schedule, but be sure that all will not be lost if you have to deviate from the schedule.

Possible auxiliary measurements: time of day, day of week, amount of sleep, psychological perception of well being, temperature and humidity.

Possible intervention: warmup.

Stair climbing is an extension of the step test that involves the possibility of experimenting with technique (e.g., use of railing).

Concluding Remarks

My central thesis is that statistics can best be learned and appreciated when instruction is combined with application in which the student undertakes real -- as opposed to contrived -- statistical projects. Student projects are necessarily limited in scope, but they can nonetheless answer interesting and important questions. They can even lead to improvement of important processes, including personal processes.

My own experience with student projects has been so encouraging that I cannot imagine teaching a statistics course without them. To add projects to an existing course that has not employed them, an instructor must make some adjustments in course coverage and invest some time in giving individual or small group guidance and feedback to students. The potential reward, however, is great: the stigma of "dullness" can be removed. The students will come to see statistics as an exciting and useful way to look at the world, and they will acquire -- and retain -- a much better understanding of statistical thinking.

Instructors, too, will find that dullness is removed. Each student project is a new adventure for them as well as for the student.

References

1. Grayson, C. Jackson and Carla O'Dell, *American Business: A Two Minute Warning*. Free Press, 1988, pages 200-201.

2. Walton, Mary, *The Deming Management Method*, Dodd, Mead & Company, New York, 1986, (now distributed by Putnam.)

Harry Roberts is Sigmund E. Edelstone Professor of Statistics and Quality Management at the Graduate School of Business of the University of Chicago. He is a Fellow of both ASA and AAAS. He has worked on statistical aspects of Quality Management since the 1950s with emphasis on applications to business (including manufacturing), academia, health care and defense. His teaching has involved hundreds of students in actual quality improvement projects in business and other areas. He is the author of numerous articles on these themes, as well as several textbooks.

Hands-on Activities in Introductory Statistics

M. A. Tanner   and   **Robert Wardrop**
University of Rochester Medical Center   **University of Wisconsin**

Why Use Investigations?

In this article we discuss the use of investigations in developing students' understanding of concepts and methods in statistics and probability. The basic goal of such investigations in the introductory course in statistics is to illustrate the role of statistics in the process of *doing* science. Investigations provide a means for the students to *participate* in the act of investigation and to *experience* statistics in action. The traditional textbook is not directed toward creating such an *active* environment. Traditional approaches introduce, reinforce, and extend concepts with examples and exercises. Yet it is important that individuals embarking on a career in science view statistics as a natural component of the dynamic process of investigation. To develop this outlook, they must see *statistics in action*.

A basic problem with introductory courses in statistics is that little or no time is devoted to addressing a scientific question by designing an experiment. Even a narrow question which lends itself to a simple experiment typically leaves students in a fog. Moreover, it has been our experience that students, even very good students, leave a standard course in statistics without the ability to analyze data. Even when an experiment has been conducted and the student is given the data, there is a certain amount of confusion because traditional classes focus on methods, not data. Having been told to perform a two sample *t*-test, most students can proceed. Having been told to analyze a data set, most students are lost.

Investigations highlight the interaction between science and statistics in a number of ways. First, participants have the opportunity to act as scientist, as well as statistician. As scientist, one participates actively in the experimental process. This experience illustrates how the design of the experiment and the analysis of the data interact. As statistician, examination of the data can be used to illustrate specific statistical concepts and methods, as well as to allow one to experience chance variation.

Chaddock [1] makes the following statement regarding statistics:

> *"It is not merely technique or a branch of mathematical science. It is concerned with a body of principles and methods developed to guide the student in assembling and in handling quantitative data. It is a point of view, a method of attack. It involves measurements, countings and estimates, a careful record, intelligent groupings, discriminating analysis, logical comparisons, clear presentations, and cautious weighing of evidence."*

By performing investigations, students experience the "point of view, the method of attack". Investigations focus the students' attention on experimental design (data collection), data (organization and description), and statistical reasoning (including inference).

122

As in any experimental situation, an activity may be subject to such common problems as outliers and/or missing data. While a rigorous treatment of these issues is beyond the scope of an introductory course, it is important for students to understand that ignoring why the observations are missing or ignoring outliers may lead to biased conclusions. These are important issues which are typically not illustrated in standard homework problems. Investigations enable the participants to see how real data really look.

In addition, investigations make the course more interesting and lively. As noted by Pool [2]: "Science must be taught in a way that will catch the interest of the student." Investigations not only catch the students' attention, but also involve the students because they are working with their own data. Investigations help to make learning an active rather than a passive experience. Investigations provide the student with a *context* in which to place the statistical concept or method, thereby facilitating the learning process.

Structured classroom investigations offer several advantages over one "big" project or even several smaller projects on topics selected by the student. First, especially at the beginning of a course, the students typically do not have the background to plan, conduct and analyze a study that is needed for a successful project. Structured investigations serve as "training wheels" to move students in this direction. Moreover, projects require a greater expenditure of time for grading since each student chooses a different topic. This is particularly troublesome in large lecture classes. A corollary is that a "model" solution is not available for project grading; thus teaching assistants may be unable to grade projects. In addition, with investigations, the entire class can participate in an examination of the data set looking for coding errors, implementation errors or other surprises in the data.

This preference for structured investigations at an early stage in the student's development parallels that of a creative writing class where the first assignments are paragraphs and short essays, rather than novels. In this regard, the first outing in a beginning hiking class is not to the summit of Mt. Everest. A first experience in doing science should not be a semester project. Finally, a carefully selected series of investigations can key on most of the important topics in a course. A simple large scale project typically focuses on at most a handful of topics.

Example: Counting Ease

The investigations should illustrate basic paradigms for scientific research. A question is posed, an experiment is performed to address this question, the results are summarized and conclusions are drawn based on an examination of the data. While the examples may be quite simple, the paradigms that the examples illustrate should be used in a wide range of scientific activities. For instance, consider the question: does the ability to recognize a letter of the English alphabet in any way depend on the letter?

To address this question, each student is asked to flip a coin to determine whether they are to count the number of times the letter "*e*" occurs in a passage or the number of times the letter "*r*" occurs in the same passage. In addition, basic demographic data are collected from the students. It is necessary to provide a data sheet for collecting the demographic information (as shown in Appendix 1), the passage and the line to record the count (Appendix 2). We suppose the students have previously seen the notions of average and median as numerical summaries of location, as well as the stem-and-leaf plot as a quick way to visualize quantitative data.

The analysis of the data would focus on a number of important points. One of the basic reasons for randomization is the hope that the randomization process will result in balanced groups. The first set of questions would ask the students to check whether, in fact, the "*e*" and "*r*" groups appear to be balanced with respect to basic demographic variables such as height, weight, and sex. While formal statistical tests are not requested, these questions focus the students' attention on an important role of randomization.

Further questions would ask the students to consider the measurement errors observed in counting the "*e*"s and in counting the "*r*"s. How are the measurement errors distributed within each group? How do the two groups compare with regard to the distribution of measurement errors? These questions encourage the students to begin to think about how to describe a set of numbers, as well as how to compare two sets of numbers with regard to shape, location and spread of the distribution.

Of particular interest are the techniques used by the students to enumerate the "*e*"s or the "*r*"s. Did they cross-out the letter as they read through the paragraph and then tally the cross-outs? Did the students keep a running count of the letter? Did the students keep a running count on each line and then tally the line totals? It is important for the students to understand that the method of enumeration impacts on the data and to consider just how it may impact on the data. Should this component of variation be controlled?

Other questions would ask the students to consider various types of experimental designs (e.g., nonrandomized, crossover, etc.) for addressing the question of interest. What benefits/drawbacks do these other designs have?

The results of this investigation can be considered at later points in the course after having introduced such concepts and techniques as the reference distribution, the scatter plot and the correlation coefficient.

Example: On The Face of It

Common to the design of the Great Pyramid of Cheops in Egypt, the Parthenon in Greece, as well as the human form is the number

$$\tfrac{1}{2}(1 + \sqrt{5}) = 1.618....$$

known as the Golden Ratio.

The Great Pyramid of Cheops was built according to right triangular forms where the ratio of the hypotenuse of the triangle to the base is 1.618.... . Similarly, the facade of the Parthenon was constructed so that the ratio of the height to the length of the front columns is 1.618.... .

In this investigation, the object is to explore whether the "typical" human face follows the magic number 1.618.... . In particular, the students are asked to measure the distance from their hairline to their chin and the distance from the right-most end of their right eyebrow to the left-most end of their left eyebrow. These distances are then remeasured by a partner (blinded to the self-measured value). The primary question of interest is whether "on average" the ratio of the larger to the smaller distance is approximately equal to 1.618.... .

To analyze these data, the students must be familiar with hypothesis tests in the one sample case, including the ideas of the null hypothesis, alternative hypothesis, test statistic and *p*-value. They would need a ruler to measure the distances and a data sheet to record the data.

The analysis of these data would begin by asking the students to look at the distribution of the data. Is the distribution of the ratios symmetric or is it skewed? Are there any outliers? Can these "outliers" be explained by such covariate information as beards or balding? Also of interest is to look specifically at the men's data and specifically at the women's data.

The next series of questions would ask the students to perform a formal hypothesis test. What is the appropriate null hypothesis in this case? What is the appropriate alternative hypothesis? What is the appropriate test statistic? What is the p-value? Is it necessary to transform the data to satisfy the assumptions of the test? In a final series of questions, the students would be asked to investigate whether different results are obtained for men and for women, as well as for self-measured versus partner-measured data and to consider why different results may have been obtained. In particular, do the self-measured data appear to be more variable than the partner-measured data? If so, is this to be expected? If so, how does this affect the test statistic?

The results of this investigation can be further considered at later points in the course when confidence intervals, paired analysis, and two-sample analysis have been presented.

Implementing Statistical Investigations

The type of investigations suggested here may be performed in the classroom, in the discussion section, or at home. Such investigations should not require special laboratory equipment to implement the activity. The results of the investigations may be examined in class, as homework, or as an extra-credit project. The particular activity being conducted often suggests the best place to collect and analyze the data.

Access to a computer, whether micro, mini, supermini, or a mainframe, is recommended for conducting the statistical analyses in this type of investigation. The students benefit from having access to the common data base (generated in the investigation) and by having the computer do the number crunching. These features will allow the students to bypass the tedious entry of data, the tedious evaluation of formulas and to spend more time thinking about *what to do*, rather than *how to do it*. Moreover, access to a graphics program will give the students the flexibility to visually explore the data from a number of different perspectives. Whether a histogram, stem-and-leaf display, or a scatter plot, graphs allow the eye to detect structure in the data.

Structure of ICS

One of the present authors (Tanner [3]) has developed a complete series of structured investigations such as the ones described above that can be used throughout a course in statistics. These investigations have been published as a book "Investigations for a Course in Statistics" (ICS). For those readers who may want to develop their own series of such statistical investigations, we outline the approach that has proved successful using ICS.

Each of the fifteen chapters begins with an introduction which includes a list of key words and a description of the problem of interest. The chapters continue with a survey of the technical background necessary to analyze the data and a list of the material needed for the investigation. It is useful to provide a common notation and to sketch briefly the statistical methods and concepts which are highlighted in the investigation. The Experimental Protocol section includes data sheets which are used to record the data and which are then detached and handed in so that the class results may be pooled. The Analysis Section provides a structured set of questions to guide the students in an analysis of the data. It is also desirable to provide illustrations to show how to use Minitab or some other statistical package to do the computations or graphs discussed in the body of the chapter.

The chapters included in ICS follow the general outline of a course in introductory statistics, as well as a course in statistical methods. The general topics addressed in the investigations include: elementary experimental design, descriptive and inferential statistics, elements of probability, discrete distributions, the normal distribution, sample surveys, one and two sample problems, analysis of categorical data, and regression.

Conclusions

It is important for the students in the introductory course to see statistics in action. Investigations direct attention to experimental design, data analysis and statistical reasoning. They make the course more interesting and lively. Investigations establish a context for the concept or method being studied. Moreover, a series of structured classroom investigations provides the "training wheels" by which students move toward learning how to do science and promotes a hands-on approach to teaching statistics. In this way, it is hoped that the introductory statistics course will direct students to a methodology for scientific investigation and understanding.

References

1. Chaddock, R. E., "The Function of Statistics in Undergraduate Training", *Journal of the American Statistical Association*, **21**, (1926), pp 1-8.

2. Pool, R., "Freshman Chemistry was Never Like This", *Science*, **248**, (13 April, 1990), pp 157-158.

3. Tanner, M.A., *Investigations for a Course in Statistics*, Macmillan, 1990.

Martin Tanner is chair of the department of biostatistics at the University of Rochester. He has been an associate editor for the *Journal of the American Statistical Association* since 1987. He is interested in Markov chain algorithms for Bayesian, likelihood and conditional inference.

Robert Wardrop is associate professor and associate chair in the department of statistics at the University of Wisconsin at Madison. His major areas of interest are statistical education, sequential analysis, experimental design, and exact inference. He has served as an expert witness in several discrimination cases. He has written a textbook in statistics which will be published in 1993.

Appendix 1

EXPERIMENTAL PROTOCOL   NAME _____

DATA SHEET

1. Record your height in inches   _____

2. Record your weight in pounds   _____

3. Record your GPA (A = 4.0)   _____

4. Record your sex   _____ MALE   _____ FEMALE

5. How would you categorize your major? Check one of the following categories.
   - 1. Physical Science   _____
   - 2. Social Science   _____
   - 3. Biological Science   _____
   - 4. Humanities   _____
   - 5. Engineering   _____
   - 6. Business   _____
   - 7. Other or Undecided   _____

6. Do you require glasses or contact lenses for reading? _____ YES   _____ NO

7. Flip the coin. Record the result.   _____ HEAD   _____ TAIL

8. If the toss resulted in a HEAD, then count the number of **e**'s between the two black lines. (See the selection below.) You are only allowed to count the *e*'s once. Please skim the selection at the speed in which you would read a newspaper or magazine article. Do not collaborate with your friends.

   My estimate of the number of *e*'s between the two black lines is _____

If the toss resulted in a TAIL, then count the number of **r**'s between the two black lines. (See the selection below.) You are only allowed to count the *r*'s once. Please skim the selection at the speed in which you would read a newspaper or magazine article. Do not collaborate with your friends.

   My estimate of the number of *r*'s between the two black lines is _____

9. Hand in your responses to questions 1-8, so that the class results can be pooled.

Appendix 2

THE JOURNAL OF

NERVOUS AND
MENTAL DISEASE

VOL. 170, NO. 8

August 1982

SERIAL NO. 1197

Operational Definitions of Schizophrenia

What Do They Identify?

MICHAEL A. YOUNG, Ph.D.,[1] MARTIN A. TANNER, Ph.D.,[2] AND HERBERT Y. MELTZER, M.D.[3]

Several studies have shown that different systems for diagnosing schizophrenia produce diagnoses with relatively low agreement. This, however, does not necessarily imply that the different systems are identifying conceptually different diagnostic groups. They may, in fact, identify a single entity, but with different degrees of accuracy. One-hundred ninety-six inpatients were diagnosed by multiple diagnostic systems. The pattern of relationship among the diagnoses was studied with latent class analysis, a statistical methodology which has not previously been applied to this type of data. Results indicated that the diagnoses of the Research Diagnostic Criteria, Flexible 6, Schneider, and 1978 Taylor and Abrams diagnostic systems all estimated a single underlying diagnosis. The Taylor and Abrams system identified this core diagnosis with high accuracy. The results also suggest that blunted affect and the absence of an affective syndrome are related to latent class schizophrenia, while the presence of only nonspecific delusions and/or hallucinations is not related. The appropriateness of the latent class definition of schizophrenia for selecting patients for research is discussed.

An Introductory Statistics Course:
The Nonparametric Way

Gottfried E. Noether
University of Connecticut

The present volume is a welcome expression of the long-delayed recognition by the mathematical community that statistics occupies an important place in our daily lives. The author has long believed that for a great majority of students, probability and statistics represent very desirable alternatives to a calculus-centered curriculum. Until quite recently, this view has not been widely shared by members of the mathematical community. Only during the last few years has there been a growing appreciation that probability and statistics are not only useful in their own right, but are helpful in motivating and stimulating quantitative thinking. An increasing number of schools teach statistical ideas as early as the primary grades and, according to a CBMS study, statistics is the fastest growing offering in the undergraduate mathematics curriculum.

Unfortunately, in spite of considerable progress in statistics instruction at the precollege level, the great majority of students in introductory college statistics still approach the course with misgivings and fear. For most, statistics is just another math course, a subject that they would rather avoid. The fact that many elementary statistics courses are taught by mathematicians with a minimum of statistical training does not help matters. For the mathematically oriented instructor, it is much simpler to emphasize mathematical details rather than statistical thinking. In particular, there is the strong temptation to spend much more time on probability theorems than is warranted in an introductory statistics course. The *nonparametric* approach to elementary statistics advocated in this paper presents statistical ideas simply and straightforwardly without overwhelming the student with mathematical derivations. As to probability, all the student needs is some understanding of the frequency interpretation of probability and of equally likely cases.

Actually, the term *nonparametric* is rather unfortunate. As presently used in statistics, it defies precise definition. Historically, it referred to hypothesis testing situations that went beyond the classical parametric population models. But over the years, statisticians have developed methods for estimating population parameters "nonparametrically" and testing hypotheses involving parameters. One- and two-sample *t*-tests are replaced by one- and two-sample Wilcoxon tests; population means are often replaced by population medians; product moment correlation is replaced by rank correlation. These nonparametric procedures are sometimes mentioned in introductory statistics texts as alternatives for the classical normal-theory methods. This paper proposes that we develop the basic concepts of statistical inference around nonparametric procedures, in particular, the Wilcoxon two-sample test.

When teaching the introductory statistics course, I spend about three or four weeks on data analysis and the simplest aspects of probability. Since, however, these two topics are discussed by other contributors to this volume, in my paper I am dealing only with how to teach concepts and methods of statistical inference.

Two-Sample Procedures

In most standard statistics texts, the basic ideas of hypothesis testing and estimation are developed for the one-sample problem involving the mean of a (supposedly) normally distributed population. However, from a practical point of view, the two-sample problem is both more realistic and conceptually simpler than the one-sample problem (which in its classical form, among other things, presupposes that we have defined the concept of mean for a continuous population.)

For our discussion, let us consider the following example: A weather modification project wants to evaluate the effectiveness of a particular cloud seeding procedure. In a randomized experiment, the following amounts of rainfall (in inches) have been recorded:

No seeding (x): 0.26 0.37 0.43 0.17 0.60 0.67
Seeding   (y): 0.75 0.55 1.26 0.28 0.77 1.02 0.50 0.71

We want answers to the following questions:

(i)   Does cloud seeding increase the amount of precipitation?
(ii) How large an increase in precipitation can be expected as a result of cloud seeding?

The first question calls for a test of the hypothesis that amount of precipitation is unaffected by cloud seeding against the alternative that cloud seeding increases precipitation. The second question calls for an estimate of the amount of increase in precipitation. We shall find both a point estimate and a confidence interval.

Before deriving the Wilcoxon two-sample test in answer to (i), we should note the greater simplicity and generality of the problem answered by the Wilcoxon test. The hypothesis tested by the Wilcoxon test simply states that x- and y-observations do not differ (they have come from the same population) against the alternative that y-observations tend to be larger than x-observations. The classical two-sample t-test tests a much more specific and more technical null hypothesis: $\mu_y = \mu_x$, where μ_y and μ_x are the means of two (normally distributed) populations with identical variances.

The formulation of the problem underlying the Wilcoxon test practically suggests an appropriate test statistic. Since we want to find out whether y-observations tend to be larger than x-observations, a natural quantity to compute is:

$$U_+ = \#(y_j > x_i), \ i = 1,\ldots,m; \ j = 1,\ldots,n,$$

the number of times y-observations are greater than x-observations. We also define a second quantity:

$$U_- = \#(y_j < x_i) = \#(x_i > y_j),$$

the number of times x-observations are greater than y-observations.[1] By comparing each of the six x-observations with each of the eight y-observations, we eventually find that $U_- = 8$ and

---

$U_+ = (6)(8) - 8 = 40$. (More generally, if there are n y-observations and m x-observations and if none of the y-observations equals an x-observation, $U_+ + U_- = mn$. For a discussion of how to handle ties among the observations, the reader is referred to more extensive treatments such as Hollander and Wolfe [1] or Noether [2].)

We would reject the hypothesis being tested in favor of the alternative under consideration, if U_+ is sufficiently large, or, what amounts to the same thing, if U_- is sufficiently small. Before we answer the question of how small U_- has to be for rejection, it seems worthwhile to remind the reader of the greater complexity of the two-sample t-statistic, which is the standard tool for answering problem (i):

$$ t = \frac{\overline{y} - \overline{x}}{\sqrt{\dfrac{\sum (x-\overline{x})^2 + \sum (y-\overline{y})^2}{m + n - 2}}} \sqrt{\frac{mn}{m + n}} $$

In this age of computers, the greater computational complexity of t compared to U is not very important. But what about an intuitive justification for t? The numerator is simple. But the same can hardly be said of the denominator. Only an authoritative assertion or hours of probability study will take care of the denominator.

There is then the question of how large, or small, U and t have to be to reject the null hypothesis at a given significance level. An answer has to refer to the sampling distributions of the two test statistics. There is no way of explaining the derivation of the sampling distribution of t in an introductory statistics course. On the other hand, the sampling distribution of a nonparametric test statistic such as U can be obtained by simple enumeration. Thus for the two-sample problem with $m = 2$ and $n = 3$, there are just 10 equally likely arrangements of the two x- and three y-observations:

arrangement	U_+	U_-
xxyyy	6	0
xyxyy	5	1
. . . . .	.	.
yyyxx	0	6

The resulting common distribution of U_+ and U_- is:

$$ P(0) = P(1) = P(5) = P(6) = 1/10, \quad P(2) = P(3) = P(4) = 2/10. $$

Computer enthusiasts among the students may want to try writing a program for larger values of m and n.

For the cloud seeding example, what does the value $U_- = 8$ tell us? According to Table U in Noether [2], the (one-sided) P-value equals 0.021. For the t-test, we find $t = 2.19$ with a P-value of 0.024.

Estimation in the two-sample shift model

As a first step in finding a point estimate and a confidence interval for the increase in precipitation due to cloud seeding, we look for a suitable model. We have described the cloud seeding experiment as a randomized experiment. Randomization of the decisions to seed or not to seed a given cloud formation is necessary to avoid bias. On a day when randomization results in cloud seeding, we write y for the amount of precipitation; on non-seeded days, we write x. If δ denotes the amount of increase due to cloud seeding, we have

$$y = x + \delta. \tag{1}$$

Model (1) is known as the *shift model*, since y-values have been shifted relative to x-values by the amount δ. If both y and x could be observed at the same time, an obvious estimate of δ would be $y - x$. In the absence of paired observations, given m x-observations and n y-observations, we compute all mn possible estimates

$$y_j - x_i, \quad i = 1, \ldots, m; j = 1, \ldots, n. \tag{2}$$

Because of their simplicity, we refer to these estimates as *elementary* estimates. Which particular one should we use as *the* (nonparametric) estimate $\hat{\delta}$ of δ? Students who have had some instruction in data analysis (including five-number summaries) will not hesitate to suggest the median as the most representative of the elementary estimates. (Another justification for the median is offered after the discussion of confidence intervals for δ.) We define the nonparametric estimate $\hat{\delta}$ of δ to be the median of the elementary estimates (2). For the cloud seeding data, $\hat{\delta} = 0.325$.[2] For comparison, the normal-theory estimate $\bar{y} - \bar{x}$ equals .313.

The elementary estimates (2) are also useful in finding confidence intervals for δ. The simplest interval is bounded by the smallest and largest of the elementary estimates. But this interval has two disadvantages: unless m and n are quite small, the length of this interval is impracticably large and the associated confidence coefficient equals one for all practical purposes. Both objections can be overcome by considering intervals of the type

$$e_{(d)} \leq \delta \leq e_{(d)}, \tag{3}$$

where $e_{(d)}$ and $e_{(d)}$ denote the dth smallest and largest of the elementary estimates (2). We shall refer to the interval (3) as the I_d-interval. Table U in Noether [2] lists d-values for confidence coefficients close to .99, .95 and .90. (In general, these nominal confidence coefficients cannot be achieved exactly, since there are only $mn/2$ choices for d.) As d increases from 1 to $mn/2$, the associated confidence coefficients decrease towards 0. The point estimate $\hat{\delta}$ of δ defined earlier can be looked upon as the limit of the midpoints for this sequence of confidence intervals.

For the cloud seeding data, the 48 elementary estimates arranged from the smallest to the largest are:

-0.39	-0.32	-0.17	-0.15	-0.12	-0.10	-0.09	-0.05	0.02	0.04	0.07	0.08
0.10	0.11	0.11	0.12	0.13	0.15	0.17	0.18	0.24	0.28	0.29	0.32
0.33	0.34	0.34	0.35	0.38	0.38	0.40	0.42	0.45	0.49	0.51	0.54
0.58	0.59	0.59	0.60	0.65	0.66	0.76	0.83	0.85	0.89	1.00	1.09

---

[2] Note on computing: Noether [2] describes a graphical method for finding $\hat{\delta}$, but computer programs are preferable. For example, in MINITAB, the command "WDIFF for Y-and X-observations" computes all mn elementary estimates. Additional commands arrange the elementary estimates according to size and find their median. There are corresponding commands for the elementary estimates associated with other problems.

The closest we can get to the nominal confidence coefficient .95 is .957 for d = 9. By simple counting, we find the confidence interval

$$0.02 \leq \delta \leq 0.60.$$

The normal-theory t-interval with the same confidence coefficient .957 extends from -0.01 to 0.64, slightly longer than the I_d-interval.

The I_d-interval bounded by the 10th smallest and largest elementary estimates has confidence coefficient .941. How can we compute the confidence coefficient associated with a given value d? If in the shift model (1), the shift parameter δ takes the value 0, x- and y-observations are identically distributed. Let us then consider the two-sample test that **rejects** the hypothesis of identical x- and y-populations, if the confidence interval (3) does not contain the value 0. This can happen only if either the number of negative elementary estimates or the number of positive elementary estimates $y - x$ is smaller than d:

$$\#(y - x < 0) = \#(y < x) = U_- < d$$

or

$$\#(y - x > 0) = \#(y > x) = U_+ < d.$$

This is the Wilcoxon two-sample test for a *two-sided* alternative, where we reject the null hypothesis if either U_- or U_+ is too small, namely, smaller than d. Since probability levels for the Wilcoxon test can be computed by enumeration, so can confidence coefficients for I_d intervals.

For the shift model (1), the hypothesis $\delta = \delta_0$ can be tested by redefining U_- and U_+ as the number of elementary estimates (2) that are, respectively, smaller than δ_0 or greater than δ_0.

The Role of Elementary Estimates

In the remainder of the paper, we will use the concept of elementary estimates to find nonparametric solutions for the one-sample problem and for linear regression and rank correlation. It is helpful to summarize the use of elementary estimates in the two-sample case. Given the shift model $y = x + \delta$, we may ask: what is the minimum information that would enable us to estimate the shift parameter δ? The answer is clearly one x- and one y-observation, for which the difference $y - x$ is the only possible estimate of δ. For m x- and n y-observations, we then define the set of elementary estimates as the collection of all possible differences $y - x$. For this set of elementary estimates

(a) the point estimate equals the median of all mn elementary estimates;
(b) confidence intervals are bounded by the dth smallest and largest elementary estimates; and
(c) tests of the hypothesis $\delta = \delta_0$ are performed with the help of two statistics equal to the number of elementary estimates that are, respectively, smaller than or greater than δ_0.

In subsequent applications, we start with an appropriate set of elementary estimates and then apply rules (a) through (c) to define point estimates, confidence intervals and test statistics.

The One-Sample Problem

We define the one-sample problem as follows. We are interested in a population that is "centered" at some value η. We want to estimate η, or test hypotheses about η, on the basis of a sample of N observations x_i, $i = 1, ..., N$. In classical statistics, the *mean* is usually taken as the population center. But for a nonparametric approach, the *median* has definite advantages, not the least of which is its greater conceptual simplicity. Thus, we shall usually refer to η as the population median, even if, as in the case of symmetrically distributed populations, the median and the mean coincide.

Our discussion will be in terms of the following example. In the 1989 income tax instruction booklet, the IRS stated that the typical tax payer requires 225 minutes to fill out Form 1040 (3 hours and 10 minutes for preparation and 35 minutes for copying the information.) Many tax payers may have felt that it took them considerably more than 225 minutes.

Assume that 25 tax payers have agreed to keep track of the amount of time (to the nearest 10 minutes) it took them to complete Form 1040. The following are the 25 times, rearranged from the smallest to the largest:

$$100 \quad 120 \quad 140 \quad 150 \quad 150 \quad 180 \quad 190 \quad 230 \quad 230 \quad 240 \quad 240 \quad 250 \quad 250$$
$$250 \quad 250 \quad 270 \quad 270 \quad 280 \quad 280 \quad 290 \quad 320 \quad 340 \quad 450 \quad 540 \quad 720$$

We want answers to two questions:

(i)   What is an estimate of the median time for completing Form 1040?
(ii) Do the observations support the IRS claim?

As indicated in the previous section, our first task is to define an appropriate set of elementary estimates for η. Since we can estimate η with as few as *one* observation, let us first consider the set of elementary estimates consisting of the individual observations. For this set of elementary estimates, according to (a) the point estimate of η is the sample median, 250, in the example. According to (b), confidence intervals are bounded by the dth smallest and largest observations. In particular, according to Table S in Noether [2], the confidence interval with confidence coefficient .957 (the closest we can get to .95) is bounded by the 8th smallest and largest observations and thus extends from 230 to 280.

For testing the hypothesis $\eta = 225$ against the alternative that η is greater than 225, we use the test statistic $S_- = \#(x_i < 225) = 7$ with a (one-sided) P-value of .022. Since we can write S_- as the number of negative differences $x_i - 225$, this test is known as the *sign test*, and correspondingly, we refer to the estimates we discussed earlier as S-estimates and S-intervals. S-procedures, that is, the S-estimate, the S-interval, and the sign test, are the most general solutions available for the one-sample problem. They make no assumptions about the population under investigation.

Statisticians who use the t-test and related t-estimates for solving the one-sample problem assume, at least implicitly, that the population under investigation follows a near-normal distribution. Another set of nonparametric procedures, the so-called W-procedures are less general than S-procedures but considerably more general than t-procedures. The W-procedures assume symmetry, or at least near symmetry of the underlying population, but not normality, nor any other specific distribution form.

The earlier elementary estimates $x_1, ..., x_N$ remain valid estimates of a center of symmetry η, but they do not specifically utilize the assumption of symmetry. A simple way to utilize symmetry is to supplement the set of actual observations by the set of averages of pairs of observations. The resulting set of $\frac{1}{2}N(N + 1)$ elementary estimates

$$\tfrac{1}{2}(x_i + x_j), \; 1 \leq i \leq j \leq N,$$

known as the set of *Walsh averages*, forms the basis for the *W*-procedures.

Let us illustrate the *W*-procedures with the tax data. Some readers will question the assumption of symmetry implicit in the use of *W*-procedures. The largest three sample observations would seem to suggest some degree of population skewness to the right. But the three observations can also be explained as outliers. By applying the *W*-procedures, and then the *t*-procedures, to the tax data, we get an indication of how these two sets of procedures stand up to outliers among the observations.

We find the *W*-estimate = median of Walsh averages = 250, the same as the earlier *S*-estimate. Both estimates are quite unaffected by the outliers. The same is not true of the classical estimate $\bar{x} = 269.2$.

According to Table W in Noether [2], a confidence interval for η with confidence coefficient .952 is bounded by the 90th smallest and largest Walsh averages, 210 and 290, respectively. For the same confidence coefficient, the *t*-interval extends from 213.1 to 325.3. While the *S*-interval (230 to 280) is quite unaffected by the outliers, the *t*-interval is greatly affected, with the *W*-interval taking its place somewhere in between.

The *W*-test, known in the statistical literature as the *Wilcoxon One-Sample or Signed Rank Test*, for testing the hypothesis $\eta = 225$ is based on the following test statistics[3]:

$$W_- = \#(\text{Walsh averages} < 225) = 113$$

and

$$W_+ = \#(\text{Walsh averages} > 225) = \tfrac{1}{2}N(N + 1) - W_- = 212.$$

The corresponding one-sided *P*-value equals .09. For the *t*-test, we find $t = 1.64$ with a *P*-value of .06. Compared to the *S*-test, both the *W*-test and the *t*-test show the effect of the outliers.

Linear Regression and Rank Correlation

As a last example of the nonparametric approach, we discuss briefly linear regression and correlation. In classical statistics, these two topics rely on the principle of least squares and product moment correlation. While the principle of least squares can be satisfactorily motivated at the introductory statistics level, its utilization requires either calculus or rather complex algebraic manipulations. Further, any attempt to provide an intuitive explanation of the *meaning* of the product moment or Pearsonian correlation coefficient is bound to end in disaster. The following nonparametric approach offers a simple alternative.

We consider the following example. For five successive years, we have observed the median family income x (in thousands of dollars) and the number of housing starts y (in millions of houses) one year later:

---

[3] We have allotted one of the two elementary estimates that are equal to the hypothetical value 225 to W_- and the other to W_+.

The *W*-statistics can also be computed as

$$W_- = \Sigma_{x_i < 225} \text{ rank } |x_i - 225|$$
$$W_+ = \Sigma_{x_i > 225} \text{ rank } |x_i - 225|,$$

explaining the name *signed rank* test.

Year	x	y
1	29.1	1.0
2	27.9	0.9
3	30.2	1.5
4	32.4	1.2
5	32.3	1.1

A scatter diagram presents a graphical picture:

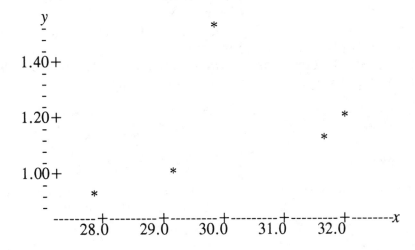

How can we fit a line $y = a + bx$ to a scatter diagram? In more formal terms, we assume that for each value of the x-variable, the center of y-values equals $\alpha + \beta x$. We want to find estimates of the regression coefficients α and β and, possibly, test the hypothesis of independence, $\beta = 0$, using N observed sample pairs (x_i, y_i).

We need two sets of elementary estimates: one set for estimating the slope β of the regression line; a second set for estimating its intercept α.

We start with the slope β. Given two sample pairs (x_i, y_i) and (x_j, y_j) with $x_i < x_j$, we can estimate β as

$$s_{ij} = (y_j - y_i)/(x_j - x_i). \tag{4}$$

If there are no ties among the x-observations, there are $\frac{1}{2}N(N - 1)$ slope estimates. We then estimate β as

$$b = \text{median of all } s_{ij}. \tag{5}$$

If β were known, according to the regression model, each sample observation would furnish an estimate $y_i - \beta x_i$ of the intercept α. Since β is unknown, we use the estimate (5) instead and estimate α as

$$a = \text{median of all } (y_i - bx_i). \tag{6}$$

For our example, we find $b = 0.0636$ and $a = -0.8618$, giving the nonparametric estimate of the regression line $y = 0.06x - 0.86$. The least square regression estimate is $y = 0.05x - 0.33$.

For the regression model $y = \alpha + \beta x$, the value $\beta = 0$ indicates that a change in the variable x does not produce a change of location for the variable y. The hypothesis $\beta = 0$ can be tested with the help of the following two test statistics:

$$C = \#(\text{positive } s_{ij}) \quad \text{and} \quad D = \#(\text{negative } s_{ij}).$$

The statistics are known as the number of *concordant* and *discordant* pairs of sample observations. Table K in Noether [2] furnishes critical values for this *Kendall test of independence*.

Rank Correlation

The quantities C and D are basic for the *Kendall rank correlation coefficient*. When we talk of the correlation between two variables x and y, we are interested in measuring to what degree the two variables tend to move in the same or in opposite directions. The greater the difference between the two quantities C and D, the stronger the relationship between x and y. $C > D$ indicates a positive relationship (x and y tend to move in the same direction); $C < D$ indicates a negative relationship (x and y tend to move in opposite directions). It is then possible to look at the quantity $S = C - D$ as a measure of the strength of relationship between x and y. However, since it is customary to standardize correlation coefficients to take on values between -1 and +1, we divide S by its maximum value $\frac{1}{2}N(N - 1)$. In the statistical literature, the quantity

$$r_K = \frac{2(C-D)}{N(N-1)}$$

is generally known as *Kendall tau*. It is better to reserve this term for the population equivalent of r_K. Since $C/[\frac{1}{2}N(N - 1)]$ is an estimate of the probability that two observation sets (x_i, y_i) and (x_j, y_j) are concordant and $D/[\frac{1}{2}N(N - 1)]$ is an estimate of the probability that two observation sets are discordant, the population equivalent of r_K is

$$\tau = \pi_c - \pi_d, \tag{7}$$

where π_c and π_d are the probabilities of concordance and discordance, respectively. In contrast to the product moment correlation coefficient, the right side of (7) provides a simple operational interpretation for the Kendall rank correlation coefficient. Further, if we set $q = \pi_c/\pi_d$, then q equals $(1 + \tau)/(1 - \tau)$.

For the income-housing data, $C = 8$ and $D = 2$, so that $r_K = .60$. Further, it follows that $(1 + r_K)/(1 - r_K) = 4$, suggesting that family income and housing starts are 4 times as likely to move in the same direction as in opposite directions. The product moment correlation coefficient r equals .41.

Final Comment

In addition to the material indicated in this paper, a one-semester introductory nonparametric statistics course may well take up the Kruskal-Wallis and Friedman tests which extend the Wilcoxon two-sample test and the sign test for paired observations to more than two populations.

References

1. Hollander, M. and D. A. Wolfe, *Nonparametric Statistical Methods*, 1973, John Wiley & Sons, New York.

2. Noether, G. E., *Introduction to Statistics: The Nonparametric Way*, 1990, Springer-Verlag, New York.

Gottfried E. Noether, emeritus professor of statistics at the University of Connecticut, died on August 22, 1991. His loss is one which is deeply felt by both the mathematics and the statistics communities.

   Professor Noether was a leading authority in the field of non-parametric statistics and had published widely. He was a fellow of both ASA and the Institute of Mathematical Statistics. He was also an active member of the MAA and a strong advocate of the importance of statistical education.

The Quest for Randomness and its Statistical Applications

S. L. Zabell
Northwestern University

Randomness is a word tinged with paradox. Strictly speaking, truly random events cannot occur within the context of deterministic classical physics; but the ordinary uses of randomization are far from quantum-mechanical in nature. The concept itself resists easy or precise definition, and its successful implementation is often surprisingly difficult to achieve in practice, yet the use of chance mechanisms can be found throughout human history, at all times and in all societies. In this paper we discuss the concept of randomness; the means that may be employed to generate random outcomes; the general uses to which randomization has been put in the past; and the specific statistical applications of randomization to be found today in subjects as diverse as opinion polling and the conduct of the modern medical clinical trial. There is an extensive literature on the above topics; the final section of the paper discusses further sources of information for those interested.

I. The Concept of Randomness

"Nothing occurs at random, but everything for a reason and by necessity."
-- Leucippus [Greek pre-Socrateser philosopher, c. 5th century B. C.]

The concept of randomness is one of the most elusive in mathematics. This is in part because it contains two very distinct (and sometimes conflictling) ideas, involving *pattern* and *process*. Consider, for example, the sequence

32823066470938446095.

To all intents and purposes, this appears to be a completely random sequence of numbers. But it is in fact decidedly nonrandom: it is the decimal expansion for π from the 111th to 130th digit. Or consider the sequence: 414344259125. This too appears to be random, but is not, although for a very different reason from the first sequence: it is the series of stops on the IND subway in New York City! (Note that two different senses of "pattern" are at play here: one mathematical, one involving human agency.)

On the other hand, consider the sequence:

01010101010101010101.

Our first reaction is that this is not a random sequence; it does not exhibit a random pattern. But it could well arise from a random process. (For example, if you toss a coin ten million times, recording a "1" when a head comes up, and a "0" when a tail comes, and then scan the

139

entire sequence for such a subsequence, you would be more likely than not to find one.) These two aspects of "randomness" are of course closely related: we ordinarily expect outcomes generated by a random process to display a random pattern; a sequence whose pattern is random may often be used as a surrogate for a sequence generated by a random process; and sequences whose pattern is not random cast doubt on the random nature of the process which generated them. These themes will arise many times in the sequel.

Example: Laplace and "Constantinople." A classic example is provided by Laplace in his justly celebrated *Essai philosophique sur les probabilités* of 1814:

> *On a table we see letters arranged in the order "C o n s t a n t i n o p l e",
> and we judge that this arrangement is not the result of chance, not because
> it is less possible than the others, for if this word were not employed in any
> language, we should not suspect it came from any particular cause, but this
> word being in use among us, it is incomparably more probable that some
> person has thus arranged the aforesaid letters than that this arrangement is
> due to chance.*

II. Generating a Random Sequence

> *"Alea jacta est" -- Julius Caesar, on crossing the Rubicon.*

How do we generate random sequences of numbers? There are essentially three basic methods: physical randomization, the use of random number tables, and pseudo-random number generators.

1. Physical randomization

The resort to a physical process of some sort to generate random outcomes is at once the oldest, most familiar, and most natural method. In the ancient world randomization was extensively employed in games of chance, religious ceremonies, selection of individuals for military or political service, and for purposes of fair or impartial allocation (a number of these are discussed in the next section).

The key point to emphasize, however, is just how difficult successful implementation of physical randomization is. The next example provides a dramatic illustration of this.

Example: The 1970 Draft Lottery

In 1969 the order of call for compulsory induction into the armed forces was determined by a lottery. Capsules containing the dates from January 1 to December 31 were placed in a container, the capsules in the container were then mixed, and finally the capsules were drawn out one at a time. Since September 14 was the first date drawn, that was first in the order of call, and persons born on that day would receive induction notices first.

This was clearly a serious matter, affecting the lives of millions of young Americans, and one would expect that the lottery was conducted in a fair and impartial manner. And indeed, if one inspects the graph of order-of-call versus date-of-birth, a purely visual inspection suggests that the allocation of numbers was indeed random. Nevertheless, subsequent statistical analysis revealed a clear nonrandom element: days towards the end of the year were more likely to be selected first, and thus persons born towards the end of the year were more likely to be selected first. In hindsight, it was relatively easy to come up with an explanation for this surprising phenomenon: the capsules were initially placed in the container in order, starting at the beginning of the year, so that the capsules on top corresponded to days at the end of the year. Although the capsules were mixed, the amount and difficulty of mixing necessary to disorder the capsules was underestimated, and the tendency of the person selecting the capsules not to reach all the way down to the bottom did the rest.

The example illustrates the difficulties and pitfalls that attend attempts at physical randomization. Experienced card players know that it is necessary to shuffle a deck of 52 cards as many as 7 times before it can be regarded as adequately shuffled, although in the real world of recreational penny-ante poker this seldom happens. (Recent studies in the mathematical literature have verified that under plausible models of shuffling, 7 shuffles is necessary.)

True randomness, of course, would entail that the outcome of a trial be indeterminate even in principle; that there are no "hidden variables" whose specification would determine the outcome. Stochastic outcomes in quantum mechanics are thought to be of this nature; this is the thrust of the so-called EPR (Einstein-Podolsky-Rosen) paradox, and the experimental verification of Bell's inequalities.

2. Random number tables

Key to the systematic application of randomization was the development of reliable and publically accessible tables of random numbers; for, as we have just seen, true physical randomization is difficult to carry out. Lord Kelvin, due perhaps to his background in experimental physics, was very sensitive to such problems, when he carried out his Monte Carlo experiments to test the equipartition theorem, writing:

> *I had tried numbered billets (small squares of paper) drawn from a bowl, but found this very unsatisfactory. The best mixing we could make in the bowl seemed to be quite insufficient to secure equal chances for all the billets. Full-sized cards like ordinary playing-cards, well shuffled, seemed to give a very fairly equal chance to every card. Even with the full-sized cards, electric attraction sometimes intervenes and causes two of them to stick together. In using one's fingers to mix dry billets of card, or of paper, in a bowl, very considerable disturbance may be expected from electrification.*

Tippett encountered similar problems while preparing tables of the mean range for Karl Pearson in the 1920's. Tippett had decided to "test and confirm" his results for the distribution of the mean range by simulation, using a normal population of 10,000 numbers, divided into subranges of 0.1 standard deviations and written on 10,000 tickets. Tippett found that, despite careful shuffling and sampling without replacement, the resulting distributions did not agree with theory. Pearson suggested the idea of using random numbers, tables of which Tippett then began to compile, using the areas of parishes in the United Kingdom census returns. He did this by discarding the first two and last two digits in order to avoid biases due to rounding, and then copying the remaining digits in the order in which they appeared in the returns.

Other tables, notably those of Fisher and Yates (1938), and Kendall and Babington-Smith (1939), soon followed, eventually culminating in the publication of the RAND Corporation's *A Million Random Digits* in 1955.

3. Pseudo-random number generators

A final important method of generating random numbers is by the use of a formula. In a *congruential psuedo random number generator*, for example, one starts out with a "modulus" p, fixed integers a and b, an initial value (or "seed") x_0, and solves the successive congruences $x_1 =: ax_0 + b \pmod{p}$, $x_2 =: ax_1 + b \pmod{p}$, ..., and in general

$$x_{n+1} =: ax_n + b \pmod{p}.$$

If p is large, and the integers a and b chosen appropriately, then the resulting sequence of integers x_0, x_1, x_2, ..., x_n, will, although not strictly random, appear so. (For example, $x_0 = 0$, and $x_{n+1} =: 3141592653x_n + 2718281829 \pmod{2^{35}}$ turns out to be a good choice.)

III. Early Uses of Randomization

The early uses of randomization were primarily recreational, religious, allocational, and political, although these catetgories are neither exclusive nor exhaustive.

1. Recreational. Games involving a chance element are the commonest, most familiar, and perhaps oldest instance of the use of physical randomization. Although Herodotus (*Histories* 1.94) reports that the Lydians claimed to have invented all games known to themselves and the Greeks, including "dice, knuckle-bones [i.e., *astragali*], and ball games," the astragalus is known to have been used in Egyptian board games dating back to the 1st Dynasty (i.e., c. 3500 B.C.). Gambling was endemic throughout the ancient world and by no means confined to the lower classes; Suetonius, for example, reports of the Roman emperors that Augustus was very fond of gambling, that "when Caligula would play at dice he would always cheat and steal," and that "so fervent was [Claudius's] devotion to dice that he wrote a book on the subject" (*Augustus* 71, *Gaius* 41, *Claudius* 33).

Some aleatory games are purely *stochastic*, whereas others involve some element of *skill*. Common examples of purely stochastic games include the use of coins, dice (hazard and craps), the astragalus or talus (strictly speaking, the knuckle-bone, but also commonly used in a generic sense for any type of bone), and throw-sticks. The most primitive of these, the astragalus, was in use throughout the ancient world, and remained popular until at least the Renaissance. Homer, who lived sometime before 700 B.C.), relates that Patroclus killed a companion in anger while playing with astragali (*Iliad* 23:87-88), while Cardano (1501-1576) thought them still important enough to analyze in his *Liber de Ludo Alea*. Irregular in shape, the astragalus has four long faces - one flat, one convex, one concave, and one irregular -- traditionally thought to be numbered 1,3,4,6. Usually they were tossed in groups of four and various throws were given names such as the "Venus" (all four sides different), the "dog", and the "vulture".

When the transition to dice began is unclear, but it must have been quite early. Cubical dice have been found in excavation sites as scattered as northern Iraq, Mohenjo-Daro in India, and 18th Dynasty Egypt (1400 B.C.). Typically the faces of these have values from 1 to 6 indicated by a corresponding number of engraved circular pips. An intriguing exception is a pair of Etruscan dice, discovered in a tomb at Toscanelli in 1848, which are inscribed with what are assumed to be the Etruscan symbols for 1 to 6. Other types of regularly shaped randomizers which developed include throwsticks and the medieval Jewish *dreidel*, a four-sided top.

Such devices are often used to introduce an element of chance into a game which also permits a variety of strategies, and therefore skill. Some of the major families encountered are the *race games* (including backgammon, pachisi, and probably the ancient Egyptian game of "hounds and jackels"), *domino games* (including the oriental *mah jongg*), and *card games* (including primero, cribbage, whist, poker, and bridge). The birth of the probability calculus was closely connected with determining the odds or fair prices for a variety of such games, and even today examples involving coins, dice, and cards are commonly used to introduce and illustrate the subject.

2. Religious

Equally ancient is the use of sortition (the use of lots) for a variety of religious or divinatory purposes. Although certain types of divination, such as astrology and the interpretation of dreams, do not employ chance mechanisms, many other forms do. The most common examples include the use of lots (*cleromancy*), and the random consultation of books (*bibliomancy*), favorite sources for the later being the Bible (*sortes Biblicae*), and poets such as Homer (*sortes Homerica*) and Virgil (*sortes Virgilianae*).

One type of divination which combines the two methods is based on the ancient Chinese *I Ching* or *Book of Changes*. The text of the *I Ching* is divided into 64 sections, each corresponding to a hexagram of six lines, each line either broken or unbroken. The two standard methods of consulting the book involve tossing either yarrow stalks or coins to randomly select a hexagram. The consultation of the Sibylline books, a collection of oracles brought to Rome at about the time of the founding of the Republic, may also have involved some form of randomization. The Germanic tribes of the 1st century A.D. would throw wooden strips inscribed with runes and, after "an intent gaze heavenward," sequentially pick up three and interpret their meaning (Tacitus, *Germania* 10). After the invention of playing cards in Europe c. A.D. 1350, the *Tarot cards* became a popular method of fortunetelling.

The use of lots in ancient Israel for divinatory puposes was quite common. Although the story of Jonah and the whale (*Jonah* 1:7) is the best known example, the Bible also relates several other episodes where persons who had offended God were discovered by the resort to sortition, e.g., Joshua's use of the lot to identify Aachen (*Joshua* 7:16-18), and Saul's use of the lot to identify Jonathan (I *Samuel* 14:40-42). (During the middle ages this practice continued in the form of the *ordeal by lot*.) The biblical justification for such practices was that "The lot is cast into the lap; but the whole disposing thereof is of the Lord" (Proverbs 16:33).

The religious use of lots persisted in Europe throughout the Middle Ages, although their legitimacy was often challenged. Thomas Aquinas, for example, distinguished between the *sors divinitoria*, or use of the lot to forecast the future, which he viewed as superstition and therefore prohibited, and both the *sors divisoria*, or use of the lot for allocation, and the *sors consultoria*, or use of the lot for making decisions, which he considered permissible (*Summa Theologica* II-2, Q.95, a.8). Use of the lot continued into the Protestant period; John Wesley learned the practice of drawing lots from the Moravians, with whom he became acquainted en route to America in the Fall of 1735, and he records a number of occasions when he employed it when in doubt as to a course of action.

3. Allocation

Although the Bible portrays the use of sortition in the story of Jonah as a means of identification, lots have also been used in similar situations as a means of fair or impartial selection. Homer relates that the Achaeans besieging Troy shook lots in a helmet to decide who would fight Hector (*Iliad* 7:170-191). Another, idiosyncratically Roman application of random selection was the practice of *decimation*: if a large body of men broke ranks during battle, a tenth of their number were chosen randomly by lot and executed. Polybius (*History*, 6.38) described this as an effective means of military discipline because "the danger and the fear of drawing the fatal lot threatens every man equally, and ... there is no certainty on whom it may fall." A similar motivation underlies the use of random audits by the U.S. Internal Revenue Service.

The use of the lot for military conscription goes back to Biblical times (*Judges* 20:9-10). In modern times lotteries have also been used for the draft, e.g., in Austria-Hungary between 1889 and World War I, and in the United States during both World Wars and the Vietnam War. Due to the widespread impact of and interest in such lotteries, their results have been closely scrutinized on several occasions. Both the 1940 and 1970 U.S. draft lotteries attempted to achieve physical randomization of numbered capsules by stirring or mixing them in a container, but in both cases later statistical analysis revealed clustering or trend incompatible with truly random selection (the case of the 1970 draft lottery was discussed earlier). Whether such lotteries were "fair" is primarily a philosophical question. An important goal of the American lotteries, however, was to present the appearance of fairness, and this, publicly executed as they were, they seem to have achieved.

4. Political

A particularly important use of random selection in the ancient world was its employment as a means of selecting public officials and jurors. In Sparta, members of the *gerousia* (council of elders) were elected by voice vote, candidates being presented sequentially in a random order determined by lot (Plutarch, *Life of Lycurgus* 25). In Athens, most ordinary magistrates and, from 487 B.C., the archons were chosen by lot, as were the Council of 500 and the juries, the latter by a very complicated, multistage procedure.

In Rome, sortition was used for a wide variety of major and minor political and judicial functions: e.g., juror selection, allocation of duties for consuls, censors, and praetors, registration of freedmen in one of the four urban tribes, and choice of senatorial ambassadors. (The outcome of the last was not always binding: when the lot fell on Cicero and Pompey in 60 B.C. for a legation to be sent to Gaul, the Senate voided the result, declaring that both were too important to leave Rome).

Although in general the ancient sources provide little information about the manner in which the lot was carried out, Athenian juror selection is a rare instance where a remarkable amount is known. This is partly owing to the happy accident of a papyrus copy of Aristotle's previously lost *Constitution of Athens* being dug out of the sands of Egypt in the 19th century, partly to excavations in the Athenian Agora which have unearthed *kleroteria*, the allotment machines used to implement sortition.

In both Greece and Rome, precautions were taken to ensure the absence of fraud in political sortition. The very complexity of the Athenian juror selection process was presumably intended to make tampering with the composition of a jury impossible. The second act of Plautus's *Casina* has a parody of Roman electoral sortition which suggests that the outcome could be affected by using lots of slightly smaller size or lighter wood. Augustus's first Cyrene edict of 7/6 B.C. requires that lots in jury selection be "as equal in size and weight as possible."

IV. Statistical Applications

Despite the wide variety of uses and applications surveyed above, a common thread which runs through many of the examples just discussed is the implicit notion of *fairness*: virtually all exemplify systems of selection or decision that are impartial or avoid responsibility because the outcomes are unknown in advance. In contrast, the uses characteristic of twentieth century statistics lie in a very different direction: the somewhat paradoxical notion that by imposing disorder in a structured way, it is possible to extract information -- by sampling, experimentation, or simulation.

1. Random sampling

Perhaps the most common statistical application of randomization lies in the selection of a random sample from a population. Inference based on the use of *random* (rather than *systematic*) sampling is a distinctively 20th century development, although earlier anticipations can be found in such procedures as the trial of the Pyx, an English test of coinage that may go as far back as the 12th century, and Laplace's use of a ratio estimator to establish the size of the French population in 1802.

A key initial step was A. N. Kaier's advocacy of the "representative method" in succesive meetings of the International Statistical Institute at Berne (1895), St. Petersburg (1897), Budapest (1901), and Berlin (1903); it is perhaps not entirely coincidental that documented instances of the use of physical randomization to draw samples begin to appear shortly after. By 1925 the outlook of the statistical profession had so shifted that discussions at the ISI meeting in Rome that year no longer concerned whether, but how to draw samples.

Despite famous snafus as the *Literary Digest*'s prediction of FDR's defeat in 1936 and Gallup's prediction of Truman's defeat in 1948, public opinion polls eventually gained general acceptance and today are perhaps the most publically visible example of a pervasive survey sampling methodology.

2. Experimental Design

L. J. Savage wrote in his famous Fisher lecture that "randomized design, and perhaps even the notion of a rigorously random sample, seems to originate with Fisher .... though this technique is so fundamental to modern statistics that to credit Fisher with it sounds like attributing the wheel to Mr. So-and-So." Savage's statement pinpoints the difficulty in untangling the historical development of this area. The use of randomization in experimentation certainly predates Fisher, notably in the Peirce-Jastrow experiment (1885) on the existence of a "least perceptible difference" in sensation (which employed a balanced randomization scheme carried out using shuffled decks of cards). Nevertheless, it was Fisher who first systematically forged and successfully popularized randomization as a powerful tool of scientific research, both in his books -- *Statistical Methods for Research Workers* (1st ed. 1925) and *The Design of Experiments* (1st ed. 1935) -- and in numerous papers.

Example: The Lady Tasting Tea

Fisher had a genius for choosing simple yet instructive examples to illustrate statistical methods and concepts. One of the most celebrated of these appears in his book *The Design of Experiments* (Chapter 2):

> *A lady declares that by tasting a cup of tea made with milk she can discriminate whether the milk or the tea infusion was added first to the cup. We will consider the problem of designing an experiment by means of which this assertion can be tested.*
>
> *Our experiment consists in mixing eight cups of tea, four in one way and four in the other, and presenting them to the subject for judgment in a random order. The subject has been told in advance of what the test will consist, namely that she will be asked to taste eight cups, that these shall be four of each kind, and that they shall be presented to her in a random order, that is in an order not determined arbitrarily by human choice, but by the actual manipulation of the physical apparatus used in games of chance ... . Her task is to divide the 8 cups into two sets of 4, agreeing, if possible, with the treatments received.*

If the subject of the experiment does not possess the claimed ability, the probability that she correctly guesses the order in which the ingredients were added to all 8 cups is $1/70 = 0.014$, or about $1\frac{1}{2}\%$ (because ${}_8C_4 = 8!/[4! \cdot 4!] = 70$). Note that because of the use of physical randomization in the design of the experiment, this probability statement is valid whatever method of selection the subject employs. (A substantial literature has sprung up around this example, concerning the issues of experimental design it raises.)

Fisher described his experiment as "psycho-physical"; from time to time there have in fact been attempts to demonstrate the existence of extra-sensory perception by statistical means. This has resulted in a considerable literature, some of it not without interest.

The extensive use of random assignment of treatment and placebo in the large-scale 1954 Salk polio vaccine field trial (several hundred thousand school-children participated) marks what may be described as the coming of age of the modern clinical trial.

3. Monte Carlo methods

Monte Carlo methods are techniques involving the use of random numbers to solve physical or mathematical problems that may be entirely nonstochastic in nature. Although primarily a 20th century development, examples of their use may be found earlier, notably in several 18th and 19th century experiments involving the estimation of π by Buffon's method of tossing a needle on a grid of equidistant parallel straight lines, and Lord Kelvin's tests of the equipartition theorem in statistical mechanics. Nevertheless, it was only during World War II that the potential of Monte Carlo methods was first fully realized, when Ulam and von Neumann employed simulation techniques to solve problems involving neutron diffusion in fissile material.

The most important statistical application of the Monte Carlo method involves the use of simulation to approximate sampling distributions. In 1908 W. S. Gosset ("Student") simulated the distribution of the sample correlation coefficient by drawing 750 samples of size 4, using shuffled slips of cardboard, from W. R. Macdonell's 1901 correlation table giving heights and middle-finger lengths of 3,000 criminals. Bradley Efron's "bootstrap" technique takes such methods one step further, simulating the sampling distribution of a statistic by drawing random samples with replacement from a miniature population which is itself a random sample drawn from an original population of interest.

V. Examples and Exercises

Employing any of the methods of randomization discussed earlier, it is possible to devise a great variety of exercises illustrating various aspects of chance phenomena and statistical inference. The follow are two examples.

1. Experimental determination of the arc-sine law

Using a random number generator provided as part of the statistical or other software for a personal computer, one can generate 100 replications of a symmetric random walk having n = 10, 000 steps in order to investigate the typical sample path behavior of this stochastic process. Let $X_1, X_2, \ldots, X_{10,000}$ represent the outcome of tossing a fair coin 10,000 times, let

$$S_n = X_1 + X_2 + \ldots + X_n,$$

let $(X_0, Y_0) = (0, 0)$, and let $(X_n, Y_n) = (n, S_n)$ for $n \geq 1$. The *sample path* of the random walk corresponding to this particular realization is the set of points $\{(X_j, Y_j) : 0 \leq j \leq 10,000\}$; or better, the curve that arises from connecting adjacent points by straight-line segments. For many it is a considerable surprise to discover that the typical sample path of such a symmetric random walk does not evenly divide its time above and below the horizontal axis, but rather spends its time either mostly above or mostly below this axis.

The precise mathematical description of such behavior is the content of the *arc-sine law*; using the 100 replications of the random walk and tabulating the percentage of the resulting sample paths which spend 0% of their time above the horizontal axis, 10% of their time, 20% of their time, etc., one can discover experimentally the qualitative shape and characteristics of the arc-sine law.

2. Monte Carlo estimation of π

Let $(X_1, X_2, \ldots, X_n)$ and $(Y_1, Y_2, \ldots, Y_n)$ denote two independent sequences of numbers drawn at random according to a uniform distribution from the the unit interval [0, 1]. Then (X_j, Y_j), $1 \leq j \leq n$, represents a point selected at random from the unit square [0, 1] x [0, 1]; in consequence, the frequency with which such a point falls in any region A is equal to the area of that region. Thus, in particular, if A $=: \{(x, y) : x^2 + y^2 \leq 1\}$, then $P[(X_j, Y_j) \in A] = \pi/4$. This simple observation enables us to estimate π by Monte Carlo: count the number m of pairs (X_j, Y_j) for which $X_j^2 + Y_j^2 \leq 1$, and let $\pi =: 4m/n$.

VI. Bibliography

The literature on the various uses of randomization is extensive and fascinating. The following references are intended in part to document statements made in the text, in part as an entry and guide into this literature.

Random Number Generation

The history of the Tippett, Kendall and Babington-Smith, and RAND tables is discussed in M. Muller, "Random Numbers," *International Encyclopedia of Statistics* (New York: Free Press, 1972), W. Kruskal and J. Tanur, eds., pp.841-842. For a brilliant discussion of pseudo-random number generation, see D. Knuth, *The Art of Computer Programming, Vol. 2: Seminumerical Algorithms* (New York: Addison-Wesley, 1969), Chapter 3 ("Random Numbers"). Adequate discussion of randomness in quantum mechanics would require an article in itself. R. P. Feynman, *QED: The Strange Theory of Light and Matter* (Princeton: Princeton University Press, 1985) and A. Fine, *The Shaky Game: Einstein, Realism, and the Quantum Theory* (Chicago: University of Chicago Press, 1986) are two very readable places to start.

Recreational and other non-statistical applications

Games of Chance

Two standard treatments are E. Falkener, *Games Ancient and Oriental and How to Play Them*, (London, 1892; reprinted New York: Dover Publications, 1961) and R. C. Bell, *Board and Table Games from Many Civilizations* (Oxford University Press, 1969). For classical antiquity, an exhaustive and definitive treatment is Lamer's article "Lusoria Tabulae" in A. Pauly, G. Wissowa, and W. Kroll, *Real-Encyclopadie der klassischen Altertumswissenschaft* (Stuttgart, 1893-), vol. 13, 1900ff. See also F. N. David, "Dicing and Gaming (A Note on the History of Probability)," *Biometrika* 42 (1955), *Games, Gods and Gambling* (Hafner: New York, 1962), Chapter 1 ("The Development of the Random Event"); L. E. Maistrov, *Probability Theory: A Historical Sketch* (New York: Academic Press, 1974), 7-15. For the history of playing cards, see Michael Dummett's *The Game of Tarot* (London: Duckworth, 1980).

The first book to mathematically analyze games of chance was the 16th century *Liber de Ludo Alea* of Gerolamo Cardano (1501-1576), although it remained unpublished until several years after the appearance of Huygen's 1657 treatise. Oyestein Ore's *Cardano: the Gambling Scholar* (Princeton University Press, 1953; reprinted New York: Dover Publications, 1965) is an excellent biography that also contains a translation of the *Liber*. An important predecessor of Hoyle that gives a picture of the commonest games of chance played at the time of the birth of the probability calculus is Charles Cotton's *The Compleat Gamester* (1st ed. 1674). Pierre-Remond de Montmort (16780-1719), however, deserves the credit for the first truly modern analysis, in his *Essai d'Analyse sur les Jeux de Hasard* (1st ed. Paris: 1708; 2nd ed. 1714; reprinted by Chelsea), although this was soon surpassed by Abraham de Moivre's *The Doctrine of Chances* (1st ed. 1718; 3rd ed. 1756; reprinted New York: Chelsea Publications, 1967), one of the great classics of probability theory.

For a probabilistic analysis of the *I Ching*, see D. Robinson, "Divination in Ancient China," *Applied Statistics* 24 (1975), 329-332.

In Ancient Israel

Useful accounts include R. de Vaux, *Ancient Israel* (New York: McGraw-Hill, 1961), 349-355; the *Encyclopedia Judaica* 11 (1971), 510-513, s.v. *Lots*; ibid., 16 (1971), 8-9, s.v. *Urim and Thummim*; N. L. Rabinovitch, *Probability and Statistical Inference in Ancient and Medieval Jewish Literature* (1973), 21-35; and A. M. Hasofer, "Random mechanisms in Talmudic Literature," *Biometrika* 54 (1966), 316-321. For information about games of chance among the

Jews in both ancient and medieval times, see the *Encyclopedia Judaica*, s.v. "Gambling," "Games," and "Cards and Card-Playing."

In Greece

For detailed information concerning the use of randomization in jury allotment and election of officials, see: T. W. Headlam, *Election by Lot at Athens* (1891, 2nd ed. 1933); S. Dow, "Aristotle, the Kleroteria, and the Courts," *Harvard Studies in Classical Philology* 50 (1939), 1-34; C. Hignett, A History of the Athenian Constitution to the End of the Fifth Century B.C. (Oxford: Clarendon Press, 1952), index s.v. "Sortition;" E. S. Staveley, *Greek and Roman Voting and Elections* (Cornell University Press: Ithaca, 1972), Chapter III, "The Operation of the Lot," 61-72; J. M. Moore, *Aristotle and Xenophon on Democracy and Oligarchy* (1975), 302-308; D. M. MacDowell, *The Law in Classical Athens* (1978), 35-40.

In Rome

For the use of the lot in Roman assemblies, see: L. R. Taylor, *Roman Voting Assemblies* (Ann Arbor: University of Michigan Press, 1966), 70-74; Staveley, (cited earlier), pp. 230-232. For decimation, see Fiebiger's entry in the *Real-Encyclopadie*, s.v. *decimatio*.

The Middle Ages

For the ordeal by lot, see H. C. Lee, *The Ordeal* (Philadelphia: University of Pennsylvania Press, 1973), 106-112. Rabelais has an amusing discussion of Homeric and Virgillian bibliomancy in his *Gargantua and Pantagruel*, Book 3, Chapters 10-12.

Statistical applications.

Sampling

For the history of sampling, see F. F. Stephan, "History of the Uses of Modern Sampling Procedures," *Journal of the American Statistical Association* 43 (1948) 12-39. Kaier's contributions and the later evolution of the notion of representative sample are discussed in W. Kruskal and F. Mosteller, "Representative Sampling, IV: the History of the Concept in Statistics, 1895-1939," *International Statistical Review* 48 (1980) 169-195. For the trial of the Pyx, see S. M. Stigler, " Eight centuries of sampling inspection: the trial of the Pyx," *Journal of the American Statistical Association* 72 (1977) 493-500; for Laplace's estimate of the size of the French population, W. G. Cochran, "Laplace's ratio estimator," *Contributions to Survey Sampling and Applied Statistics*, H. A. David, ed. (New York: Academic Press, 1978), 3-10.

Experimental Design

For details of the Peirce-Jastrow experiment, see C. S. Peirce and J. Jastrow, "On Small Differences in Sensation," *Memoirs of the National Academy of Sciences for 1884* 3 (1885) 75-83; for a brief modern discussion, see S. M. Stigler, "Mathematical Statistics in the Early States," *Annals of Statistics* 6 (1978), pp.247-249. No really satisfactory assessment of Fisher's contributions exists to date, but for a useful summary, see the articles by Cochran, Holschuh, and Picard in S. Fienberg *et al.*, eds., *R. A. Fisher: An Appreciation* (New York: Springer-Verlag, 1980). Fisher's example involving the lady tasting tea has spawned a substantial literature of commentary: see, e.g., J. Neyman, *First Course in Probability and Statistics* (New York: Holt, Rinehart, and Winston, 1950), 272-289; N. T. Gridgeman, "The Lady Tasting Tea and Allied Topics," *J. Amer. Statist. Assoc* 54 (1959), 776-783; C. I. Bliss, *Statistics in Biology*, vol. 1 (New York: McGraw-Hill, 1967), Chapter 1 ("A Taste Experiment").

The earliest probabilistic analysis of guessing card-type in a randomly shuffled deck would appear to be C. Richet, "La suggestion mentale et le calcul des probabilités," *Revue Philosophique* 18 (1884), pp.608-671; see also the discussion by Sorel in the same journal, 23 (1887), pp.50-66. At about the same time, F. Y. Edgeworth analyzed similar experiments in Great Britain: "The Calculus of Probabilities Applied to Psychical Research," *Proceedings of the Society for Psychical Research*, 3 (1885) 190-199, 4 (1886) 189-208. For Feller's criticisms of the parapsychological literature, see his article "Statistical Aspects of ESP," *Journal of Parapsychology* 4 (1940), 271-298, and the reply by J. A. Greenwood and C. E. Stuart immediately following. For a valuable survey and discussion, see P. Diaconis, "Statistics and ESP," *Science* (1977).

Monte Carlo Methods

There is an extensive literature on Buffon's needle problem and its generalizations: particularly noteworthy is M. D. Perlman and M. J. Wichura, "Sharpening Buffon's Needle," *American Statistician* 29 (1975), 157-163; see generally H. Solomon, *Geometric Probability* (Philadelphia: SIAM, 1978), Chapter 1, for a survey and further references. For details of Kelvin's Monte Carlo calculation of an integral, as well as further history and references, see J. M. Hammersley and D. C. Handscomb, *Monte Carlo Methods* (London: Methuen, 1964); W. Thompson, "19th century clouds over the dynamical theory of heat and light," *Phil. Mag.* Ser 6, 2 (1901), pp. 1-40. S. M. Ulam relates his discovery of the Monte Carlo method in *Adventures of a Mathematician* (New York: Charles Scribner's Sons, 1976), pp.196-200. For the bootstrap, see B. Efron, *The Jackknife, the Bootstrap and Other Resampling Plans* (Philadelphia: SIAM, 1982).

Sandy Zabell is professor of mathematics and statistics at Northwestern University where he is also a member of the Program in the History and Philosophy of Science. He has also taught at the University of Chicago, the University of California at Berkeley, and Stanford University. His interests include mathematical probability, and the history, philosophy and legal applications of probability and statistics. He is a Fellow of both the ASA and the Institute of Mathematical Statistics, and has twice served on panels of the National Academy of Sciences dealing with the legal applications of statistics.

The Psychology of Learning Probability[1]

Ruma Falk[2] and **Clifford Konold**
Department of Psychology Scientific Reasoning
and School of Education Research Institute
Hebrew University of Jerusalem University of Massachusetts, Amherst

Today our vision of the world is permeated by probability, while in 1800 it was not. Probability is the great philosophical success story of the period.

Hacking, in [8]

1. Introduction

Probability is too frequently regarded as a subsidiary topic of statistics. Consistent with this view, statistics educators often teach the minimum amount of probability they regard as sufficient for learning statistics. We regard this as a grave mistake. *Probability is a way of thinking.* It should be learned for its own sake. In this century probability has become an integral component of virtually every area of thought. We expect that understanding probability will be as important in the 21st century as mastering elementary arithmetic is in the present century.

The term "probabilistic revolution" [8 and 9] broadly suggests a shift in world-view from a deterministic description of reality, phrased in terms of universal laws of stern necessity, to one in which probabilistic ideas have become central and indispensable. Concepts of uncertainty are introduced into science partly because *we are ignorant* of the multiplicity of variables affecting our data, and because there is error in our measurements. Hence, even those who believe in the ultimate determinism of nature, nevertheless use probabilistic language to account for human shortcomings. The revolution goes further, however, viewing chance as an irreducible part of natural phenomena. That conception is typically represented by the *inherently indeterminate* view of quantum physics. An equally dramatic example of a probabilistic revolution that has changed our thinking about our own existence is found in evolutionary biology.

In learning probability, we believe the student must undergo a similar revolution in his or her own thinking. There are good reasons to expect that this breakthrough cannot be easy. There might have been a firm psychological basis for the historical persistence of the deter-

[1]Received: November 1990; Revised: August 1991

[2]This article was written while Ruma Falk was on sabbatical leave at the Department of Psychology, University of Massachusetts, Amherst. Preparation of the chapter was supported in part by the Sturman Center for Human Development, the Hebrew University, Jerusalem, and by the National Science Foundation, Grant MDR-8954626. We wish to thank Raphael Falk for his helpful comments.

151

ministic outlook in science. As we shall show, students (and some teachers) are apriori widely inclined to think in causal, deterministic terms and to overlook the effects of chance.

In our statistical age, *public* discourse is cast in the language of statistics and probability. However, our *private* musings and observations may still be subject to the old deterministic habits of thought. Whatever our psychological leanings and untutored intuitions, "for certain public purposes we shift our gaze from individuals to averages, from deep-felt certainties to probabilities, from impressions to numbers" [4, p. 291]. The teaching of probability should strive not only to provide students with the framework necessary for carrying on such a public discourse, but also to transform their private "musings."

We first focus on some major psychological obstacles inherent in making the switch to probabilistic reasoning. Next, we illustrate a few prevalent misconceptions of chance, adding a caveat for the teacher not to overdo confronting students with their own erroneous intuitions. Finally, we advocate starting the process of probabilistic education by building on the firm basis of students' sound intuitions, endeavoring to strengthen students' conceptual grip of probability theory without getting too technical. We address issues that should be relevant to a wide range of levels of introductory courses.

2. The Search for Certainty

> *All that happens around children is interwoven with probabilistic elements, and all they hear and say, with logical ones.*
>
> T. Varga

2.1 Decisions versus outcomes

Some years ago the first author devised a probability game in which young children's ability to compare probabilities [3] could be practiced: each of two players is required to choose which of two discs to spin before moving on a game board. Each of the discs is divided into sectors of two colors, one favorable and one unfavorable. The discs are divided in different proportions, and it is to a player's advantage to select and spin the disc with the greater probability of a favorable outcome. The game was sharply criticized by parents and educators as being 'uneducational.' They objected to the notion of a game in which one might make a correct choice (of the disc with a higher probability of success) and yet obtain an unfavorable outcome, while on the other hand, an incorrect decision may be rewarded. Obviously, they wished for a consistent, 'just' system. Implied in their criticism was the expectation that good decisions would always be reinforced, while bad ones would never be. Apparently, the two concepts of a *correct choice* and a *favorable outcome* were not clearly differentiated in these critics' minds.

Learning and acquiring knowledge are typically associated with increasing certainty and dispelling doubt. Hence, the primary objective in teaching probability is to reconcile students' habitual search for (certain) knowledge with the understanding that learning probability will not enable us to predict the immediate outcome of a random process. Even the most probable event, which is expected to occur most frequently in the long run, is not guaranteed to occur on the next opportunity.

2.2 Underrating the role of chance

The tendency of students of all ages either to reject uncertainty altogether or to underestimate its impact has been documented in a variety of educational and psychological studies [e.g., 3]. Young children of about age 6 said of a spinning top with a 2/3 probability of success "Now I am sure to win." Probabilities greater than 1/2 were all "sure to win," while those below 1/2 were "sure to lose." Children playing a game of chance in which they repeatedly observe the different possible outcomes, often change their minds about their correct choices when they are not promptly rewarded. They also stick to incorrect choices which happen to result in a favorable outcome. Similarly, some adults [7] consider a weather forecast stating that the probability of rain on a particular day is 70% to be wrong if it does not rain that day.

Many of the biases and fallacies found in the research on judgment under uncertainty [6] result from employing various heuristics which may roughly be characterized as minimizing the role of chance. There is widespread confusion between the terms 'representative sample' and 'random sample.' People tend to identify a random sample with a *true cross section* of the population, neglecting the ever-present variation due to chance sampling. A variety of studies conducted in the last decades offer a fairly coherent picture of people's beliefs about randomness: disregarding chance fluctuations, people expect population proportions to be represented not only globally in the entire random sequence, but also locally in each of its parts. They regard chance as a self-correcting mechanism in which a deviation in one direction is *promptly* balanced by a deviation in the opposite direction. That view clearly ignores the most significant characteristic of chance, namely, its *blindness*. Unlike the human observer, a chance process does not review its productions and feels no obligation to restore equilibrium in a sample of any size.

Nonregressive predictions, typically obtained in judgment research [6], can also be construed as a form of denial of chance's contribution. To expect sons' heights to be as extreme as those of their fathers is to ignore the imperfect validity of father's height as a predictor. The overlooked variables, such as mother's genes and pre- and post-natal environmental factors, are either weakly correlated with, or completely independent of, father's height. It is because these other multiple factors are nonsystematic, or *random*, that we get regression toward the population mean. An extreme reliance on the input as predictor is, in fact, a typical manifestation of shunning chance.

One of the first goals in teaching probability should be to help the student recognize the fact that chance cannot be 'driven out of the system.' No matter what progress we make in our study of probability or of the phenomena of interest, uncertainty cannot be eliminated. The best we can hope for is to sharpen our tools for accurately quantifying our long-run expectations (or our degrees of uncertainty, depending on the interpretation of probability that we endorse).

2.3 Short versus long run

When randomness is present, probability answers the question, "How often in the long run?" and expected value answers the question, "How much in the long run?"

D. S. Moore

Probability is one of many mathematical and physical concepts that have been studied by cognitive-developmental psychologists using the binary-choice paradigm. In these experiments, two stimuli are presented in each trial, and the subject is required to judge in which of them the studied variable assumes a greater value. Concepts studied using this paradigm include number of objects, liquid quantity, equilibrium in a balance-scale apparatus, proportionality of orange-juice concentration, and sweetness of sugar solutions. Piaget and Inhelder [11] were first to present binary comparisons of probabilities to children, and many studies have since introduced modifications and improvements to this paradigm. One significant feature, however (the educational implications of which are far-reaching), distinguishes probability from the group of studies of other mathematical-physical concepts: in the latter, a correct judgment is invariably accompanied by positive feedback. The correctness of choice between the two sides of the balance scale, or between two cups of sugar solution, can be promptly verified by a single experimental trial. In the probabilistic situation, however, the role of immediate experimental feedback is equivocal.

If success is not guaranteed even though one chooses the alternative with the highest probability of success and, moreover, choosing the option with a lower success-probability may nevertheless prove successful, in what sense is the former choice correct and the latter incorrect? The answer is to be found in the *long run*. According to the frequency interpretation of probability, the probability of success is the limiting proportion of successful trials when the number of repeated trials grows indefinitely. A test of the correctness-of-choice between two urns with different proportions of beads of the winning color requires multiple draws. Furthermore, regardless of the sample size, the conclusion is still an inference with some, though small, degree of uncertainty.[3]

Suppose that, in accord with the short-run expectations of the critics of the game described above, we arrange that a choice with 75% probability of success is 100% rewarded. In that case we would be teaching children something incorrect! The real (uncertain) world is so structured that even wise decisions are only probabilistically reinforced. It is only in the long run that we should expect good decisions to pay off. Thus, choice of the disc constructed of the highest proportion of the winning color is justified, not because it ensures an immediate reward, but because it will turn up most favorably over *many repetitions*. Simple as that statement must seem, it is not easily internalized and it defies many students' intuition.

The concept of probability can be decomposed into two subconcepts: proportion and chance. It is because of chance that a correct decision proves true only in the long run. The distinction between a correct probabilistic choice and a favorable outcome is apparently quite elusive since, in most of the studies of choice between probabilities, investigators have overlooked the distinction, analyzing the task as if it involved only comparison of proportions.

The complicated status of feedback as a vehicle of instruction poses a serious challenge to the teacher of probability. In discouraging students from regarding *one* experimental result (despite its evidential value) as an ultimate criterion of correctness, we do not want to diminish

---

[3]It should be noted that on a deeper level of analysis (in accord with the probabilistic world-view) the difference in the role of feedback between 'deterministic' experiments and a random experiment can be construed as just a difference in degree. It is not absolutely certain that a sip from the cup with the stronger sugar solution will indeed taste sweeter. Even when the solutions have been properly stirred, there is always a non-zero (though practically negligible) probability that a sample of that cup may taste less sweet than a sample of the other cup. It is because of the gigantic number of particles in any small sample of the sugar solution that success in a single tasting from the correct cup is virtually guaranteed. This is simply a manifestation of the "law of large numbers." Were we to devise a mechanism of drawing single molecules from the cup, the situation would reduce to that of drawing beads from an urn.

in their minds the importance of observing relevant data; we simply want to shift their perspective from the very short to the long run.

2.4 Prediction versus quantification

I don't believe in probability, because even if there is a 20% chance, it could happen. Even 1%, it could happen. I don't believe in probability.
                                                                    Student Quotation, July, 1987

While most students entering their first course in probability do not come with an alternative discipline of thought, they do speak a language different from that of the teacher (or at least that part of the teacher who stands ready with an urn and a pocket full of formulae). This is because students, as inhabitants of an uncertain world, have learned to think and converse in everyday language about 'chance,' 'probability,' 'luck,' 'randomness,' and have developed a rich vocabulary with which to communicate degrees of belief from 'I'm certain' to 'it's impossible.' Underlying these terms exists a perspective on uncertainty that is, at points, at odds with formal theory.

A very basic difference between formal and informal views of probability concerns the perceived objective in reasoning about uncertainty. Formal probability is mostly concerned with deriving measures of uncertainty, or answering the question *'How often will event A occur in the long run?'* On the other hand, what most people want is to *predict* what will occur in *a single instance* — to answer the question *'Will A occur or not?'*

Konold [7] has referred to this latter perspective as the "outcome approach." People reasoning via the outcome approach tend to interpret a request for a probability of some event as a request to predict whether or not that event will occur on the *next* trial. For example, subjects were given an irregularly shaped bone that could rest (with unequal probabilities) on six different surfaces [7]. They were asked, "If you were to roll this, which side do you think would most likely land upright?" After answering this question, subjects rolled the bone. Several spontaneously judged their answer as being wrong (or right) after that single roll. And as mentioned previously, subjects evaluated the forecast "70% chance of rain tomorrow" as being incorrect based on the information that it did not rain. Furthermore, given records of what happened on 10 days for which a 70% forecast of rain had been made on each day, subjects judged the forecaster as performing suboptimally when, in fact, it rained on 70% of those days. According to these subjects, a perfect performance would have entailed rain on all 10 days.

Understanding that many students view the objective of probability as making predictions can help the teacher of probability make sense of what otherwise are incomprehensible statements. The statement at the beginning of this section was made by a student about half-way through a two-week workshop on probability. Our interpretation of this student's claim is that she is in the process of giving up her outcome orientation. She is questioning her ability to make accurate single-trial predictions. Note that if the phrase 'making single-trial predictions' is substituted for the word 'probability,' the claim is understandable and normative. What this student has not yet come to realize, however, is that the word 'probability' refers to something other than single-trial predictions.

Given the desire for predictions, outcome-oriented subjects translate probability values into yes/no decisions. A value of 50% is interpreted as total lack of knowledge about the outcome, leaving one without justification for making a prediction. For example, rather than interpreting the 50% probability of heads in flipping a fair coin as the expected percentage of heads in the long run, many subjects interpret it as "You can't tell...it could be anything." Similarly, a forecast of 50% chance of rain is taken as an admission of total ignorance: "If he said 50/50 chance, I'd kind of think that was strange ... that he didn't really know what he was talking about ... ." ([7] p. 68).

Probability values sufficiently above or below 50% are coded as "yes" or "no" predictions, respectively, as illustrated below. (*I* and *S* stand for Interviewer and Subject, respectively.)

I: What does the number, in this case the 70%, tell you?

S: Well, it tells me that it's over 50%, and so, that's the first thing I think of. And, well, I think of the half-way mark between 50% and say 100% to be like, well, 75%. And it's almost that, and I think that's a pretty good chance that there'll be rain. ([7] p. 66.)

Konold [7] also found a correlation between the tendency to interpret questions about probability as requests for single-trial predictions and a preference for analyzing probabilistic situations from a causal perspective. Indeed, the belief that one could successfully predict single instances would seem to require an understanding of the causal mechanisms involved. One component of this causal orientation is the preference for basing predictions on information that seems causally related to the outcome of interest rather than on frequency information. Subjects who tended to evaluate predictions as right or wrong after a single trial also tended to give little weight to the results of rolling the bone. For example, after rolling the bone 7 times and getting 3 Ds, 2 Cs and 2 Bs, the subject below judged C to be the most likely based on an analysis of the 'shape' of the bone. She was then asked:

I: Would rolling that a lot more times help you, in any way, decide which side is most likely to come up?

S: I don't think so.

She was then given a list showing the results of 1,000 rolls.

I: Would you feel safe in concluding, looking at this table, that D is, in fact, more likely than B? [Frequencies were A-50, B-279, C-244, D-375, E-52, F-0]

S: I don't know. Not really. I think it could just have something to do with your luck, or your chance. If I did the same number of rolls, I don't know if it would come out the same.

I: What is your belief about which side is most likely to come up if you rolled it?

S: I have to look at this [bone] again. It changed shape in the past few minutes. [Laughter] I still think C.

In addition to underweighting frequency data, a few subjects suggested that the number in the statement "70% chance of rain" was a measure of some factor causally related to rain -- "Maybe it means that there's 70% humidity in the air... ."

Most instruction in probability ignores problems associated with interpretation, so that disagreements are allowed to remain unexamined. These ultimately reveal themselves, however, in students' inability to transfer or to show other signs of understanding. Students who reason according to the outcome approach cannot be treated as if they are suffering from a simple mis-understanding. It does not do, for example, for the teacher to keep repeating, "What I mean by probability is... ." Each student will have to struggle with various contradictions that the outcome approach (and judgment heuristics) entails before they are able to adopt a more normative view.

The difficulties for students of probability are not over, however, once they have accepted uncertainty and adopted the convention of representing this uncertainty quantitatively. The problems we discuss below are characteristic of those that continue to plague students even after they have gotten past the more rudimentary aspects that we have discussed to this point.

3. Difficulties in Applying the Probabilistic Model

Probability is especially rich in counterintuitive examples, which often entail fallacies and paradoxical conclusions. Some of these examples played an important role in the development of probability theory. Students may likewise benefit from comparing their intuitions concerning puzzles and paradoxes with normative solutions. This activity requires increased awareness of one's own thought processes. Knowledge of one's own thinking (metacognition) is no less important than learning the right solution, and reflective thinking is a vital step toward achieving abstract mathematical ability. A paradox usually triggers a conflict. Such conflicts may encourage students to critically examine their intuitive theories. In parallel to the historical development of the theory, this examination might reveal to students deficiencies in their understanding and so promote the development of normative concepts. In addition, many difficulties in probability arise from ambiguities and complexities involved in translating real-world situations into formal ones. Solving an intriguing puzzle requires an unequivocal definition of the problem at hand. The problem is often solved once the assumptions involved in modeling are made explicit. These possibilities will be demonstrated via a few examples.

3.1 The gambler's fallacy

Many people seem to believe that there is something akin to the law of atone-ment in probability. If a coin thought to have no built-in bias produces nine heads in a row, they reason, one would be wise to bet on tails the next flip. After all, if the ratio of tails to total flips is ½ in the long run, then, in the aforementioned long run, there should be as many tails as heads and that means tails have some catching up to do. The fallacy in this line of reasoning lies in the fact that the coin has no memory nor conscience. The coin simply isn't aware that tails are lagging behind heads.

S. K. Campbell

Events in a random sequence are statistically independent of each other. The concept of *statistical independence*, though easy to formally define via simple equalities of probabilities, is not easily internalized. Observers find it difficult to avoid the perception of interdependence among events that are in fact unrelated. That fallacy is revealed not only in the gambling set-ting, but in diverse tasks, such as generation and judgment of randomness and assessment of likelihoods. The gambler's fallacy appears in varied disguises. People typically regard long runs that turn up by chance as non-random occurrences. Thus, when they observe sequences of hits and misses in a basketball game, they attribute the random runs, which *seem* too long for the production of chance, to a mysterious, non-random agent [5]. This agent has been dubbed the 'hot hand' in the case of long runs of hits. In complete analogy, the attention of casino gamblers is drawn to random runs of wins or losses that are subjectively too long. They similarly invent a non-chance agent, 'luck,' to account for these apparent deviations from randomness [12].

Students should be encouraged to view statistical independence from different perspectives in the hope that they will assimilate its meaning to the point of eliminating these notorious biases. But overcoming these biases will not be easy; they are quite compelling. Nor are they limited to naive subjects; even experts can fall prey to the fallacies. Consider the following problem,[4] which is addressed to a large class of students.

> Suppose each of you writes down on a note how many brothers and how many sisters you have. We collect the notes and separate them into those written by women and those written by men. Should we expect men to have more sisters than women have? and would men have more sisters or more brothers? or are no differences expected?

Despite agreeing to assume equal probabilities for male and female births and independence among births, most people expect men to have both more sisters than women have and more sisters than brothers. They reason that because, 'on the average,' families have an equal number of sons and daughters, the set of siblings of men, who are themselves not counted, should comprise an excess of women (sisters). This compelling reasoning is nevertheless wrong! Men are expected to have the same number of sisters as women have, and the same number of sisters as brothers.

Overcoming the gambler's fallacy in this case requires a full realization of the significance of statistical independence. Asking a random child from a family of n children how many brothers and how many sisters he or she has, is equivalent (mathematically) to picking a random family with $n-1$ children and asking how many sons and how many daughters it has. In the latter case we expect, however, the same number of sons and daughters. Hence, irrespective of whether the questioned persons are males or females, they would have, on the average, the same numbers of brothers and sisters.

Even though the above reasoning seems right, some doubt may linger in students' minds. The feeling may persist that men ust have more sisters since the men are not counted in the statistics of their families. That doubt may be enhanced by the apparent similarity to sampling *without* replacement from an urn with half Ms and half Fs. It is important, therefore, to dwell further on this problem until one clearly sees that asking a man about the sexes of his siblings is -- statistically speaking -- identical to asking a woman the same question. This is the essence of the concept of statistical independence: the probabilities of male or female births in the respondent's family are *not affected* by knowledge of the respondent's sex.

Some students may nevertheless experience some conflict between the long-run expectation of equal proportions of males and females and the exclusion of the respondents from consideration in a finite population. The following is a representative argument that was given

---

[4]We advise the readers to try to solve this, as well as the other problems, before reading our exposition of the solution.

by a male student in one of our classes: "Suppose there were 1,000 men in our class. Consider all 1,000 of us together with our brothers and sisters. That large group should comprise approximately equal proportions of males and females. Now, exclude us, 1,000 men, and there would remain more sisters than brothers." This reasoning fails to take into account the fact that all-daughter families are not represented among those 1,000 families, and, moreover, that families with many sons are overrepresented in that large group. The group will therefore contain an excess of males.[5] Only after removing the original 1,000 men should the expected number of males (brothers) equal that of females (sisters).

3.2 Overlooking the chance mechanism

Problems about sex distributions of children in families provide a useful context for applying basic probability theorems. The example below involves only two children, yet it is challenging even to those who accept the conventional assumptions concerning sex determination.

> Mrs. F. is known to be the mother of two children. You meet her in town with a boy whom she introduces as her son. What is the probability that Mrs. F. has two sons? One possible answer is that the probability is 1/2: you now have seen one boy, and hence the question is whether the other child is a male. Another possibility is that the probability is 1/3: you learned that Mrs. F. has 'at least one boy,' and hence three equiprobable family structures are possible (BB, BG, GB), of which the target event (BB) is but one.

Both arguments make sense. No further discussion can decide between the two, because each may be true depending on the assumptions one makes. The explication of the assumptions, however, is missing in the problem's statement. The solution should depend on the exact method by which the observation has been obtained, namely, on the chance mechanism that has generated the datum. If the woman in question typically chooses at random one of her two children to accompany her on her outings, then your observation is 'a randomly selected child of the two-children-family was found to be a boy.' In such a case, a family with two sons is twice as likely to yield that observation than a mixed family. A simple Bayesian calculation shows that the posterior probability that the woman has two sons is 1/2 (see [1]). If, however, being a part of a male-chauvinistic society, she always prefers taking a son along with her, then your observation becomes 'the family has at least one son' and the three families of that kind are equally likely to bring about the above meeting. Consequently, the probability of two sons should be one third.

There is obviously no point in arguing about the correct answer to a problem that is under-defined. We could only *favor* one or the other of the assumptions as more reasonable. Lacking any other information, the most sensible assumption would be that chance determines the child to accompany the mother on a given trip. Indeed, most real-life situations, from which you learn that a given woman has a son, are similarly structured: you overhear the woman mentioning a son, or you pay a visit and catch a glimpse of a boy. In each of these cases, the greater the proportion of sons among the children in the family, the more likely is a chance meeting with one of its sons. Inventing a scenario in which such an encounter is equally likely

---

[5]The example 'inadvertently' provides an interesting opportunity to point out the sampling bias introduced by sampling families via their sons.

under all three family compositions -- BB, BG, and GB -- is considerably more difficult. The answer 1/2 is thus favored over 1/3 only in the sense of regarding the assumptions it is based on to be more realistic.

Textbooks of probability theory often begin with the concept of a *statistical experiment*, namely, an idealized, well-defined chance process whose elementary outcomes comprise the sample space. Indeed, probability theory is concerned with the outcomes of statistical experiments. As demonstrated in the analysis of the above family problem, the probabilistic conclusions depend on the exact definition of that random process. Describing the habits of the mother in choosing the child to accompany her to town amounts to modeling her behavior as formally equivalent to one urn model or another. The statistical experiment should clarify whether one child is randomly sampled from a (random) two-children family, or whether one family is randomly sampled from two-children families having at least one son.

Assumptions play a central role in reasoning and problem solving in all areas. Challenging one's assumptions is especially important when solving probability problems because of the subtle intricacies of modeling real-life situations. The psychologists Nisbett and Ross advise teachers to offer useful slogans for didactic purposes. One of their slogans, *Which hat did you draw that sample out of?* may be instrumental in alerting the students to reflect on the chance mechanism underlying the problem. A more general reminder would be, *Make implicit assumptions explicit*! Keeping that advice in mind, let us turn to another problem.

The "Flaws, Fallacies, and Flimflam" column of the *College Mathematics Journal* (January, 1990, p. 35) offered the following teaser from the folklore of probability (suggested by R. Guy):

> *Where the Grass is Greener.* There are two cards on the table. One of them has written on it a positive number; the other, half that number. One of the cards, selected by a coin flip, is revealed to you. You may get in dollars either the number on this card or the number on the other card. Which should you choose?

> Suppose that the number revealed to you is A. Then the other card has the number $2A$ or $0.5A$, each with equal probability. If you stick with the card shown, your expected winnings are A. If you switch, your expected winnings are $0.5(2A) + 0.5(0.5A) = 1.25A$. Thus, you should always select the card other than the one revealed to you.

Moreover, even without having seen the number on the first card you should switch, but then you should also switch back, and so on. One is caught in a never-ending pendulum swing. But is the argument indeed valid for *any positive A*? Let us examine what we know about the positive numbers written on these two cards. How did these numbers come about? We are uncertain about their values, but we ought to know the (chance) mechanism that has generated them. Nothing, however, is said about the selection process in the problem.

Suppose you were told that a person had drawn at random one of the integers 1, 2, 3, ..., 10; the outcome of that draw was written on one card, and twice that number on the other card. Now, considering that experimental procedure, if the number revealed to you were, say, 18, you would know that you should stick with it. If, however, you were shown a 7, you would switch. Different considerations could be applied, depending on the number revealed to you, and on the description of the number-generating process. Lacking any specification of the chance procedure in the problem statement, your considerations must be based on some *assumed* distribution of the pairs of numbers.

The crux of the puzzle was based on the assertion that given any $A > 0$, the probabilities that the other card bears either $0.5A$ or $2A$ are equal. The one assumption that is compatible with that requirement for *each* positive A, at least in the discrete case, is that the first number was randomly drawn from the set of all, say, integer powers of 2, and then it was doubled. Put differently, your expected winnings upon switching would be $1.25A$, *for every A*, only if you assume the first number was randomly drawn from an infinite uniform distribution. A uniform distribution, however, *cannot* be infinite. If a discrete random variable is uniformly distributed, it can assume only a finite number of values. On first reading, the reader is not aware of the fact that he or she is accepting an impossible statistical experiment. The paradoxical conclusion can, therefore, be traced to contradictory implicit assumptions.

Is there a chance set-up according to which it is always reasonable to switch? Suppose the game is conducted so that you are first shown some positive number W. Then, out of your sight, somebody flips a coin to decide whether the other card will bear the number $0.5W$ or $2W$. You are now to decide whether to accept W dollars as your winning or to switch. In this case, you would do better to switch. The lesson from this can be summed up by the admonition, *Inquire about the chance mechanism!*

4. Probabilistic Reasoning as an Extension of Commonsense Thinking

4.1 Grounding instruction on students' valid intuitions

Puzzles are often piquant. They may challenge students and capture their interest. It is tempting to bring some of the more devious problems to the classroom to demonstrate to students their erroneous tendencies and perhaps enlighten them. However, if a teacher persists in pointing out to students how prone they are to inferential errors, they may become so convinced of their incapacities that they despair of ever mastering more appropriate techniques. A balance needs to be struck between illustrating some misconceptions and biases and reassuring students of their existing capacity. Furthermore, it seems reasonable to begin instruction in probability by building on students' sound intuitions. Valid probabilistic intuitions are not hard to come by. Despite the abundance of studies describing people's inferential biases and shortcomings, many of the rules prescribed by probability theory are compatible with commonsense. Indeed, much of what is known as Bayesian inference is intuitively sensible. Everybody will agree, for example, that when a sick child develops a rash, the probability of measles should increase relative to that of flu. That is so because, no matter what the prior probabilities of the two diseases, a rash is more likely assuming measles than assuming flu.

In probability, as in physics, "not all preconceptions are misconceptions." That expression is borrowed from the title of a paper by Clement, Brown, and Zietsman [2]. They propose that in teaching physics it is desirable to ground new material in students' intuitions that are in agreement with accepted theory. Likewise, in teaching probability, one expedient strategy is to offer nontrivial probability problems for which students can guess whether the answer is greater or smaller than a given number. This would allow the student to experience the satisfaction of having the prediction borne out by the results of the probabilistic calculation. They may realize that commonsense can still be a good guide though it should be exercised with caution.

4.2 Bayesian inference and commonsense

Many people's sound intuitions in learning from experience and revising their beliefs are consistent with Bayesian analysis. Furthermore, in specific limiting cases the results of Bayesian computations coincide with those of deductive inference. Thus, Bayesian reasoning can be viewed as an extension of logical inference into the gray area of uncertainty. A clear and comprehensive exposition of Bayesian statistics can be found in [10].

We give one simple example that can be presented in introductory courses and may appeal as commonsensical even to children.

A man was arrested as a suspect of murder. The investigating officer summed up his impressions of the suspect and all the relevant information at his disposal and arrived at the conclusion that the suspect's probability of guilt was .60.

a. As the investigation went on, it was found (beyond any doubt) that the murderer's blood-type was O. The relative frequency of blood-type O in that population is .33 (i.e., this is the probability that a 'random person' in that population has blood-type O). The suspect's blood was tested and found to be of type O. What is the posterior probability of the suspect's guilt (from the officer's point of view) considering all the data?

b. Suppose both the murderer's and the suspect's blood-types were found (with certainty) to be A. The relative frequency of blood-type A in the population is .42. How would the posterior probability in that case compare with the same probability in question (a)?

c. Suppose everything is as in question (a) above. The murderer's blood-type is O, but the suspect's blood-type is B. What is the suspect's posterior probability of guilt?

Let us denote the event "the suspect is guilty" by G. Guessing the direction of change in the probability of G in the three cases would pose no difficulty to most students. It would be gratifying, however, to see that formal computations yield the same conclusions.

a. Let "O" denote the evidence that the suspect's blood-type was found to be O. We know that $P(O|G) = 1$ and $P(O|\bar{G}) = .33$, and we are interested in the probability of guilt, given that evidence, namely, in $P(G|O)$. Now, employing Bayes' formula, one obtains:

$$P(G|O) = \frac{P(O|G)\,P(G)}{P(O|G)P(G) + P(O|\bar{G})P(\bar{G})}$$

$$= \frac{1 \times .60}{1 \times .60 + .33 \times .40} = .82$$

verifying the intuition that the result of the blood test should be incriminating for the suspect.

b. Commonsense clearly indicates that blood-type A would make the probability of the suspect being guilty smaller since there are more people with that blood type. When substituting A for O as the evidence in Bayes' formula, the only change would be to replace .33 in the denominator by .42, thus increasing the denominator and decreasing the resulting probability of guilt. Students may speculate about the limiting cases where either 100% of the population share the murderer's and suspect's blood-type, leaving the suspect's probability of guilt unchanged, or, there is known to be just one person in the population with the blood-type found to be the murderer's and the suspect's, causing our suspicion to jump to certainty. An important insight that students ought to distill from these various examples is that the lower the rate of the blood-type shared by the murderer and suspect, the more incriminating the evidence.

c. If the murderer's blood-type is O and that of the suspect is B, the latter is obviously exonerated. Formally, one has to replace $P(O|G) = 1$ in the formula by $P(B|G) = 0$, resulting in $P(G|B) = 0$. That logical result is obtained as an extreme case of the probabilistic model.

In comparing parts (a) and (c), we note that although $P(O|G) = 1$, observing O [in (a)] did not prove G; it only raised its probability (from .60 to .82). On the other hand, because $P(B|G) = 0$, observing B [in (c)] disproved G, thus proving the suspect's innocence. This is a demonstration of the conclusiveness of negative evidence compared with the relative corroboration affected by positive evidence.

5. Conclusion

The teacher of probability is bound to encounter some initial resistance that may originate from the deeply-rooted human quest for certainty. Every student must undergo a 'probabilistic revolution.' This involves understanding that questions of probability do not require prediction of immediate outcomes — that they refer to long-run relative frequencies. Teachers should be aware of the prevalence of misconceptions of chance and help students confront difficulties in assimilating the concept of statistical independence. Explication of the random mechanism that has generated the data given in a probability problem is one of the most important steps toward solving the problem. Despite the counterintuitive nature of some probabilistic results, many of probability theory's rules are compatible with everyday logic and are psychologically acceptable. We advocate capitalizing on such commonsensical inferences to establish students' confidence in their ability to reason probabilistically, and to help them see that reasoning as an extension of common-sense reasoning.

References

1. Bar-Hillel, M., and R. Falk (1982). Some teasers concerning conditional probabilities. *Cognition*, **11**, 109-122.
2. Clement, J., D. E. Brown, and A. Zietsman, (1989). Not all preconceptions are misconceptions: Finding 'anchoring conceptions' for grounding instruction on students' intuitions. *International Journal of Science Education*, **11** (special issue), 554-565.
3. Falk, R., R. Falk, and I. Levin, (1980). A potential for learning probability in young children. *Educational Studies in Mathematics*, **11**, 181-204.
4. Gigerenzer, G., Z. Swijtink, T. Porter, L. Daston, J. Beatty, and L. Krüger, (1989). *The empire of chance*. Cambridge: Cambridge University Press.
5. Gilovich, T., R. Vallone, and A. Tversky, (1985). The hot hand in basketball: On the misperception of random sequences. *Cognitive Psychology*, **17**, 295-314.
6. Kahneman, D., P. Slovic, and A. Tversky, (Eds.). (1982). *Judgment under uncertainty: Heuristics and biases*. Cambridge: Cambridge University Press.
7. Konold, C. (1989). Informal conceptions of probability. *Cognition and Instruction*, **6**, 59-98.

8. Krüger, L., L. Daston, and M. Heidelberger, (Eds.). (1987). *The probabilistic revolution: Vol. 1. Ideas in history*. Cambridge, MA: MIT Press.
9. Krüger, L., G. Gigerenzer, and M. S. Morgan, (Eds.). (1987). *The probabilistic revolution: Vol. 2. Ideas in the sciences*. Cambridge, MA: MIT Press.
10. Phillips, L. D. (1973). *Bayesian statistics for social scientists*. London: Nelson.
11. Piaget, J., and B. Inhelder, (1975). *The origin of the idea of chance in children* (L. Leake, Jr., P. Burrell, and H. D. Fishbein, Trans.). New York: Norton (Original work published 1951).
12. Wagenaar W. A., and G. B. Keren, (1988). Chance and luck are not the same. *Journal of Behavioral Decision Making*, **1**, 65-75.

Ruma Falk is a psychologist at the Hebrew University of Jerusalem. She has taught probabilistic and statistical thinking for many years to students of psychology, social work and education. Her research deals with the judgment of randomness and coincidences, inference via conditional probabilities, and the development of the concepts of probability and infinity in children.

Cliff Konold is at the Scientific Reasoning Research Institute in the Hasbrouck Laboratory of the University of Massachusetts at Amherst. He is Principal Investigator of the NSF-funded project, "A Computer-Based Curriculum for Probability and Statistics". As part of this project, he has designed interactive software for use in introductory high school and college courses.

Part III

Technology

in

Statistical Education

Does the Computer Help Us Understand Statistics?[1]

J. Laurie Snell and **William P. Peterson**
Dartmouth College   Middlebury College

The computer can be used profitably in almost any elementary mathematics course, but it is especially helpful in teaching probability and statistics. A general discussion of computing in a probability course can be found in Snell and Finn [10]; Florence Gordon [6] gives such a discussion for a statistics course. Statistical software packages are regularly reviewed in the *American Statistician*. With these references in mind, the present paper is not intended as a comprehensive discussion of how computing fits into the curriculum, nor do we attempt a complete analysis of the software resources available. Instead, we proceed by means of examples to illustrate what can be accomplished in an environment where students have access to a statistical computing package and also where they are familiar with a simple computing language.

For demonstration purposes, we have used the Minitab and Statview statistical packages, and the True BASIC programming language. Our choices have admittedly been influenced by our own familiarity with them. Nevertheless, we feel that these options give us enough flexibility to illustrate our main points, and it should be clear how to adapt our examples to other environments. Minitab, of course, is widely used in the teaching of statistics. We appreciate the interactive environment and the excellent on-line help. Another extremely useful feature is the ability to write small executable loops of commands, which amounts to simple programming. However, by today's standard for graphical displays, Minitab sometimes falls short. Statview is a package for the Apple Macintosh computer that takes full advantage of the Macintosh interface and graphics capabilities. Finally, we consider programming exercises that go beyond the standard calculations found in statistical packages. We have used True BASIC to demonstrate the power of having students explore new kinds of problems and calculations.

We have identified three broad areas in which the computer is helpful: reducing the need for lengthy manual calculations, facilitating graphical data analysis, and illustrating statistical concepts by means of simulation experiments. The first two categories tend to be well supported in the available packages, and we present here some representative examples. The last category seems not to be well developed in the standard texts, and our treatment here is more extensive, including both manipulations with standard packages and programming exercises.

---

[1] This work was supported by the New England Consortium for Undergraduate Sciences Education through a grant from the Pew Charitable Trust.

167

Numerical Calculations

Probably the most obvious use of the computer is to relieve the student of time-consuming numerical operations traditionally associated with statistics. Many textbooks now include exercises intended to be worked with a computer or a statistical calculator, and provide descriptions of the relevant commands for some of the more widely used software packages, such as Minitab and SAS. Some authors have even developed their own software to accompany their texts.

There is no doubt that tedious manual calculations have discouraged many students of statistics. A statistical computing package has the advantage that is hides a great number of technicalities that are not necessary for understanding of the analysis of data. This allows both designers of the package and students to concentrate on user-friendly manipulations of the data. The disadvantage is that the package often hides some of the things that are useful to the understanding. In almost all cases, it is still important to understand what a formula means conceptually, and how the components of the formula can be expected to influence a computed value.

As an example, consider the sample correlation coefficient r, computed from a set of data points (x_i, y_i), $i = 1,\dots,n$. By the time this concept is introduced, the students will already be familiar with the single variable mean and standard deviation calculations:

$$\bar{x} = \frac{1}{n}\sum_{i=1}^{n} x_i, \qquad s_x = \sqrt{\frac{1}{n-1}\sum_{i=1}^{n}(x_i - \bar{x})^2},$$

$$\bar{y} = \frac{1}{n}\sum_{i=1}^{n} y_i, \qquad s_y = \sqrt{\frac{1}{n-1}\sum_{i=1}^{n}(y_i - \bar{y})^2}.$$

They will also have seen how to express variables in terms of standard units via the transformations $(x_i - \bar{x})/s_x$ and $(y_i - \bar{y})/s_y$. The *correlation coefficient* r can then be defined as:

$$r = \frac{1}{n-1}\sum_{i=1}^{n}\left[\frac{x_i - \bar{x}}{s_x}\right]\left[\frac{y_i - \bar{y}}{s_y}\right]$$

In this form, it is easy to see that r is invariant under changes in unit of measurement for either variable. It can also be explained that pairs (x_i, y_i) with both observations above their respective sample means, or both below, will contribute negative terms. Thus as a measure or linear association on a scatterplot, a positive value of r indicates positive association and a negative value indicates negative association.

The above is not the most convenient expression for computer (or manual) evaluation of r. The algebraically equivalent formula

$$r = \frac{n\sum x_i y_i - \left(\sum x_i\right)\left(\sum y_i\right)}{\sqrt{\left[n\sum x_i^2 - \left(\sum x_i\right)^2\right]\left[n\sum y_i^2 - \left(\sum y_i\right)^2\right]}}$$

is more suitable for direct computation, but contributes much less to the conceptual understanding. It is an example of a computational technicality that we can avoid discussing in an introductory course, given the availability of a statistical package that will perform the computation behind the scenes. However, our philosophy is that the discussion of the first form for r should definitely not be dispensed with.

When using the sample correlation coefficient (along with the means and standard deviations) as a summary statistic for a scatterplot, attention must be paid to outliers. It is important for students to see that r is not resistant to outliers; a few extreme observations can severely affect its value. The effects, which can be particularly pronounced in small data sets, are illustrated by the following example (adapted from Kimble [7, Chapter 9].

x	1	2	3	4	5	6	7	8	9	10	11
y	6	11	3	4	2	7	9	1	10	8	4

These data give $r = 0.000$ to three decimal places. However, successively changing the last y value from 4 to -10, 10, and 20 results in values for r of -0.360, 0.258, and 0.452, respectively. Students should learn to inspect their data with a critical eye before calculating. The value -10 immediately stands out as the only negative number. Closer examination suggests that 20 is a possible outlier in the distribution of y values. Such observations should be checked for possible mistakes in data entry, or conditions in their measurement that may have somehow differed from the others. It is not reasonable to ask that very many calculations of this sort be performed by hand, but with a software package the required manipulations are obviously quite simple.

It must also be stressed that r is a measure of linear relationship only, and can be quite misleading when calculated in other settings. As an example, consider the data points:

x	0	1	2	3	4	5	6	7	8	9
y	0	4	7	9	10	10	9	7	4	0

These fall on the parabola $y = 4.5x - 0.5x^2$, in pairs around its axis of symmetry. Calculation shows that $r = 0$ exactly for these data. But, of course, we would not want to say that there is no relationship between x and y.

We have focused in this section on the correlation coefficient, but it should be clear that similar comments apply to other statistical computations as well. The availability of a statistical software package frees up a great deal of time that would otherwise be spent teaching computational details. There may be a temptation to devote the time saved to covering an ever-greater number of statistical procedures. There is no doubt that this is one of the benefits of software. However, we feel that a substantial portion of the time gained should be spent on understanding exercises designed to promote conceptual understanding of the material.

Graphical Data Analysis

Early lectures in an introductory course are typically devoted to descriptive statistics. Graphical displays for exploratory data analysis, as pioneered by Tukey [11], have gained increased importance in applied statistics. Students need to gain familiarity with standard graphical displays of data, such as histograms, boxplots, and stem-and-leaf diagrams. Constructing these plots by hand quickly becomes tedious. Nevertheless, it is useful to explore their behavior for a number of different shapes of distributions. There is no guarantee that the first attempt will produce an informative plot. There are, for example, questions of where to split stems in a stem-and-leaf diagram, or how to choose class intervals for a histogram. The computer becomes the ideal tool, because of the ease with which plots can be created and modified. To develop an intuitive feel for the correlation coefficient, discussed in the last section, students can practice "guessing" r for a variety of scatterplots and comparing with the computed values. By varying the ranges on the axes for a fixed data set, they can explore the phenomenon that variables tend to look more highly correlated when there is more blank space around the point cloud. The role of software in promoting understanding here is arguably as important as its role in relieving the burden of numerical calculations.

Graphical displays, of course, have many uses in data analysis and presentation. The remainder of our discussion here will be devoted to one important use of graphics that arises from the need to validate the assumptions on which various statistical calculations are based. We highlighted this need earlier. The problems with outliers and nonlinearity in our correlation coefficient analysis would have been more easily diagnosed graphically. A scatterplot can often provide a key insight that a blind calculation would not. Anscombe [1] constructed a classic collection of four data sets to illustrate these ideas in the context of fitting a regression line.

I		II		III		IV	
x	y	x	y	x	y	x	y
10.0	8.04	10.0	9.14	10.0	7.46	8.0	6.58
8.0	6.95	8.0	8.14	8.0	6.77	8.0	5.76
13.0	7.58	13.0	8.74	13.0	12.74	8.0	7.71
9.0	8.81	9.0	8.77	9.0	7.11	8.0	8.84
11.0	8.33	11.0	9.26	11.0	7.81	8.0	8.47
14.0	9.96	14.0	8.10	14.0	8.84	8.0	7.04
6.0	7.24	6.0	6.13	6.0	6.08	8.0	5.25
4.0	4.26	4.0	3.10	4.0	5.39	19.0	12.50
12.0	10.84	12.0	9.13	12.0	8.15	8.0	5.56
7.0	4.82	7.0	7.26	7.0	6.42	8.0	7.91
5.0	5.68	5.0	4.74	5.0	5.73	8.0	6.89

Each of these sets of $n = 11$ data points has $\bar{x} = 9.0$ and $\bar{y} = 7.5$. Each gives rise to the least squares equation $y = 3.0 + 0.5x$ with $r^2 = 0.67$. However, the differences in the scatterplots of the original data, shown in Figure 1, are dramatic. Only Set I is a straightforward candidate for fitting a regression line. Set II suggests a curvilinear relationship, for which a simple regression line is not appropriate. In the last two sets, a single observation has critical influence, and additional discussion is called for. In Set III, a different regression line would fit perfectly if one outlying observation were removed. In Set IV, only one observation shows

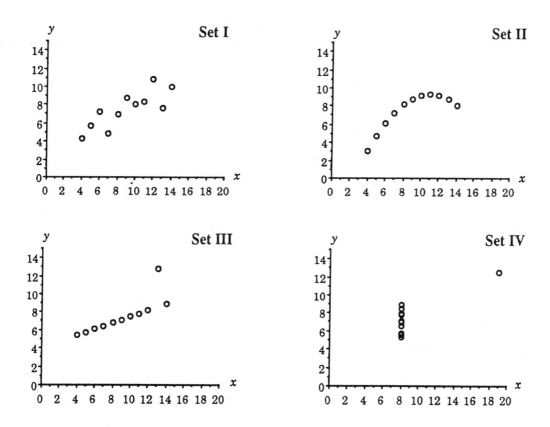

Figure 1. Statview Plots of Anscombe Data Sets I-IV

any variation in the x coordinate. Without this point, no regression slope could be calculated.

For more general data models, for example multiple regression, it is more difficult to get an initial impression from graphing the data. This should not discourage us from plotting along any dimensions that mighty be of interest (it should probably be noted that higher dimensional graphics routines are becoming available as well). Furthermore, most models can be characterized in the form

$$\text{Data} = \text{Fitted Values} + \text{Residuals}.$$

Thus after fitting the model, the residuals can be plotted against the fitted values in a two-dimensional scatterplot. The results for the Anscombe data are shown in Figure 2. For Set I, the lack of any pattern or trend indicates that the data do not differ from the fitted model in a systematic way. This is ideally what we hope to observe. For Set II, the nonlinearity of the original data shows up in the pattern of the residuals, which are first negative, then positive, and then negative again. In Set III, one residual is nearly three times as large in magnitude as any other, indicating a possible outlier. The systematic linear trend in the remaining data indicates that a better fit would be obtained if the suspected outlier were removed. In Set IV, it is clear that a single value has dominated the fitting of the regression equation: all other observations have the same fitted value.

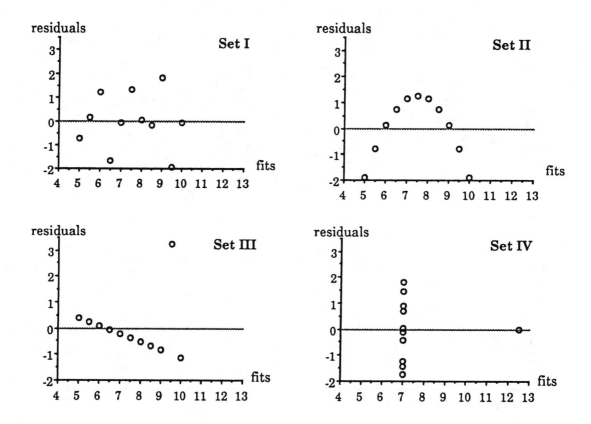

Figure 2. Residuals vs. Fits for Anscombe Data Sets I-IV

Observe that the residual plots amplify our perception of the features we noted earlier. Analysis of the residuals is always a good idea, even when plots of the original data have been examined. An illustration of this is provided by the following problem, which is adapted from Freedman, Pisani and Purves [5]. It was created to demonstrate the pitfalls of assuming we know more than we do about a relationship just because a scatterplot exhibits and apparently linear trend. Consider an investigator who is unfamiliar with basic geometry, and does not know the area formula for a rectangle. He suspects that area must be related to perimeter, so he takes a sample of "typical" rectangles and measures both perimeter and area. (We need to use a little imagination here. Perhaps perimeter is measured by wrapping a string around the figure and then counting off unit lengths against a standard. Area might be measured by counting the number of unit squares needed to cover the figure. We're trying to imagine a scenario in which the investigator is completely ignorant of the basic relationships.)

We now simulate the experiment, by supposing that the length and width for typical rectangles are independently chosen from a uniform distribution on the interval [0,1]. (We will have more to say about the uses of simulation in the next section.) Using a random number generator routine to compute 20 length-width pairs, we then calculate perimeters and areas, using the actual formulas from geometry. These data are shown below.

Length	Width	Perimeter	Area
.215	.409	1.248	.088
.725	.602	2.655	.437
.814	.32	2.267	.26
.682	.818	3.000	.558
.011	.376	.773	.004
.588	.96	3.097	.565
.437	.091	1.057	.04
.931	.462	2.786	.43
.408	.254	1.324	.104
.328	.451	1.559	.148
.705	.908	3.226	.64
.964	.215	2.358	.207
.567	.543	2.221	.308
.492	.288	1.559	.141
.766	.743	3.019	.569
.055	.222	.552	.012
.349	.325	1.348	.113
.334	.211	1.09	.07
.527	.808	2.67	.426
.233	.118	.703	.028

Recall that our hypothetical investigator only has the perimeter and area data. A plot of area vs. perimeter, with the fitted regression line, is shown in Figure 3. The trend is deceptively linear in appearance, and the regression gives $r^2 = .938$!

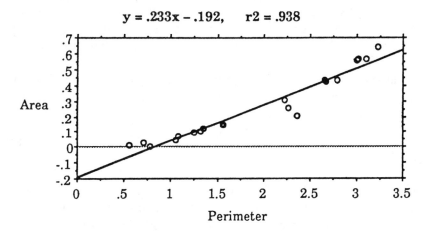

Figure 3. Scatterplot and Regression Line for Rectangle Data

A plot of residuals against fits makes patterns in the residuals more readily discernable, as seen in Figure 4. The trend that emerges here is that the residuals are positive for the smaller fitted values, negative for medium values, and the positive again for the larger values. This is evidence that fitting a line was a mistake.

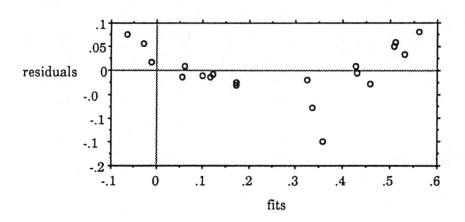

Figure 4. Residuals vs. Fits for Rectangle Data

Geometrically speaking, the negative area values fitted to perimeters close to zero don't make sense, either. The lesson here is that we can be seriously misled when important factors are ignored. In the present case, we know that the length and width dimensions determine both area and perimeter. The model the investigator was trying to fit, predicting area on the basis of perimeter, simply isn't meaningful.

We conclude this section with one more important use of graphics for validating assumptions. Many statistical procedures are based on the assumption that some underlying variable is normally distributed. A quick visual check of this assumption is provided by a normal quantile plot, which is available in many packages. The idea is to sort the values in the sample data and then plot the corresponding quantiles of the standard normal distribution against them (some adjustment is necessary because the 1.0 quantile of the normal distribution is $+\infty$, and implementations differ slightly in their details). A straight line pattern indicates that the data are approximately normal. Miller [8] gives a nice discussion of these plots for preliminary data screening, and recommends them for this purpose over more formal tests (such as Kolmogorov-Smirnov or chi-squared).

For introductory students, it is useful to create some examples to illustrate how departures from normality will be revealed in a quantile plot. Consider the following data set, consisting of $n = 15$ observations:

5.41	9.34	7.72	9.68	4.19	6.31	9.99	3.88
2.31	1.74	9.61	6.67	7.12	6.73	2.92	

The Minitab output presented in Figure 5 shows a histogram of the data, which have been entered into column C1. The histogram itself shows departure from a normal bell curve shape. There is no distinct peak or tails -- the data drop off abruptly at the end of the range, rather than tapering off.

The first two lines of Figure 6 show the Minitab commands required to produce a normal quantile plot for the data. The resulting output follows.

The tail behavior is shown by the bending upwards at the high values in the sample, and downward at the low values, which shows that the tails of the sample are less elongated than would be expected for a sample from a normal distribution. In contrast to this, the normal quantile plot for a distribution whose tails are more elongated than a normal would bend to the right on the high end, and to the left on the low end, flattening away from a straight line pattern. For a positively skewed distribution, the plot would bend down at the low end and to the right at the high end. Negative skewness would be shown by bending to the left at the low end and upward at the high end.

```
MTB > histogram c1

Histogram of C1    N = 15

Midpoint    Count
       2        2   **
       3        1   *
       4        2   **
       5        1   *
       6        1   *
       7        3   ***
       8        1   *
       9        1   *
      10        3   ***
```

Figure 5. Minitab Histogram of Sample Data

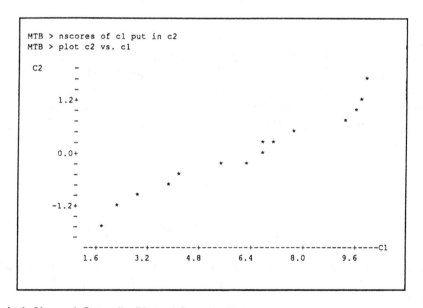

Figure 6. Minitab Normal Quantile Plot of Sample Data

In the particular example shown here, the departure from normality was already evident in the histogram. In the next section, we present an example where the distinction is harder to pick up without the quantile plot.

Simulation

Pseudo-random number generation routines are available in most scientific computing environments, and can be used to drive simulation experiments that illustrate the behavior of statistical techniques. For example, one can generate a large number of samples of a given size from a known distribution and observe the distribution of the sample means. A related idea is to compute from each sample a confidence interval for the population mean, and count the number of intervals that cover the true value. By a "large number" of samples, we mean tens or hundreds, so this type of exercise is critically dependent on the availability of a computer. In more advanced courses, mathematical techniques expressions for the density function or the moments of a sampling distribution. At the introductory level, simulation provides an alternative way to investigate the form of the distribution. It has the added benefit of making the sampling procedure itself more tangible. Furthermore, it should be noted that simulation is sometimes the method of choice in advanced statistical applications. In a later section, we illustrate the use of simulation in "bootstrap" methods.

```
MTB > random 500 observations put in c1-c4;
SUBC> normal mu=10, sigma=2.
MTB > rmean c1-c4 put in c5
MTB > rstdev c1-c4 put in c6
MTB > let c7 = (c5-10)/(c6/2)
```

Figure 7. Minitab Computation of *t* Statistics from Simulated Normal Samples

As a first example, consider how the Student's *t* distribution arises in sampling from a normal population. In a more advanced course, students would study the distribution theory of the sample mean and standard deviation, and the construction of the *t* statistic from them. The following Minitab experiment uses simulation to investigate these ideas. The commands shown in Figure 7 simulate 500 samples of size $n = 4$ from a normal population with $\mu = 10$ and $\sigma = 2$. The samples are represented by the rows of columns C1 through C4. For each of these, we compute the *t* statistic

$$t = \frac{\bar{x} - \mu}{s/\sqrt{n}}$$

where $\bar{x}$ is the sample mean and s the sample standard deviation (in this experiment, our *t* statistic has 3 degrees of freedom). The sample means are computed using the RMEAN (row mean) command, and are put in C5. The standard deviations are computed using the RSTDEV (row standard deviation) command, and are stored in C6. Finally, the *t* statistics are computed and stored in C7.

Figure 8 shows a dotplot of the *t* statistics. The distribution does seem to have a bell shape, and it would be difficult to distinguish it from normal data on this basis alone.

One concrete way to make the comparison is to use the normal quantile plotting ideas introduced in the last section. Figure 9 shows the plot for our data.

Knowing how to interpret quantile plots, we can see that our simulated distribution has more elongated tails than the normal. This is indicated by the flattening away from the straight line trend.

A final note on normal quantile plots is in order here. It is important to distinguish random fluctuations from systematic deviations from a straight line pattern. As an exercise, students can simulate a number of samples of different sizes ($n = 10, 20$ etc.) from a normal distribution, and construct normal quantile plots. This will help develop a feeling for the kinds of fluctuations that can occur in data that does come from a normal population.

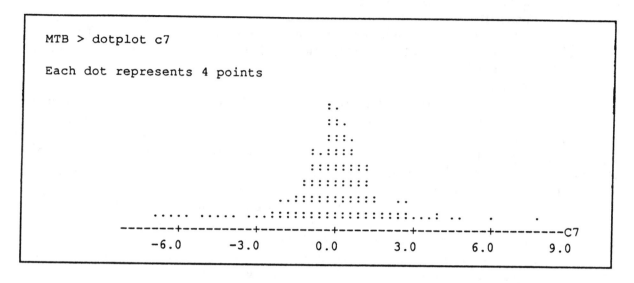

Figure 8. Dotplot of Simulated *t* Distribution

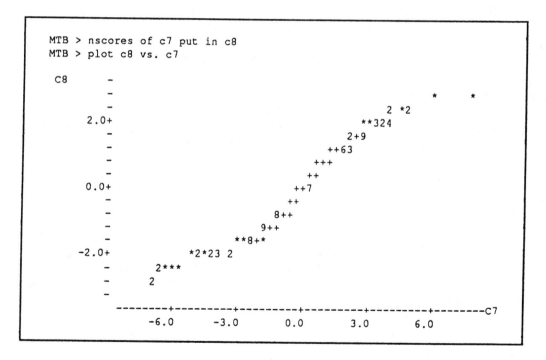

Figure 9: Normal Quantile Plot of Simulated *t* Distribution

Examples using Programming

There are situations where we feel that the statistical package hides more from the students than is desirable for the learning process. Seeing the way that probabilities are computed often reinforces the understanding of statistics. This is even more true when a student implements the algorithm for computing the probabilities in a computer program. In addition, in the typical statistical package, it is awkward to introduce new experiments and new kinds of problems than would be the case if he or she were using a statistical package in conjunction with the ability to write simple programs.

Of course, we pay a price by requiring students to know how to write programs and so it is important to keep the technicalities of programming at a minimum. We want to show that this is possible by using an easy to learn computer language that supports the use of libraries of subroutines. Once a simple concept, such as computing binomial coefficients, has been developed, understood and implemented by a subroutine, the subroutine can be put into the library and used from that time on by a simple call to the library. Also, technical programming difficulties such as labeling the axes when plotting histograms can be relegated to a subroutine that the student never needs to see. Similarly, all the familiar tables that appear in the back of a statistics book can be made available to the students in a much more convenient form and for use in their programs. The particular numerical techniques needed to most efficiently calculate the tables can again be buried in a subroutine not normally seen by the student. Of course, the interested student is free to look at this if he or she wished to and in some cases this is instructive.

Students at Dartmouth College learn to write simple Basic programs in their first mathematics course, usually calculus. Over 80 percent of the students have Macintosh computers and the others have easy access to these computers in the clusters around campus. The language currently used is a structured form of Basic called True BASIC. Thus at Dartmouth, it is possible to assume that students can write simple programs in any of their mathematics and probability courses that can be implemented in such an environment. We have found that the True BASIC language is ideal because students can write programs after a very brief introduction to the language and it has very easy to use graphics. If the students were already familiar with Pascal, there are implementations such as Turbo Pascal that allow libraries of subroutines and could be used just as well if the students were familiar with the language. The difficulty with Pascal is that it is not reasonable to expect students to learn this language pretty much on their own as we do with Basic. It is encouraging that other simple languages such as ISETL are being developed and used in the teaching of mathematics.

We begin with an example chosen from a course that we are currently developing called CHANCE. This course will be based primarily upon the articles in *Chance* magazine published by Springer-Verlag. The magazine *Chance* provides articles on applications of probability and statistics that are of current interest to the public. These include topics such as: the undercount in the census, DNA fingerprinting, statistical problems in AIDS, streaks in sports, etc. These articles will be supplemented by other general science journal and newspaper articles. We plan to teach the course *CHANCE* as a case studies course and to develop the necessary probability and statistics in the context of the example being studied. A bibliography of typical sources currently being used is included in Part IV of this volume. Such an approach has been very successful in the new course *Case Studies in Quantitative Reasoning* currently being taught at Mount Holyoke College.

We will show how we plan to use the computer in the discussion of the article *The Cold Facts about the "Hot Hand" in Basketball* by Amos Tversky and Thomas Gilovich that appeared in the Winter 1989 issue of *Chance*. As the title suggests, the authors are interested in seeing if there is any validity to the belief that basketball players are more successful with their shots after they have made several shots and less successful after times that they have failed in several previous shots. The authors observe that "the belief in the hot hand is really one version of a wider concept that 'success breeds success' and that 'failure breeds failure' in many walks of life."

Tversky and Gilovich begin their paper by observing that part of the problem in judging whether players shots behave like "random sequences" comes from the fact that people are poor judges of what a "random sequence" should look like. To show this, they carried out experiments with basketball fans with data that they created. We can generate the kind of data they used by simulation. We write a subroutine **shoot** to simulate a sequence of shots for a player whose outcome on a shot is the same as the last outcome with probability p and the opposite of this outcome with probability $q = 1 - p$. We use this subroutine in a program to present the subject with a sequence randomly chosen to be either an *alternating sequence* ($p = 1/4$), a *chance sequence* ($p = 1/2$), or a *streak sequence* ($p = 3/4$) and ask the student to guess which it is. A typical outcome for this program is shown in Figure 10.

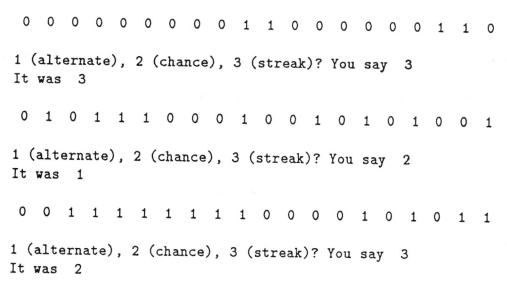

```
0  0  0  0  0  0  0  0  0  1  1  0  0  0  0  0  0  0  1  1  0

1 (alternate), 2 (chance), 3 (streak)? You say  3
It was  3


0  1  0  1  1  1  0  0  0  1  0  0  1  0  1  0  1  0  0  1

1 (alternate), 2 (chance), 3 (streak)? You say  2
It was  1


0  0  1  1  1  1  1  1  1  1  0  0  0  0  1  0  1  0  1  1

1 (alternate), 2 (chance), 3 (streak)? You say  3
It was  2
```

Figure 10: Which Sequences are Random?

Students could generalize our experiment to the case that the proportion of successes is different. To do this, they only have to allow the probability of a success following a success to be different from the probability of a failure following a failure.

Tversky and Gilovich provide data for 9 players from the Philadelphia 76ers during the 1980-81 season. It would be an easy mater for students to obtain data from their own basketball team. Tversky and Gilovich apply a number of statistical tests to see if the data supports the hypothesis that a player has a higher probability of a hit following a hit than a hit following a miss. In particular, they use the runs test. This is a good test to illustrate the concept of test of hypothesis since the necessary calculations are quite simple.

The null hypothesis for the runs test is that the outcomes are Bernoulli trials with an unknown probability p for success. The test assumes that the number of successes is given. In this case, all sequences with the given number of successes are equally likely. If there are m successes and n failures, to have $2k$ runs beginning with a success run we have to choose k-1 division points in m-1 successes and k-1 division points in n-1 failures. Thus, if R is the number of runs,

$$P(R = 2k) = 2\frac{\binom{m-1}{k-1}\binom{n-1}{k-1}}{\binom{N}{n}}$$

The 2 in the second terms comes from the fact that the sequence could have started with either a success or a failure. If we have an odd number of runs and the first outcome is a success we have $k+1$ groups of successes, and k groups of failures and so we have to choose k division points in the m-1 successes and k-1 division points in the n-1 failures. Similar reasoning applies for the case where the first element is a miss. Thus,

$$P(R = 2k+1) = \frac{\binom{m-1}{k}\binom{n-1}{k-1} + \binom{n-1}{k}\binom{m-1}{k-1}}{\binom{N}{n}}$$

We can use this formula in a subroutine to calculate the exact probability of k runs when there are t trials and s successes. We can then use this subroutine in a program to plot the distribution for the number of runs when the number of shots and hits is specified. We illustrate the output of such a program in Figure 11.

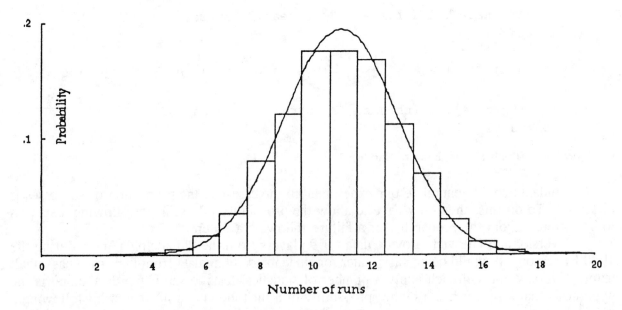

Figure 11: Distribution for Number of Runs with 20 Shots and 11 Hits

We can use the graph to illustrate the concept of upper and lower critical values for a test of hypothesis that the sequence is independent trials. The power of such a test would require knowing the distribution of runs for the Markov chain corresponding to dependent trials. This is more difficult but again easy to illustrate by simulation. For example, if our test were to involve 400 trials, we can simulate a large number of sequences of 400 trials and see how often the hypothesis of independence is rejected. The results of running such a program for a number of different values of p is show in Figure 12.

p	Proportion of times randomness is rejected.
.5	.007
.55	.317
.6	.938
.65	1
.7	1
.75	1

Figure 12: The power of the Runs Test by Simulation

We see that when we have as many as 400 shots, we are very likely (probability $> .9$) to reject the data from a streak shooter who has probability .6 of making a shot after a successful shot. The basketball players considered by Tversky and Gilovich made about 400 shots so we see that if p is greater than .5 and is not too close to .5, we have a good chance of rejecting the hypothesis of randomness. In fact, the chance hypothesis was not rejected for any of 9 players considered by Tversky and Gilovich, suggesting that, if they do improve their chances after a successful shot, the improvement is small.

We do not have simple formulas for the distribution for the number of runs when the outcome depends upon the previous outcome. However, we can use our subroutine **shoot** to simulate this situation and plot an empirical distribution for the number of runs for different values of the dependence parameter p. Doing this for the cases $p = 1/2$ and $p = 3/4$ allows us to compare a streak player with a chance player. The results are shown in Figure 13. We note that we can expect on the average only about half as many runs for the streak player as for the chance player.

As we have previously indicated, the ability to write programs allows the student to explore other approaches to a problem. An interesting project for a student would be to investigate the use of a different statistic to test randomness. For example, Tversky and Gilovich mentioned a test suggested by David Freedman that involves partitioning each record into non-overlapping sequences of four consecutive shots. Then count the number of series in which the player's performance is high (3 or 4 hits), moderate (2 hits) or low (0 or 1 hit). The student could first explore the effectiveness of this test using simulated data and then apply it to his of her own data. Another interesting test would be to consider the distribution of the longest streak. The instructor could provide a subroutine for calculating this distribution and the student could use this to obtain the critical values and compare the power of this test with that of the runs test.

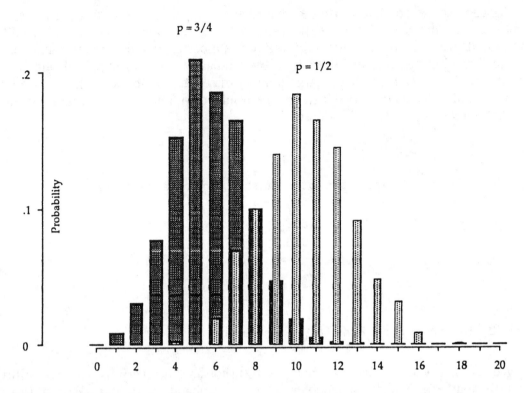

Figure 13: Runs Distribution by Simulation

The Bootstrap Method

Bradley Efron has remarked that the availability of fast computing has radically changed the way that statistics is done. His own development of the *bootstrap* methods is an excellent example of this. The bootstrap is a very simple but powerful new idea in statistics and it can easily be illustrated by writing a few very simple programs. A popular account of the bootstrap method is given by Persi Diaconis and Bradley Efron in [3] and a survey of the method can be found in Efron and Tibshirani [4].

A typical problem in statistics is to obtain some descriptive quantity relating to a large population of data. The data determine a density $f(x)$ equal to the proportion of the population data that have the value x. We assume that f is unknown to us, and we want to estimate some descriptive property of the population. For example, the population might be the cholesterol counts of all women in a particular age group, and we want to estimate the average of these counts, that is, the mean of the density f. We choose a sample of size n with replacement $x = (x_1, x_2,...,x_n)$ from the population and use the sample mean

$$\overline{x} = \frac{x_1 + x_2 + \ldots + x_n}{n}$$

as an estimate for the population mean. In assessing the value of this estimate, it is important to know how much variability there is in the sample mean. If we knew f, we could compute its variance σ^2 and our sample would have variance σ^2/n. We could then use the standard deviation of the sample mean $\sigma/\sqrt{n}$ to estimate the variability in our sample mean. However, we do not know f and so must base our estimate of the variation of the sample mean on the only information we have, namely, the sample itself.

In some sense, the best estimate for f is the empirical density $\hat{f}$ determined by the sample. Then $\hat{f}(x)$ is the proportion of data elements with value x. Now, as an approximation, we can assume that our sample is determined by $\hat{f}$ rather than f. Then $\bar{x}$ is the mean of $\hat{f}$ and the variance is

$$\overline{\sigma}^2 = \frac{\sum_{i=1}^{n} (x_i - \overline{x})^2}{n}$$

This is $n/(n-1)s^2$ where s^2 is the sample standard deviation for the sample x. Then the standard deviation for a sample of size n using $\hat{f}$ is $\overline{\sigma}^2/n$. We use this as an estimate of the variation in the sample mean for samples from our true population determined by f. How good an estimate this is will depend upon how close $\hat{f}$ it to f.

Now suppose that we are considering some other statistic of our population, say the median. Then we proceed in the same way. Now, however, there is no simple way to compute the standard deviation of the median of samples of size n determined by $\hat{f}$. Of course, we can compute this standard deviation by brute force. But, since each sample could take on n different values, there can be as many as n^n possible samples of size n and the time required to analyze these might well be prohibitive. When the exact computation is not feasible, we can use simulation to estimate the distribution of the sample median from this approximate distribution. For our original problem of estimating the variation in samples of size n determined by f, this will result in two approximations: one due to replacing f by $\hat{f}$ and another due to the simulation process. The error due to simulation can be made small by making a reasonably large number of simulations. Experience has shown that a few hundred simulations gives an excellent approximation. If the sample size n is large we would expect also the empirical density of $\hat{f}$ to be close to the population density f.

When we have to resort to simulation, using $\hat{f}$ for our samples amounts to sampling with replacement from our original sample. This seems to be pulling ourselves up by our bootstraps and, for that reason, the method of is called the *bootstrap method*. The samples used in the simulation are called *bootstrap samples*. We shall call the original sample the *true sample*.

Of course, our real interest is knowing something about the variation in the statistic of interest for samples from the population. The success of the bootstrap method can be measured in terms of how well variation in the statistic for samples from $\hat{f}$ agrees with that we would have obtained had we used f. We can use computer simulation to show that we do get reasonable results by the bootstrap method. We first consider the case of estimating the variation in a sample mean and sample median from a large population.

To see if the method works we assume that we do know the population (we do know f) and write a simple program to carry out sampling with replacement from the population (use f) and sampling with replacement from the sample (use $\hat{f}$). We then plot a histogram of the sample statistic in each case and see how similar they are. Running the program for the case of the mean, we have the typical outcome shown in Figure 14.

We notice that the variation is quite similar indicating that the bootstrap method gives us a good idea of the variation of the sample means in this case. Since our population was obtained by sampling from an exponential distribution with mean 1, the theoretical standard deviation is 1. Thus the standard deviation for the mean for a sample of size 100 is .1. Therefore, we

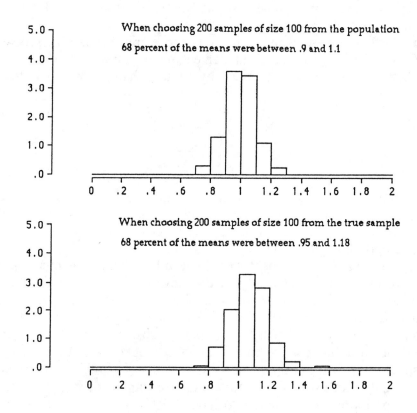

Figure 14: Compares Variation of Sample Means when Sampling from the Population and from the True Sample

expect about 68 percent of the means to fall between .1 and -.1 of the mean of 1. This happened exactly in the population samples and approximately in the bootstrap samples.

We can make the same kind of caparisons where we are interested in the variation in the sample median. We only have to replace the computation of the sample mean in our program by the corresponding computation of the median. The results of this simulation are shown in Figure 15.

In this case the population median is log 2 = .693, but we do not have an easy way to compute the standard deviation of the sample median. However, the variation of the sample medians when sampling from the population and from the sample again are comparable, suggesting that the bootstrap method gives a reasonable estimate for this variation.

Finally, we give an example of the use of the bootstrap method in a more complicated situation suggested by a standard example of Efron. Suppose that we are teaching a section of calculus with 20 freshmen randomly selected from the 238 freshmen who took calculus this year at Dartmouth. We have the mathematics SAT scores for our students and we know their total examination scores on exams which are common to all 238 students. These 20 pairs of numbers are

SAT scores	Calculus Exam Scores
640	363
630	233
630	396

700	396
660	339
580	391
730	359
630	262
690	377
700	414
690	348
650	444
690	478
630	270
720	426
620	417
730	450
580	391
770	401
690	450

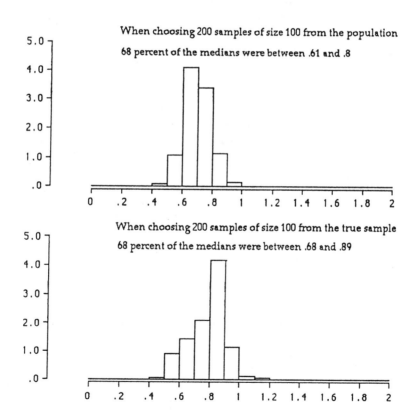

Figure 15: Compares Variation of Sample Medians when Sampling from the Population and from the True Sample

We became curious to see how much correlation there is between their math SAT scores and their calculus exam scores. We use our statistical package to provide a scatterplot and a coefficient of correlation. We obtain the plot in Figure 16.

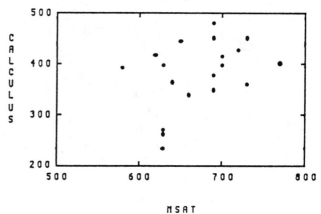

We find that the correlation coefficient is .344. This seems low to us and we wonder what the variation in the correlation coefficient is between sections and in the population as a whole. Again, it is an easy matter to estimate this variation by the bootstrap method. We just take a large number of samples from the data for our class. That is, we choose samples of size 20 with replacement from our table of pairs of values for the math SAT scores and the calculus exam scores.

Figure 16: Math SAT Scores versus Calculus Exam Scores for the Small Class

We compute the correlation coefficient for each of our samples and see how much variation there is. Again, because we happen to know the whole population in this case, we can have our program sample both from our class and from the population of all 238 students to see if the bootstrap method gives us a reasonable estimate of the variation. The result of carrying out this simulation is shown in Figure 17.

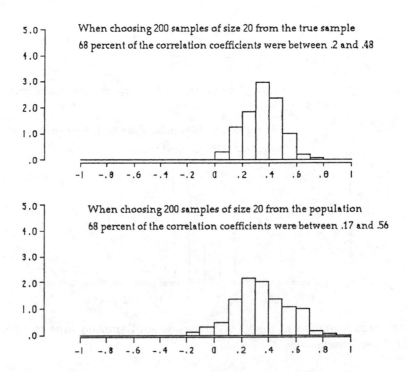

Figure 17: Bootstrap Estimate for the Variation in the Correlation Coefficient

Again, we see that the bootstrap method gives us an estimate of the variation that we might expect between classes which is quite consistent with that which we would have obtained if we were to use the entire population.

Our computer simulations allow us to explore the bootstrap method in cases where we know the population and so we can get a pretty good idea of how it works. Of course, the strength of the method lies in simplicity and applicability to a wide variety of estimation problems when we do not know the population, but must rely completely on the information contained in our sample.

In summary, we have tried to show some ways that computing can be used in teaching statistical packages with some modest programming if that is possible. The programming is less necessary with Minitab and some of the bigger packages such as S PLUS. Whatever computing methods are used, it is important that the environment is such that they are easy and natural for the students to use. The authors would be happy to provide interested readers with the programs used for illustrative purposes in this paper.

References

[1]   Anscombe, F. J., Graphs in statistical analysis. *The American Statistician* **27** (1973), 17-21.

[2]   Cleveland, William S., *The Elements of Graphing Data*, Wadsworth, Monterey, California, 1985.

[3]   Diaconis, Persi and Bradley Efron, Computer-intensive methods in statistics. *Scientific American* **248** (May, 1983), 116-130.

[4]   Efron, B. and A. Tibshirani, Bootstrap Methods for Standard Errors, Confidence Intervals, and Other Measures of Statistical Accuracy, *Statistical Science* **1** (February, 1986), 54-57.

[5]   Freedman, David, Robert Pisani, and Richard Purves. *Statistics*, W.W. Norton and Company, New York, 1978 (second edition 1991).

[6]   Gordon, Florence S., Computer use in teaching statistics. in: *Computers and Mathematics* edited by Smith, Porter, Leinbach, and Wenger, MAA Notes Number 9, The Mathematics Association of America, 1988.

[7]   Kimble, Gregory A., *How to Use (and misuse) Statistics*, Prentice-Hall, Englewood Cliffs, N.J., 1978.

[8]   Miller, Rupert G., Jr., *Beyond ANOVA, Basics of Applied Statistics*, John Wiley and Sons, New York, 1986.

[9]   Moore, David S. and George P. McCabe. *Introduction to the Practice of Statistics*, W.H. Freeman and Company, New York, 1989.

[10]   Snell, J. Laurie and John Finn. The use of the computer in a probability course, in *Computers and Mathematics* edited by Smith, Porter, Leinbach, and Wenger, MAA Notes Number 9, The Mathematical Association of America, 1988.

[11]   Tukey, John W., *Exploratory Data Analysis*, Addison-Wesley, Reading, Mass., 1977.

Laurie Snell is Benjamin Cheney Professor of Mathematics at Dartmouth College. He worked with John Kemeny and Gerald Thompson to develop the Finite Mathematics course and the textbook based on it. He has published numerous articles and books; the most recent is a Carus Monograph, *Random Walks and Electronic Networks*, with Peter Doyle. He and Doyle are currently editing a forthcoming volume in the MAA Notes series, *Topics in Contemporary Probability*.

William Peterson received his PhD in Operations Research. After a sojourn in industry, he joined the mathematics and computer science department at Middlebury College where he teaches courses in probability, statistics, and operations research. His research interests are applied probability, stochastic processes and queueing network theory.

Exploring Probability and Statistics Using Computer Graphics

Henk Tijms
Department of Econometrics
Vrije University, The Netherlands

Introduction

In the teaching of probability and statistics, microcomputers can certainly take the drudgery out of statistical calculations. More importantly:

- they can be used to teach basic concepts and ideas;
- they make it possible to perform laboratory experiments;
- they enable students to discover first principles themselves.

Many of the basic ideas in probability and statistics seem exceedingly difficult for most students to grasp. It is extremely important to give students a sound intuitive feeling for essential concepts such as randomness and the normal curve before teaching them formal statistical theory. Most people find it much easier to grasp concepts in visual terms and through direct experiences than in terms of formulas. Computer graphics are very effective in conveying an understanding of fundamental concepts through pictorial representations. For example, key ideas such as the Law of Large Numbers, random walks, and the Central Limit Theorem come alive before one's eyes through computer graphics. Direct experience and actual experimentation is the best way the student can obtain a feeling for these concepts.

It cannot be emphasized enough how important it is that students obtain a feeling for probabilistic reasoning at an early stage. People are not born with a natural feeling for probabilistic thinking, although probability reasoning is essential for solving many real-world problems. The microcomputer is the ideal tool to use to help students develop a sound probabilistic intuition. Computer graphics -- the most powerful feature of the micro -- should be exploited.

This paper discusses how we use graphical software to introduce the beginning student in both a motivating and coherent way to the following very basic concepts:

- The Law of Large Numbers. Computer animations of the experiment of coin and dice tossing is the best way to obtain insight into random fluctuations. Using an interactive simulation program, it is possible to fight misconceptions that even short runs of the coin-tossing experiment should reflect the theoretical 50:50 ratio of heads to tails.

- Random walks. The graphical display of random walks provides the student with a lot of insight into random fluctuations. The random walk showing the actual number of heads minus the expected number in the coin-tossing experiment is very instructive. Using this random walk, a natural link can be made with the Central Limit Theorem.

- The Central Limit Theorem. A visual demonstration of this very important concept in probability and statistics can be given. In an interactive way the student can generate the probability histograms for different sample sizes and see how fast the probability histogram approaches the normal curve as the sample size gets larger.

▶ Statistical graphs. Plotting the density graphs of various distributions such as the binomial, Poisson and chi-square enables the student to discover limiting relations between these distributions and the normal distribution.

Law of Large Numbers

Many people tend to expect that even short runs of the coin-tossing experiment will reflect the theoretical 50:50 ratio of heads to tails. This misconception is known as the gambler's fallacy since some gamblers feel that the probability of tails becomes larger after a run of heads. Mere exposure to the theoretical laws of probability may not be sufficient to overcome such misconceptions. However, an interactive simulation program on the microcomputer is ideally suited to correct misconceptions such as the gambler's fallacy.

Computer animation of the experiment of coin and dice tossing is the best way to obtain a feeling for randomness. Within a few seconds, students can simulate and repeat this statistical experiment on the computer. Looking at the graphical representation of the results, they see what randomness means. By doing the experiment of tossing a fair coin and observing the relative frequency at which heads appear, the student will see that this relative frequency may still significantly differ from the value ½ after a large number of tosses. As the number of tosses grows, the relative frequency will eventually approach the value ½ according to the Law of Large Numbers. To see this experimental demonstration of the Law is very in- structive. Also, it is valuable to see that the relative frequency typically approaches the value of ½ in a rather irregular way.

To make the above concepts come alive, Kalvelagen and Tijms (1990) developed a software module that simulates the experiment of rolling a die. The program gives the user the option of using either a fair die or a loaded die. For the case of a loaded die, the user has to specify the probabilities of each of the six possible outcomes of any given throw of the die. By

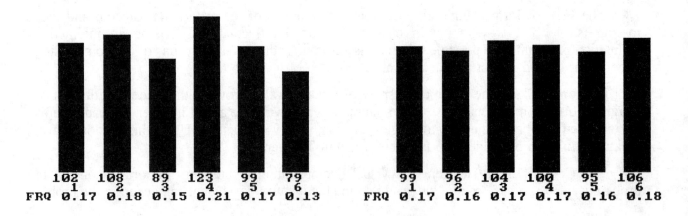

Figure 1: Two Simulations of the Die-Tossing Experiment

assigning positive probabilities to only two outcomes, the coin-tossing experiment is a special case of the die-tossing experiment. Once the data have been specified, a computer animation of the die-tossing experiment is given in real time. The final results for two computer simulations each consisting of 600 throws of a fair die are displayed in Figure 1. Such pictures give the student a good feeling for random fluctuations.

The software for the die-tossing experiment is elementary, but it is very instructive. Students like to do the experiment themselves and in this way they learn what randomness means. Much can sometimes be achieved by very simple techniques!

Random Walks

Let's consider the coin-tossing experiment using a fair coin. Many people erroneously believe that a run of heads should be followed by a run of tails so that heads and tails even out. In placing their bets, many gamblers use some system that takes into account any imbalance between the past number of heads and tails. It is illusionary to think that such a system can be of any help. The coin has no memory.

Suppose a fair coin has been tossed 100 times and has landed heads 60 times. Then, in the next 100 tosses, the absolute difference between the number of heads and tails may still increase, while the relative difference decreases. For example, this occurs when heads appears 51 times in the next 100 tosses. The only thing one can be sure of is that the relative frequencies of heads and tails will eventually be about equal, but there is no such thing as a law of averages for the absolute difference between heads and tails. In fact, the absolute difference between the number of heads and tails tends to increase as the number of tosses gets larger.

This surprising fact can be shown convincingly by simulation and a graphical display of the results on the screen. For a simulation experiment consisting of 5,000 tosses of a fair coin, Figure 2 displays the graph of the random walk giving the actual number of heads minus the

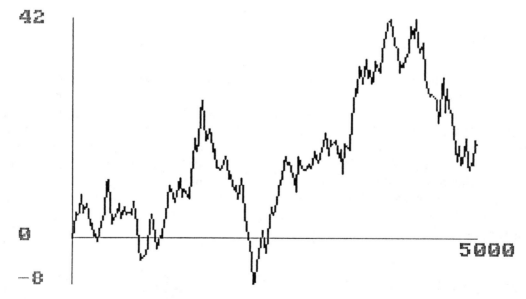

Figure 2: A random walk for the coin-tossing experiment

expected number. By trying different simulation runs of various lengths on the computer, the students will learn by experience that realizations such as those in Figure 2 are no exception, but are typical for the coin-tossing experiment. The longer the run, the bigger and bigger the waves of the random walk and the relatively more rare the crossings of the x-axis. This finding is rather counterintuitive, but can be mathematically explained from the Central Limit Theorem.

However, before turning to this basic theorem, it is important that the student first learns by direct experience that a series of independent probability trials tends to behave better and better in the average sense but wilder and wilder in the absolute sense as the number of trials increases. This is a basic lesson in probability! It is nicely demonstrated in Figure 2 and Table 1. In Table 1, the convergence of the frequency of heads to ½ is shown by giving, for a number of intermediate values of n, the relative frequency of heads after n tosses in the same simulation experiment as used for Figure 2.

n	frequency
334	0.4970
668	0.5075
1002	0.5030
1336	0.5082
1670	0.5114
2004	0.5045
2338	0.5013
2672	0.5052
3006	0.5060
3340	0.5048
3674	0.5082
4008	0.5087
4342	0.5067
4676	0.5049
5000	0.5036

Table 1   The relative frequency of heads

Once the students have learned by computer experimentation that the random walk of the actual number of heads minus the expected number tends to exhibit bigger and bigger fluctuations as the number of tosses grows, they may be ready to learn more about the mathematical explanation of this behavior. At this point, a natural link can be made with the Central Limit Theorem. Taking the Theorem for granted, how do we explain mathematically the behavior exhibited in Figure 2? Let the random variable $X_i = 1$ if the ith toss gives heads and $X_i = 0$ otherwise. It then follows that the actual number of heads minus the expected number can be represented as $X_1 + \ldots + X_n - n\mu$, where $\mu = $ ½ denotes the mean of the X_i's. Since the normal distribution has about 68% of its mass within one standard deviation of the mean, a simple consequence of the Central Limit Theorem is that

$$P\{ \mid X_i + \ldots + X_i - n\mu \mid > \sigma \sqrt{n}\} \approx 0.32 \qquad \text{for } n \text{ large enough,}$$

where $\sigma = $ ½ denotes the standard deviation of the X_i's. In words, the probability that, after n tosses, the random variable $X_1 + \ldots + X_n - n\mu$ will take on a value larger than $\sigma\sqrt{n}$ is about 32% when n is large. This gives the mathematical explanation why the absolute difference between the number of heads and tails tends to increase as the number of tosses grows. This result does not contradict the fact that the difference between the relative frequencies of heads and tails tends to zero, since the absolute difference between the number of heads and tails roughly tends to increase proportionally to the square root of the number of tosses. This important but subtle point is best understood by students through experimental studies using computer graphics.

Central Limit Theorem

The Central Limit Theorem is the most important result in probability and statistics. Using the random walk discussed in the previous section, the student's interest could be aroused in this theorem, but how do we explain it? The Central Limit Theorem is extremely difficult to prove. Moreover, the proof will not substantially help the student to understand the working of the theorem. It will not give the student a clear insight into how large n should be before the sum $X_1 + \ldots + X_n$ of n independent and identically distributed random variables $X_1, \ldots, X_n$ is approximately normally distributed. An intuitive feeling for the Central Limit Theorem is best obtained through experimental studies based on computer graphics.

A possible way to demonstrate the Central Limit Theorem graphically is to use computer simulation. In an interactive way, many samples may be drawn from a given probability distribution and a graphical display of the probability histogram of the sum may be given for different sample sizes. It is clearly evident that the probability histogram gets close to the bell-shaped form of the normal density as the sample size increases. The drawback of the simulation approach is that, for any fixed sample size n, many observations of the sum are needed before the simulated probability density of $X_1 + \ldots + X_n$ is sufficiently close to the true density. How many observations are needed is often not clear. Hence the simulation approach has the drawback that the Law of Large Numbers interferes with the Central Limit Theorem.

To avoid this complication which obscures the working of the Central Limit Theorem, we advocate an analytical approach in combination with computer graphics. Restricting our attention to discrete probability distributions allows us to compute analytically the true probability density of the sum $X_1 + \ldots + X_n$ for any fixed value of n. How to do this can be easily explained to the beginning student, as is done in Kalvelagen and Tijms [1]. An illustrative case is obtained by using the good old die and taking

$$X_k = \text{the number of points shown at the } k\text{th rolling of a die.}$$

The sum $X_1 + \ldots + X_n$ then represents the total number of points obtained when the die is rolled n times.

The software module we developed in Kalvelagen and Tijms [1] gives the student the option of using a fair die or a loaded die. For a loaded die, the student has to specify for $j = 1, \ldots, 6$ the probability $p(j)$ of getting j points at any given throw of the die. Through the specification $p(1) = 1 - p$ and $p(2) = p$ for some $0 < p < 1$, the student has the option of verifying experimentally that the binomial probability density tends to the normal curve as the number of trials gets large. For this specification, the number of points obtained in n throws minus n has a binomial distribution with parameters n and p.

Once the underlying probability distribution $\{p(j)\}$ of the die and the number n of throws have been specified, the computer program calculates the probability density of the sum $X_1 + \ldots + X_n$ and displays the graph of this density on the screen. A glance at the screen is sufficient to see whether this density has the bell-shaped form of the normal curve or not. Since the probabilities $p(j)$ of the die can be varied, students can discover themselves that how large

n should be before the probability density of the sum $X_1 + \ldots + X_n$ is close to the normal curve depends very much on the degree of asymmetry in the underlying probability distribution of the die. It is quite instructive to find out that the more the underlying distribution is asymmetric, the larger n should be.

As an illustration, Figure 3 gives the probability densities of the sum $X_1 + \ldots + X_n$ for $n = 3$ and $n = 5$ tosses of a fair die. It is remarkable how fast the probability density of the sum resembles the normal curve when the underlying distribution is symmetric. However, the situation is quite different for an asymmetric distribution. Figure 4 displays the probability densities of the sum $X_1 + \ldots + X_n$ for $n = 5$ and $n = 20$ tosses of a loaded die having the asymmetric distribution $p(1) = 0.2$, $p(2) = 0.1$, $p(3) = p(4) = 0$, $p(5) = 0.3$ and $p(6) = 0.4$. It is really fun to learn about the Central Limit Theorem in this experimental way.

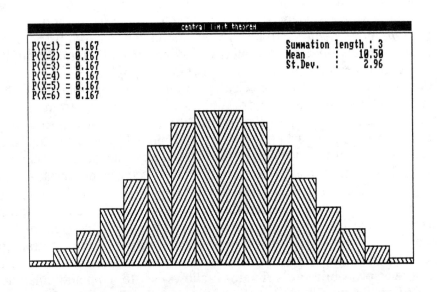

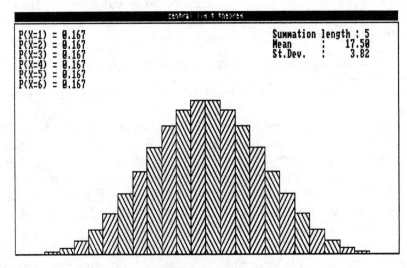

Figure 3: Probability Histograms for a Fair Die

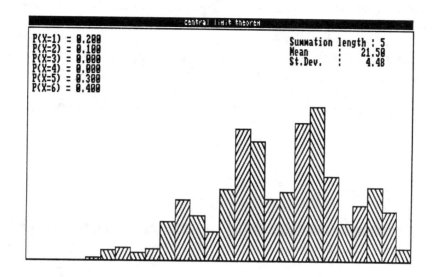

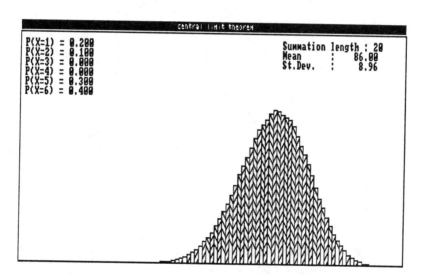

Figure 4: Probability Histograms for a Loaded Die

Statistical Graphs

The normal curve in probability theory is a law of nature. Many probability distributions are closely related to the normal distribution. Students can learn this fact not only from the Central Limit Theorem, but also by plotting the density graphs of various distributions.

The binomial distribution is the most important family of discrete probability distributions. A binomial random variable with parameters n and p can be interpreted as the total number of successes in n independent trials with probability p of success on each trial. This interpretation together with the Central Limit Theorem provide an explanation of the fact that the binomial density graph approaches the normal curve as n increases. Alternatively, students can discover this limiting relation by plotting the binomial density graphs for various values of n and p together with the corresponding graphs of the normal density. At the same time, they can experimentally verify that the normal approximation to the binomial density is quite good provided that the binomial mean is at least three (say) standard deviations away from both 0 and n, the extreme values of the binomial random variable. Assuming that students know that more than 99% of the area under the normal curve is within three standard deviations of the mean, this condition will be intuitively obvious to them; otherwise, there is not enough distance between the mean and the extreme values to be able to adopt the shape of the normal curve.

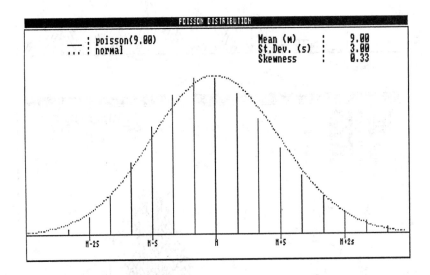

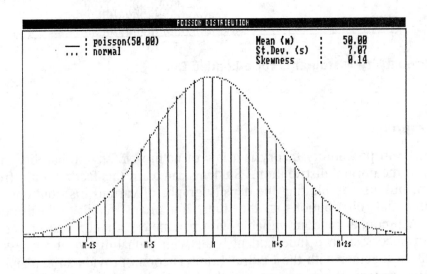

Figure 5: Poisson Densities and the Normal Curve

The approach of plotting the probability density graphs is particularly useful to discover a limiting relation between the Poisson and the normal distribution. A Poisson random variable X has a single parameter:

$$P\{X = k\} = e^{-\lambda} \lambda^k /k!, \quad k = 0,1,\dots .$$

The mean μ and the standard deviation σ of the Poisson distribution are given by $\mu = \lambda$ and $\sigma = \sqrt{\lambda}$. Students may argue that approximate normality for a Poisson distribution requires that the mean $\mu = \lambda$ is at least three deviations $\sigma = \sqrt{\lambda}$ away from 0. The condition $\lambda \geq 3\sqrt{\lambda}$ gives $\lambda \geq 9$. Using graphical plots, students can directly verify the validity of this condition. Figure 5 displays the plots for both $\lambda = 9$ and $\lambda = 50$.

A lot of insight is obtained by experimenting with graphical software that displays the density graphs of various distributions together with the corresponding density graph of the normal distribution. It is very rewarding to learn in this way about the key role of the normal distribution in probability and statistics.

Reference

Kalvelagen, E. and H. C. Tijms, *Exploring Operations Research and Statistics in the Micro Lab*, Prentice-Hall, Englewood Cliffs, NJ, 1990.

Henk Tijms is professor of operations research at the Vrije Universiteit in Amsterdam. He has published widely in applied probability and stochastic optimization, including a textbook, *Stochastic Operations Research*. He has also been extremely active in popularizing applied mathematics at Dutch high schools.

Computer Graphics and Simulations
in Teaching Statistics

Henry Krieger and **James Pinter-Lucke**
Harvey Mudd College   Claremont McKenna College

Introduction

Computer graphics and simulations can greatly enhance the teaching of statistics, especially in introductory courses. We shall illustrate this point by considering two important aspects of statistics, the limiting results of repeated sampling and the exploration of data. For reasons we shall discuss, both of these topics are either poorly taught or completely omitted from traditional basic courses. In each case, we shall show how the use of interactive computer programs, which are based on simulations and emphasize graphical displays, have made the associated concepts and methods much more accessible to our students.

This article is organized into three primary sections. In the first, we deal with the problem of teaching students the concepts of random sampling, convergence in probability, and convergence in distribution. Results such as the Law of Large Numbers and the Central Limit Theorem are neither intuitive nor easily illustrated because they involve sample size increases and often require many repetitions before the limits become evident. In a mathematical statistics course, these theorems can be proved, but there is little opportunity to do this in an introductory course. Even for students who might be expected to understand a proof, it may be more valuable to provide vivid illustrations of the results in order to instill intuitive feeling for the concepts. We shall show that it is precisely at this point that programs such as ours, featuring simulation and graphical displays, come to the fore, especially when they are designed with student interaction and experimentation in mind.

The second section considers the problem of teaching data analysis. Ordinarily, students are not exposed to a sufficient amount of data analysis for several reasons. It is difficult for teachers to produce interesting examples of data sets for their students to analyze, especially if they would like to provide many such sets. Even if data sets are available, students are often handicapped in their analysis because of a lack of tools, both for computing and visualization. We shall describe an example of a computer program that randomly assigns distributions and generates data for the students, thus reducing the effort required of the teacher. It also creates graphical displays associated with sample statistics which help the students in their investigation of the data. In addition, the program records the steps taken during each student's analysis so that teachers can discuss this process with their students and help them learn from their mistakes.

Our final section deals with the difficulties and tradeoffs involved in producing software for education in statistics. One tradeoff arises from the conflicting goals of ease of production and breadth of distribution. Complete software for statistics is easier to produce if one works on a mainframe which is typically equipped with statistical packages that provide the programmer with technical subroutines for tasks such as random variate generation. The mainframe may also come with a graphics package to facilitate the displays, but this package is often highly machine dependent. On the other hand, for broad distribution of software, machine independence and portability are desirable. The latter can be achieved on a personal computer which supports a graphics package. However the cost to the programmer is the necessity of developing one's own random variate routines. Finally, assuming that a useful package has been produced, there is the formidable task of making the rest of the statistics community aware of it.

Teaching Limit Theorems

In order to understand statistical inference, students must understand the concepts of random sampling, convergence in probability, and convergence in distribution. In particular, they must understand the Law of Large Numbers and the Central Limit Theorem, even if they are taking a less mathematical approach and do not see the proofs of these theorems. Since both of these results are concerned with the outcome of repeated random sampling, computer routines which simulate this process from a wide variety of distributions are the basis for our programs. Our goal in writing the programs, beyond just providing graphical illustrations of the mathematical concepts, was to produce programs that students could use on their own to try out their perceptions of these concepts. We felt and still feel that, if the students are active participants in the program, then they will learn more about the target concepts as well as other related ideas, such as the nature of different distributions and the effect of varying their parameters.

The Law of Large Numbers states that in independent samples of size N chosen from a population with mean μ, the sample mean will converge to μ as the sample size N increases. We illustrate graphically the weak form of this law which corresponds to convergence in probability (or to convergence in distribution to a distribution concentrated at a single point) by displaying the histograms of 100 sample means derived from random samples of increasing size. The narrower the histogram and the more centered about the population mean, the better we can say the sample means are converging to the population mean.

The program LRGNUM illustrates this process by allowing the user the choice of a distribution and the choice of a sample size. It then does the random sampling, computes the means, displays the corresponding histogram in the upper third of the screen and prompts the user for another sample size. After the sample

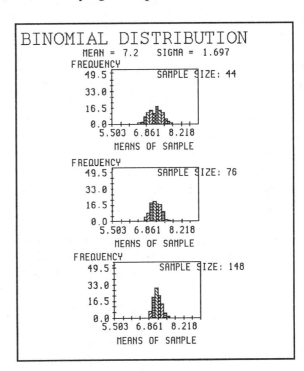

Figure 1

size is entered another histogram is produced and displayed in the middle third of the screen. Again the user is prompted for a sample size and a third histogram is displayed on the lowest third of the screen. (Figure 1)

If the user has been asking for progressively larger sample sizes, the histograms will have changed from quite broad, spread-out ones to tighter, narrower ones. As histograms for new sample sizes are produced, the oldest histogram is discarded, the other two are moved up, and the newest histogram appears in the lowest third of the screen.

LRGNUM actively involves the users in the program by having them choose distributions from one of six discrete and seven continuous distributions. The discrete choices are the Bernoulli, binomial, negative binomial, geometric, hypergeometric, and Poisson distributions. The continuous choices are the chi-square, gamma, exponential, normal, t, F and uniform distributions. Users may receive information about the distributions and their associated parameters before making their selections. Once a final choice, consisting of a distribution and parameter value(s), is made, the sample size selections are performed. Upon completion of this process, the parameter(s) of the distribution may be changed or a completely new distribution may be selected. As shown in Figure 1, the display always includes the name, mean, and standard deviation of the population distribution and each histogram is labelled with the appropriate sample size.

The Central Limit Theorem states that the distribution of the sum of N independent identically distributed standardized random variables converges to the standard normal distribution as the sample size N increases. To illustrate the concept graphically, the sample density functions of standardized means from random samples of increasing size are displayed against the graph of the standard normal density function. As sample size increases, the graph of the sample density function should lie closer to the graph of the normal density.

The program CENTRAL displays this process in a manner parallel to that of LRGNUM. The user chooses any of the same distributions, sets the parameter(s) and the sample size. The program then creates the random samples, computes 200 means, standardizes them, computes their sample density function, and graphs the result on the same graph as the standard normal density. The sample density function is computed from the sample cumulative distribution function by numerical differentiation and a smoothing function. The user is prompted for a new sample size, and the operation repeats until three sample density functions have been graphed on the same screen. As shown in Figure 2, the display indicates the sample size associated with each of the graphs.

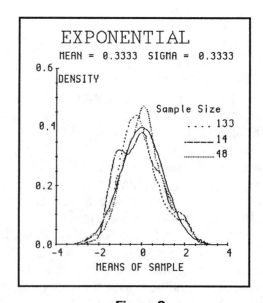

Figure 2

The Central Limit Theorem is an example of convergence in distribution. Not all such convergence is to a normal distribution, so we wanted to illustrate situations in which the limit distribution was something different. An example of this occurs in conjunction with the maximum value of a random sample, chosen from certain continuous distributions. This maximum converges to one of three types of limit distributions, called extreme value distributions, none of which is close to the normal. Our program EXTREME displays the sample density function of 200 maxima calculated from random samples of a user prescribed size, after the data has been standardized suitably. As is shown in Figure 3, this sample density function is displayed against the graph of the appropriate extreme value distribution.

Even though programs such as LRGNUM, CENTRAL, and EXTREME are designed with student experimentation in mind, use of these programs in our classes has shown that a certain amount of guidance by the teacher is extremely helpful. Once students have seen several demonstrations of a particular limit theorem, some of them may be convinced of its truth and think there is nothing further to learn about the process. Here is where the teacher can step in. For example, by suggesting specific sequences of distributions to be sampled, a student can be led to observations and conjectures regarding the relationship between smoothness of the distribution function and rate of convergence. In such a way, student experimentation can be channeled in productive directions. However, there is a definite art to giving students useful suggestions that still permit them to make their own choices and mistakes. We do not claim to have completely mastered that art.

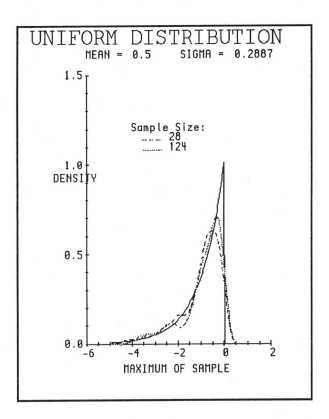

Figure 3

Teaching Data Analysis

In the actual practice of statistics, a statistician must often decide the nature of the distribution underlying a process being analyzed. In fact, he must describe it both qualitatively and quantitatively. This skill is just as important for a statistician as it is for a chemist to deduce the nature of an unknown material. Without computers there is not an easy way to give the student practice at this basic skill because it is too time-consuming to acquire data and too tedious to produce the statistics and graphics necessary to complete the analysis. The process of determining the distribution is not only good training, but is very instructive as the students must use the tools, techniques, and knowledge that they have acquired from the statistics course in a new and different manner. The computer can easily generate sample data that would otherwise require hours of field work. Additionally, we had already developed some illustrative graphics programs for statistics and felt that by combining graphics with the computational and data-manipulation power of the computer, we could produce a sophisticated program that would give students experience in data analysis.

There were two main considerations in designing our program, STATLAB. The first, and main, consideration was to produce an exercise that would stimulate the students. This is accomplished by giving them an interesting problem, with appropriate tools to solve the problem, but with enough constraints that they will have to approach the problem thoughtfully. Of course we want the program to be easy to use so the students can concentrate on the problem and not the program. In fact, the latest version was written for a Macintosh to provide the program with an intuitive interface. The second consideration was the goal of giving the professor enough information about the students' actions to allow them to continue the educational process in a "debriefing" session.

The first goal was attained by simulating a laboratory exercise in which the students are given unknown substances and expected to use their analytic skill to discover the identity of the substance. In our program, the substance is an unknown probability distribution which is randomly assigned to the student by the computer. The students are then provided with statistical tools with which to conduct their detective work. In actual practice, a statistician would have to spend a lot of time and energy doing tasks such as collecting data, computing statistics describing the data, ordering the data, graphing a histogram of the data, and conducting a goodness-of-fit test of the data against a variety of suspected answers. In our program, these tools are provided, but at a cost to the student's limited budget. This is parallel to a laboratory situation in which one has only a finite amount of a mystery substance and must finish the analysis before running out of the unknown.

The program accomplishes the second goal, allowing assessment of the student's progress, by recording the distributions assigned to the student, including the parameters of each, and all operations performed by the student. Thus the professor can recover the user's train of thought in solving the problem and can point out both any extraneous actions by the student and any concepts or constraints that the student has overlooked. Together the professor and student can see how to detect the type of distribution and the parameters that specify it.

In the initial stages of analyzing an unknown distribution, graphics can be very useful. The histogram or sample density function can point out properties such as skewness, centering about the origin, and possible normality. Since, for realism, the student can only generate a small amount of data, 150 pieces, the graphs are not going to match perfectly with a textbook graph, but they will assist in understanding the data. Finally when all this analysis has given the investigators a clue as to the nature of the distribution, they can test the hypothetical distribution through a goodness-of-fit test such as the chi-square or Kolmogorov-Smirnov tests.

The program generates and stores, for future use, random data from the distribution, calculates statistics such as the mean, standard deviation, and the extreme values, reorders the data into ascending order, and allows the user to see the histogram or sample density function. The students can also use the chi-square or Kolmogorov-Smirnov goodness-of-fit tests to see if their solutions are statistically significant. Moreover, in the latest version, the program graphically illustrates the workings of each of these tests (See Figure 4). These tools are provided so that the students can concentrate on the problem-solving. The program also contains the pool of possible distributions so that the user may review this list at any time.

The student's bank account is assigned at the beginning of the program and is updated after each operation. Each of the tools available to the student has an associated cost. Thus there is a constraint on the student trying to solve the problem with massive amounts of random computations and guessing. They can easily run out of money if they do not plan properly. This adds to the realism and challenge of the exercise.

There is another consideration about this exercise that is important for professors. Before assigning the work, they may want to be sure that the level of the exercise is correct for the class. The instructor can have the program randomly assign from one to five unknown distributions per student. The distributions may be restricted to a family of six one-parameter

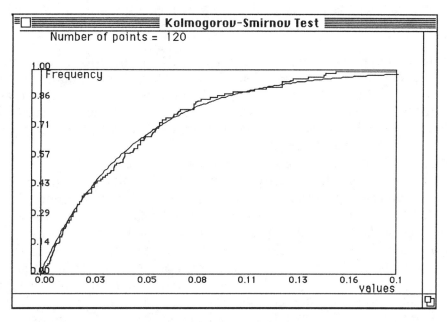

Figure 4

distributions, or to a family of twelve one- or two-parameter distributions, or even to the last family expanded to include one three-parameter distribution, the hypergeometric. The collection of distributions is the same as those used in the previously mentioned programs, only here they are organized by number of parameters. Thus, the exercise can be fine-tuned to accommodate different levels of familiarity with statistics and mathematics.

Our students found STATLAB challenging once they learned to use the program. They were running the version of STATLAB on a VAX, and when one of them found a bug in the program, (and they were good at that) it crashed the program. Then the data files for the program were invalid and no one could run the program! These initial difficulties caused some students to back away from STATLAB and illustrated again the need to make the program as foolproof and fail-safe as possible. Those students who overcame the initial hurdles found data analysis more difficult than they expected; it required a new type of thinking on their part. They had assumed that they would easily be able to determine the nature of the distribution from the shape of the histogram or sample density function. Instead they found that there was a family of potential winners and that more detailed analysis was required to find the correct choice. Many of these students became quite involved in the exercise and learned a lot from it. Based on our experience, we see STATLAB, especially the new improved versions, as a valuable addition to the statistics curriculum which provides a unique opportunity for students to test their understanding and skills in a realistic setting.

Software Development

Producing useful educational software for statistics is a highly beneficial endeavor, but it is not a simple task. To be accepted by students and faculty, the software must have both a sound educational basis and a natural, intuitive interface. In addition, it should be a solid piece of software which is impossible to break and which does not confuse or frustrate the user. Another asset in statistical simulations is access to a large number of distributions via robust random number generators. The production of such high quality software is time consuming and not to be taken lightly.

Since the developer is putting his heart and soul into the effort, he is looking for two, often conflicting, results. He wants assistance in and/or tools for writing the software to make the task as easy as possible, and he wants the results to be as widely distributed as possible. If he is truly fortunate, he may find a way to enlist a team of competent students and colleagues to assist him. Alternatively, he may find a computer with software packages that provide many of the routines he needs in the development.

In the past and, for the most part, in the present, the developer will get the most "help" on a mainframe. Such computers usually contain statistical software libraries, such as IMSL, which provide subroutines to generate many different random distributions. Additionally, mainframes often have graphics packages to assist the developer in this crucial area. Finally they contain utilities such as symbolic debuggers and language sensitive editors that greatly assist in the coding and debugging of the project. The price one pays for all this assistance is that the final program cannot be widely distributed. It can only be run on other mainframes of the same type equipped with the same packages. Even when the developer does take the time and effort to replace subroutines from the packages, thereby making the program more independent, it is not easy to promote and distribute mainframe academic software. Most professors are no longer looking for software for mainframes even though there is now some effort to facilitate the distribution process, notably the ClearingHouse[1] for VAX software.

Graphics add considerably to the impact of statistical software and may in fact be the most effective language to use in describing certain concepts. This means that mainframe computers have another handicap: there is no standard graphics terminal for any particular mainframe. Hence the mainframe developer must use the lowest common graphics or find/develop some method for the users to identify their terminal and hook the program to the correct graphics device driver. At this moment, this is a large deficit to working on a mainframe.

Another disadvantage of developing one's software on a mainframe is the lack of a standard graphical interface such as exists for the Macintosh. For educational software to be successful, it must be easily accessible to the students and require very little explanation from the professor. This is most easily achieved with a graphical interface. While there may be some window systems for mainframes that allow for the development of a graphical interface, this also means that fewer mainframes will be compatible with such software. The advantages of this type of interface are illustrated in Figure 5 where a screen from the latest Macintosh version of STATLAB is displayed. One can see that the users have much useful information at their fingertips since, by clicking in any window, they can view the information in the window. The only other choice is to write line oriented software, which will have to be carefully designed to have half the appeal of an application with the graphical interface.

If the developer chooses to use a standard personal computer as the development platform, then some things are simplified. There will be only one kind of graphics and no need to identify the terminal. There is a large base of personal computers in place for instructional purposes, so distribution will amount to sending a floppy disk to an interested colleague. (Of course that still leaves the problem of finding interested colleagues.) However, the disadvantage

---

[1] The ClearingHouse, Computational Center, 297 Durham Center, Iowa State University, Ames, Iowa 50011

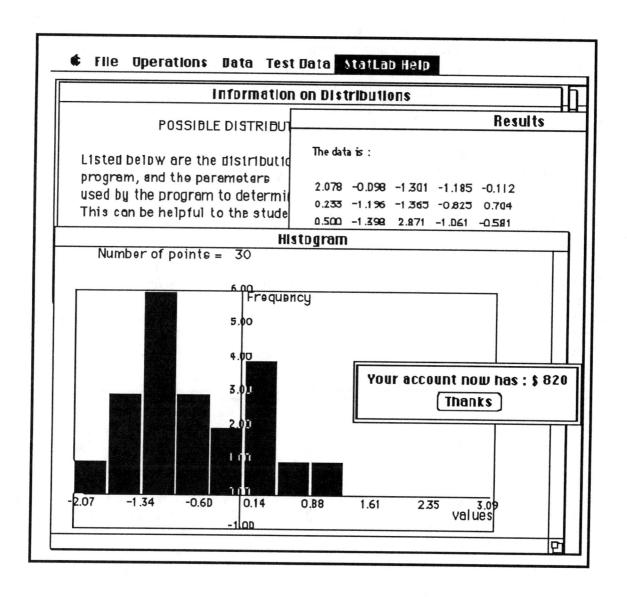

Figure 5

is that the development task is much larger since the typical personal computer has no advanced statistical package such as IMSL. While in no case is the writing of educational software a short simple task, the prospect of writing all one's own random number generators, statistical tests, and high level graphics can be a daunting one that may keep the thoughtful professor from ever getting started.

This picture is improving as more statistical packages and programming tools become available for personal computers. However students' and professors' expectations have risen. In the future, one of the large tasks for a developer will be to design a friendly, intuitive interface such as is standard on the Macintosh. In fact, this is what we have done to make

STATLAB more attractive, but not without quite a bit of work.

While we have shown that it is possible to write useful, friendly, highly interactive software to assist in teaching/learning statistics, our experience indicates that it is a highly non-trivial task. For instance, writing software to Macintosh user interface standards is much more work than writing line oriented software. At present most of the incentives in our profession pull one away from software development, primarily because the software is not considered a publication. What is needed to encourage the production of useful software for mathematics and statistical education is a national project to support developers. This may take the form of summer institutes where professors can work together and be supported both financially and by development assistance. A new peer-reviewed journal for educational software in mathematics would also promote more work in the area. In general, the mathematical community must put a higher value on this work if we wish to use it to improve our instructional efforts.

Conclusion

In this paper, we have attempted to illustrate a number of interactive computer programs, based on simulations and emphasizing graphical displays, which can be used to enhance the teaching of introductory courses in probability and statistics. While our programs offer only a glimpse of the possible uses of such techniques, we think they are representative both in their classroom uses and the difficulties associated with their development. Since one of our goals is to stimulate interest in teaching techniques which incorporate these types of programs, we hope that they will be a seed from which a great deal will grow.

Henry Krieger received his Ph.D. in Applied Mathematics from Brown University in 1964. After teaching at Cal Tech, he came to Harvey Mudd College, where he is professor of mathematics. He teaches probability, his specialty, and statistics.

James Pinter-Lucke is professor of mathematics at Claremont McKenna College. After writing simulations for research in reliability theory, he become interested in computer science and attended the Institute for Retraining in Computer Science. He now teaches both mathematics and computer science while conducting research on using computers in education. One-time tennis opponents, he and Henry Krieger have been collaborating on computer programs in statistics since 1983.

Sampling + Simulation = Statistical Understanding: Computer Graphics Simulations of Sampling Distributions

Florence S. Gordon   and   **Sheldon P. Gordon**
New York Institute of Technology   **Suffolk Community College**

Introduction

Students in introductory statistics courses are often asked to accept many statements and procedures on faith since the mathematical justification is usually far too sophisticated for them to comprehend. In large measure, this can be attributed to the nature of statistics, which is quite unlike most other undergraduate mathematics offerings. In other courses, the underlying theory applies directly to the object in question, whether it is a function being optimized or integrated, a system of linear equations being solved or a differential equation being solved.

In statistics, however, the theory applies to some unseen underlying population, not to the sample at hand. The sample is only used to make a statistical inference about the unknown population and the "answer" always involves some degree of uncertainty. Unfortunately, students see just the sample, but have no direct way to perceive the population or to develop any deep understanding of its properties.

Moreover, the fundamental notion in inferential statistics is variability based on the sample drawn from the population. In most traditional approaches, the emphasis in an introductory course has all too often been on the mechanics of performing a hypothesis test or constructing a confidence interval. As such, the students are usually given the summary statistics for a set of data (rarely the actual data values themselves) and asked to perform an inferential procedure. At best, there is some preliminary theoretical discussion in class, but it is usually very quickly forgotten in the rush to complete the mechanical solution to the problem. When students do work with the actual data, even data they have personally collected, it is still just one possible sample to which they apply the appropriate (we hope) procedures. It is therefore not surprising that many students come out of an introductory statistics course having mastered a series of computational procedures, but with relatively little statistical understanding.

It is highly desirable to consider cases where the students see many different samples and thus the resulting effects on the associated sampling distribution. However, it is usually too difficult and too time-consuming to examine large numbers of different samples in front of a class. And, in the few cases where this is done, it is much more likely that the class becomes bogged down in the details of enumerating and analyzing all possible samples, so that the students do not see the overall patterns.

Fortunately, most of the critical topics in statistical inference can be dramatically presented using computer graphics simulations which allow the students to visualize the sampling distribution and to see how it relates to the underlying population. While the *practice of statistics* is predominantly quantitative, we believe that the *learning of statistics* can often be accomplished qualitatively through recognition of visual patterns. In this way, we can significantly enhance students' understanding of the statistical concepts and methods. To paraphrase an old adage, a good simulation is worth several thousand numbers and a thousand words.

Even as simple an experiment as flipping a set of two or three coins is difficult for many students in terms of predicting the type of results which will occur. We can derive the associated probabilities theoretically, but many students are not convinced about the accuracy of the results. Computer simulations can be very valuable in conveying this conviction.

At a simplistic level, a simulation can be used to randomly generate the results of repeated coin flips (or the outcomes of other experiments). The actual outcomes might be displayed individually in the form: HH, HT, HH, TH, HT, ..., say, or a summary of the simulated results might be shown in a frequency table. Unfortunately, most students find it difficult to abstract any usable pattern from lists of individual outcomes or even from a summary table. The human mind has evolved to process information visually; processing symbolic and numerical information is not a natural mechanism and we should not expect all students to be good at it.

A computer graphics program which simulates the experiment repeatedly and displays the results visually is far more effective. Students are able, visually, to detect the resulting patterns and so develop a feel for the accuracy of the predicted results. Simultaneously, it provides them with the conviction that the theory indeed agrees with reality. Numerous articles, including [2] and other papers in this volume, describe such simulations for a variety of probability experiments.

The same type of approach can be applied throughout an introductory statistics course to simulate graphically a wide variety of discrete and continuous distributions. In this article, we discuss the use of such simulations for investigating most common sampling distributions.

Simulations of Sampling Distributions

Various authors ([1,2] as well as others in this volume) have described how the distribution of sample means can be simulated extremely effectively using computer graphics. Typically, the students can select any of a variety of underlying populations, sample size and number of samples. In most implementations, the program randomly generates repeated samples of the given size, calculates and then displays either the mean of each sample or the associated value for $(\bar{x} - \mu)/(s/\sqrt{n})$. See Figures 1 and 2 for results based on samples of sizes $n = 4$ and $n = 25$ drawn from a skewed population. Each horizontal line represents the sample mean for such a random sample; a histogram of the underlying population is superimposed for comparison. (Note that this comparison would not be evident if t-values were plotted instead of the sample means.) From such explorations, the students in an introductory statistics course are able qualitatively to predict the conclusions of the Central Limit Theorem and the properties of the sampling distribution of the mean based exclusively on the visual displays.

Of course, how large n must be to produce approximate normality depends strongly on the underlying population. Computer graphics provide an ideal tool for experimenting with different values of n and different populations. For instance, students can see that $n = 1$ is adequate if the underlying population is normal; that $n = 12$ is usually accepted as large enough if the underlying population is uniform; and that $n = 100$ may be needed if the underlying population is highly skewed.

In a previous article [3], the present authors describe the use of computer graphics simulations to investigate the properties of other sampling distributions which often arise in an introductory statistics course. These include the distribution of sample proportions, the distribution of the difference of sample means, the distribution of the difference of sample proportions and the distribution of sample variances.

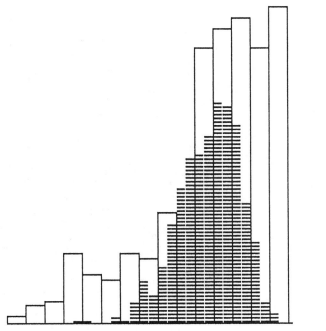

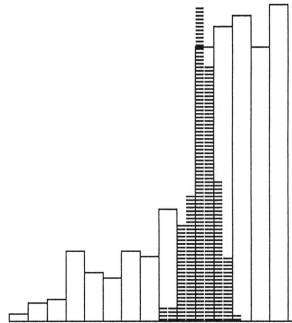

Figure 1: 500 simulated sample means with sample size *n* = 4.

Figure 2: 300 simulated sample means with sample size *n* = 25.

It is also possible to consider the sampling distributions associated with other statistics. For example, we can consider the sampling distribution of the median and use an appropriate computer graphics simulation to experiment with its properties. In Figures 3 and 4, we show the results of such a program when samples of sizes *n* = 4 and 20, respectively, are drawn from an underlying U-shaped population and the sample median for each sample is calculated and displayed. Incidentally, the occasional occurrence of an outlier, such as the one at the extreme left in Figure 3, should be pointed out in class and used as the focus for a short discussion.

A particularly effective use of such a program is to allow the students to conduct their own investigations of this and related sampling distributions on an individual or small group basis. Having seen how such an analysis proceeds with the distribution of sample means, they can ask themselves the comparable questions about the sampling distribution being studied: what is its shape? what is its mean? and what is its standard deviation? The students are then able to make some conjectures based on the visual and accompanying numerical displays. Individual students can be assigned different projects of this nature using a variety of underlying populations and a variety of sampling distributions such as those for the median, the mode, the midrange or even the variance. This type of activity is extremely desirable for giving students a feel for discovering and exploring a mathematical theory on their own. More importantly, by repeating the procedure used in developing the key ideas for the distribution of sample means in a parallel context, the students achieve a far better understanding of what the first exploration accomplished. Otherwise, the ideas presented on the Central Limit Theorem are not reinforced and hence may not make a sufficiently deep impression on all students.

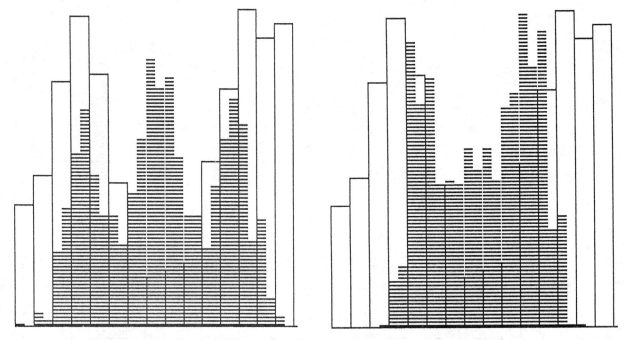

Figure 3: 1000 simulated sample medians with sample size $n = 4$.

Figure 4: 1000 simulated sample medians with sample size $n = 20$.

Estimation

The idea of graphical simulations can also be applied to many other core topics in inferential statistics to increase student understanding.

We begin with estimation of the unknown mean μ for a population. A 95% confidence interval for the mean should have a 95% chance of containing μ or, equivalently, 95% of the confidence intervals so constructed should contain μ. For most students in an introductory statistics course, this statement represents, at best, nothing more than an act of faith. They do not fully appreciate the fact that the confidence interval constructed will correctly contain μ with probability .95. There is simply no effective way to construct a large variety of different confidence intervals based on different sample data to see whether the theoretical considerations actually make sense. Instead, the students too often perform the appropriate manipulations to calculate the correct answer to any such problem in a purely mechanical fashion.

However, computer usage is ideal for this type of repeated calculation. In Figure 5, we show the results of a program which generates repeated random samples from a given underlying population, constructs the corresponding confidence intervals and displays the results visually. This graphical simulation provides an especially powerful tool to translate the statistical theory and predictions into ideas that the students can visualize and hence comprehend and remember.

The result shown in Figure 5 is based on repeated random samples drawn from a selected underlying population. Each sample is used to construct a 90% confidence interval for the mean. The program constructs and draws almost 100 successive confidence intervals on each run (until the screen is filled). Notice that the location of the sample mean $\bar{x}$ about which each confidence interval is centered is also displayed. The vertical line indicates the location of the population mean μ. Further, whenever a particular confidence interval does not contain

μ, the program displays the corresponding line in a different color for emphasis. In this particular run, the numerical results corresponding to Figure 5 show that 87 of the 96 confidence intervals, or 91%, contain μ. Repeated runs of such a program can be used to demonstrate that, in the long run, the results will more or less average out to the selected confidence level of 90%.

A useful computer laboratory exercise is to have each student run the program individually and simultaneously, and then average the percentages of intervals which contain the true mean μ. One important advantage to conducting such an activity is to reinforce the notion of variability. Too many students do not have a well-founded understanding of the word *about* as we use it in mathematics. When we say that *about 90%* of the confidence intervals will contain μ, most students interpret this to mean that *precisely 90%* of the intervals will do so. By showing them the results of repeated runs of this simulation, they can see that different sets

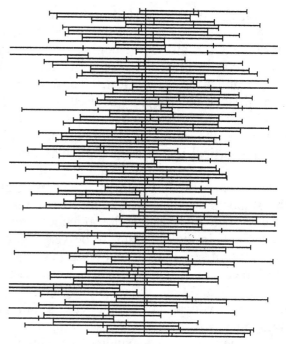

Figure 5: 96 simulated 90% confidence intervals for the mean μ.

of outcomes may, for example, include as few as 84% of successes or as many as 97% of successes, as well as the expected 90% of successes. On the other hand, if the comparable experiment is performed with a 98% confidence level, then the results may fall between 93% and 100%. One hundred repetitions is simply not large enough for the results to be as close to 90% or 98% as many students naively expect.

In addition, within each run of the program, the lines drawn for the individual confidence intervals have different lengths. This is because the length of each interval is based on the size of the sample standard deviation. Thus, the program provides a visual dimension for seeing the effects of the sample standard deviation on the outcome of an estimation problem.

Similarly, if a different confidence level, say 98%, is used, then it is visually clear that most of the confidence intervals drawn are longer than those shown in Figure 5. Moreover, very few of these confidence intervals do not contain μ. Thus, the students see that, by increasing the confidence level, we achieve a much greater likelihood of the confidence interval containing μ. Perhaps most importantly, such a program can be used to give the students a greater appreciation for the nature of inferential statistics: any statistical result is based on the data from one particular sample and the result will change if a different sample is used.

Totally comparable results can be achieved for confidence intervals for proportions. The primary difference is that the user is able to define the population proportion π of the underlying population.

Hypothesis Testing

We next consider hypothesis testing for means. Again, this is a procedure where the key ideas are, at best, usually accepted by most students on faith and the corresponding problems are handled mechanically rather than with any statistical understanding. For instance, if the significance level is .05, students often do not appreciate the fact that only 5% of all possible sample means will fall into the rejection region when H_0 is true. As with confidence intervals, these ideas can also be demonstrated very effectively with an appropriate graphics simulation.

We illustrate the results of such a program in Figure 6 using a significance level of α = .05 and a one-tailed test. The theoretical sampling distribution, which is approximately normal since the sample size used is $n = 36$, is drawn first. The corresponding critical value for the z-statistic is drawn as the tall vertical line. Finally, 100 random samples of size 36 are generated, the sample mean of each is calculated, and the associated test statistic is drawn as a shorter vertical line. In the particular run of this program displayed in Figure 6, 6 of the 100 sample values fell in the rejection region.

When this program is used repeatedly, the students see that the proportion of sample values that fall inside the rejection region comes out, in the long run, to be very close to the value of α. Further, they see that most of the "rejects" are relatively close to the critical value. In addition, by examining closely the pattern in which the sample values fall and noticing where they are dense compared with where they are sparse, students can again see how the sampling distribution is roughly normal. They also begin to appreciate the fact that they are really dealing with just one possible sample when they perform a typical hypothesis test which gives them a better understanding of the nature of hypothesis testing.

As with confidence intervals, it is also possible to treat hypothesis tests on the population proportion π in a totally analogous manner, but we will not indicate any of the details here.

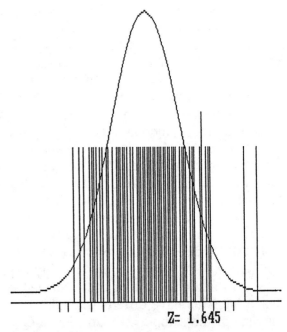

Figure 6: 100 simulated sample means for a one-tailed hypothesis test with α = .05.

Chi-Square Analysis

Our next application is chi-square analysis for contingency tables, another topic that students often end up handling in a purely manipulative manner with little real understanding. Essentially, they may memorize the fact that the chi-square statistic, generated through a certain process or formula or computer routine, follows a chi-square distribution with a certain number of degrees of freedom. Unfortunately, many students rarely gain any appreciation of what is actually happening. Using software to perform the calculations for a given contingency table frees them from the drudgery of doing the computations by hand, but does not convey any real understanding of what the procedure is all about or what is signifies.

In the authors' graphics simulation, the user is able to define the expected frequencies for a contingency table. For a given value for the sample size n, the program randomly generates samples of size n and assigns each sample entry to the appropriate cell in the table based on the individual frequencies or probabilities. Each resulting contingency table is then used to calculate the corresponding value for the chi-square statistic and the results of the repeated simulations are plotted, as shown in Figure 7. As the successive values are graphed, the students are able to see how the values fall into the pattern for a chi-square distribution. This is important for them to see visually because many students expect that everything behaves according to a normal distribution pattern. They can also see that relatively few of the simulated chi-square values actually exceed the critical value and so lead to a rejection of the null hypothesis assuming that H_0 is true. As we said before, such simulations dramatize the fact that the results of a statistical test depend on the particular sample.

The program also provides the opportunity to turn chi-square analysis into an exploratory exercise in which the students can see the effects of using different subtotals (and consequently a different set of defining probabilities) in the underlying population.

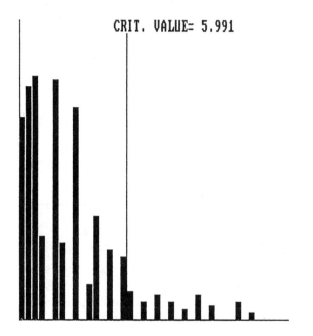

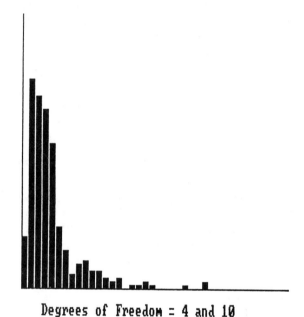

Figure 7: 400 simulated sample values of the chi-square statistic.

Figure 8: 300 simulated sample values of the F-statistic.

Analysis of Variance

Computer graphics simulations can also be applied to analysis of variance, another heavily computational topic that students end up treating in a purely mechanical way with no real statistical understanding. Again, using software that only performs the ANOVA calculations for a set of sample data does not lead to comprehension, but only to answers.

In the authors' simulation, the user can define the value of μ to be tested, the number of treatments and the number of observations per treatment. The program randomly generates repeated samples from a normally distributed population having the indicated mean μ, calculates the corresponding value of the F-statistic for each set of samples, and plots it. The results of a typical run are shown in Figure 8.

Linear Regression and Correlation

Finally, we consider the notions of linear regression and correlation. Again, students rarely appreciate the fact that the data they use to construct a regression equation or calculate the correlation coefficient is just one possible set of bivariate data. Rather, they get so involved in performing the calculations, even with computational tools, that they lose sight of the underlying statistical ideas. These procedures, though, also lend themselves to graphical simulations to enhance the concepts visually. In Figure 9, we show the results from a program which draws repeated samples of size $n = 4$ from an underlying bivariate population, calculates the regression line for each sample and displays it graphically. For comparison, the regression line for the population is also shown and the point corresponding to the two means, μ_x and μ_y, is highlighted by the circle. The students quickly see that each sample gives rise to a different regression line, though most of them remain relatively close to the population regression line and most of them pass fairly close to the indicated point. However, it is not unlikely to obtain several sample lines which lie at a sharp angle from the population line, and so regression analysis based on small samples is seen to be highly suspect. On the other hand, when larger sample sizes are used, the resulting sample regression lines usually lie very close to the population line and so support the notion that increased sample size makes for far better predictions.

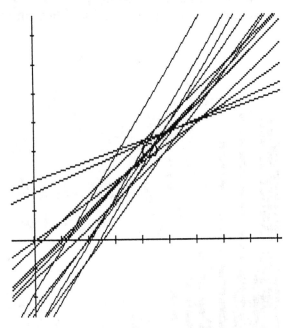

Figure 9: Simulated regression lines based on samples of size $n = 4$.

Furthermore, by examining the various sample regression lines, it is clear that they will eventually diverge from the population regression line and so the display provides an excellent argument for the dangers of extrapolating beyond the set of data values. Further, it also provides a clear demonstration of the adjustment necessary in constructing prediction intervals for the value of y based on a given value of $x = x_0$ using the regression equation:

$$y \pm t s_e \sqrt{1 + \frac{1}{n} + \frac{(x_o - \overline{x})^2}{\sum_{i=1}^{n} (x_i - \overline{x})^2}}$$

where s_e is the standard error of the estimate. Thus, rather than just citing an intimidating-looking (and, to most students, meaningless) formula, the computer simulation provides a way to communicate the significance of the different quantities in the adjustment term to the students. It is now easy to relate the fact that the further x_o is from $\overline{x}$, the larger the adjustment must be. Also, the larger that n is, the smaller the adjustment term needed to achieve a desired level of confidence in the prediction.

In a similar direction, it is also possible to perform a graphical simulation for the correlation coefficient. In Figure 10, we show the results of such a program where the vertical line corresponds to the population correlation coefficient ρ and the distribution of sample correlation coefficients is shown based on samples of size $n = 4$. For such a small value, the resulting correlation coefficients can vary significantly about the population value ρ; when the sample size is increased, it becomes visually evident that the sample values cluster ever more closely about the population value.

Implementation

The authors use the programs described here primarily on an in-class demonstration basis to motivate and explain statistical ideas. With the use of an LCD display panel, the PC graphical output is easily visible to all students in the class. In addition, several of the programs are used as the basis for individual investigation projects; students are required to conduct a study of the properties of a sampling distribution and write a formal report containing graphical output and their conclusions based on it.

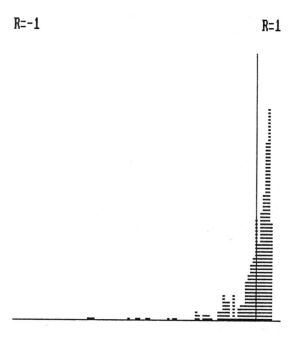

Figure 10: 400 simulated sample values of the correlation coefficient based on samples of size $n = 4$.

Moreover, the authors have found that the visual images generated by the computer tend to "stick" in students' minds. They often refer back to the computer pictures of such-and-such a simulation. Consequently, it is possible to capitalize on these solid images by referring back to them on an on-going basis in class when discussing subsequent ideas or going over particular problems.

Conclusions

The common thread running through the use of each of these graphical simulations is that they provide a visual dimension which allows a student to achieve a much firmer grasp of the statistical notions. It is no longer necessary for students to accept statistical facts purely on faith or blind memorization and worry only about mastering statistical procedures mechanically. Even in low-level introductory statistics courses, students can develop understanding of the statistical theory without actually studying the theory formally.

References

1. Dambolena, I. G., Teaching the Central Limit Theorem through Computer Simulation, *Mathematics and Computer Education*, **18**, 128-132, 1984.

2. Gordon, F. S. Computer use in teaching statistics, in *Computers in Mathematics*, David Smith, Gerald Porter, Carl Leinbach and Ronald Wenger, eds, MAA Notes #9, 79-83, 1988.

3. Gordon, F. S. and S. P. Gordon, Computer graphics simulations of sampling distributions, *Collegiate Microcomputer*, **7**, 185-189, 1989.

Florence Gordon is associate professor of mathematics at New York Institute of Technology. Her research interests include distribution theory and applications of computer graphics to statistical education. She is a member of the writing team on the NSF-funded Math Modeling/PreCalculus Reform Project and is responsible for incorporating material on probability models and data analysis into the course.

Sheldon Gordon is professor of mathematics at Suffolk Community College. He is co-project director of the NSF-funded Math Modeling/PreCalculus Reform Project and a member of the working group of the Harvard Calculus Reform Project. His interests lie primarily in the implications of technology at all levels of the undergraduate mathematics curriculum and the effects on the teaching faculty.

Computer Simulations to Motivate Understanding

Elliot A. Tanis
Hope College

Introduction

Students who are taking their first course in probability and/or mathematical statistics often lack the theoretical tools, appreciation, background, or experience for attacking the solution of a problem. And even if they are able to solve a problem theoretically, they do not always have an intuitive understanding of the result.

There are various ways in which the computer can be used to provide this understanding and motivation for further study. These include:

- ▸ Use simulation to provide insight into a theoretical solution.
- ▸ Use simulation to confirm and/or gain a better understanding of the theoretical solution.
- ▸ Simulate a plausible solution when the necessary theoretical tools are not known and perhaps motivate further study.
- ▸ Let the computer calculate the mean, variance, and standard deviation when a closed form solution is not known.

Some examples from a variety of sources are given that illustrate various uses of the computer. These examples are motivated by journal articles, textbook exercises, talks at professional meetings, and conversations with statisticians. Computer programs that you write to solve these problems will usually be quite short. They require only a decent "random number generator" such as the built in RND in IBM BASIC or GWBASIC. Graphical output will enhance several of the solutions. (See the Conclusions Section.) In addition, physical experiments can sometimes be used to help understand the problems.

Examples

1. On the average, what is the minimum number of random numbers that must be added together so that their sum exceeds 1? That is, if u_1, u_2, u_3 ... is a sequence of independent random numbers that are selected from the interval (0,1) and $X = min \{k: u_1 + u_2 + ... + u_k > 1\}$, what is the value of $E(X)$?

Using a table of "random numbers" or a "random number generator" on a calculator, it is easy to simulate observations of X. This is an interesting problem for an entire class to simulate, and in the process, a large number of observations can be found rather quickly. Writing the results obtained by individual class members on the board will help students see not only the randomness of such an experiment, but will also provide additional data for guessing the answer. Furthermore, if you tally the numbers of times that $x = 2, 3, 4, ...$ are observed,

217

the students will perhaps be able to guess that the probability function of X is

$$f(x) = P(X = x) = \frac{x-1}{x!}, \quad x = 2,3,4,\ldots$$

It is instructive to count and tally and calculate by hand, but time constraints also suggest that a computer simulation could provide a larger number of repetitions of this experiment more easily and rather quickly. Questions about the appropriate sample size can be discussed so that students begin to think about the number of repetitions that should be used.

A program to simulate this experiment is short and the simulation does not take very long. To see how a computer program can be written to solve a problem like this, consider the following program in BASIC for an IBM (compatible) computer:

```
100   CLS: RANDOMIZE TIMER: KEY OFF
110   N = 500: REM N is the number of repetitions
120   FOR K = 1 TO N
130       X = 0: SUM = 0
140       SUM = SUM + RND
150       X = X + 1
160       IF SUM < 1 THEN 140
170       PRINT X,
180       SUMOFX = SUMOFX + X
190   NEXT K
200   XBAR = SUMOFX/N
210   PRINT
220   PRINT "The sample mean of the x's is "; XBAR
230   END
```

On each of five successive runs of this program, the average numbers of random numbers needed so that their sums exceeded 1 were 2.714, 2.690, 2.736, 2.728, and 2.736.

After the students have guessed that the answer to the question is $E(X) = e$, hopefully they will become intrigued by the problem and will want to prove this result theoretically. This problem appeared on The William Lowell Putnam examination in 1958. Some of the places in which the solution appears and extensions of this problem are discussed are given in [1,3,4,5,6,7,10].

2. An urn contains n balls that are numbered from 1 to n. Take a random sample of size n from the urn, one at a time. A match occurs if ball numbered k is selected on draw k. Let A be the event that at least one match is observed. Show that

$$P(A) = 1 - \left(1 - \frac{1}{n}\right)^n$$

when sampling with replacement, and

$$P(A) = 1 - \left(1 - \frac{1}{1!} + \frac{1}{2!} - \frac{1}{3!} + \ldots + \frac{(-1)^n}{n!}\right)$$

when sampling without replacement. In addition, show that

$$\lim_{n\to\infty} P(A) = \lim_{n\to\infty}\left[1 - \left(1 - \frac{1}{n}\right)^n\right] = 1 - \frac{1}{e}$$

and

$$\lim_{n\to\infty} P(A) = \lim_{n\to\infty}\left[1 - \left(1 - \frac{1}{1!} + \frac{1}{2!} - \frac{1}{3!} + \ldots + \frac{(-1)^n}{n!}\right)\right] = 1 - \frac{1}{e}$$

Thus for "large" n, $P(A) \approx 1 - e^{-1}$ when sampling either with or without replacement.

There are several ways to simulate this problem physically. You could write consecutive integers from 1 to n on n balls or on n slips of paper, place them in a container, and select a sample of size n, sampling either with or without replacement. Show that the proportion of trials on which at least one match occurs is approximately $1 - 1/e$ when n is sufficiently large.

Another way to simulate sampling with replacement is by rolling an n sided die n times, checking whether face k is observed on roll k, and counting the number of trials on which at least one match occurs.

Sampling without replacement can be simulated by shuffling each of two identical decks of n cards and then comparing them to see whether the cards in position k match. Again calculate the proportion of shuffles that lead to at least one match.

Physically simulating this experiment in class provides an excellent opportunity to get the class involved. It also helps them to really understand the problem. However, to obtain a sufficient number of repetitions, a computer program could be written to simulate this problem. Care must be taken when simulating sampling without replacement.

The following table gives the results of several repetitions of this experiment so that you can also see the randomness in the estimates. Each of these repetitions was based on 200 trials with $n = 6$.

	Estimates Based on 200 Repetitions						Probability
Sampling with replacement	0.705	0.620	0.620	0.650	0.660	0.670	0.6651
Sampling without replacement	0.645	0.580	0.625	0.655	0.675	0.640	0.6319

Using simulation, or calculating probabilities using the formulas, the students can decide how greatly the probability is affected by whether the sampling is done with or without replacement. The students can also determine the effect of n on the probability and decide when "n is sufficiently large" for the probability of at least one match to equal, approximately,

$1 - 1/e = 1 - 1/2.71828... = 0.6321$. For a more complete discussion, see [9].

3. A modification of the "birthday problem" can be stated as follows. Consider successive rolls of a fair "M-sided die." Let the random variable X equal the minimum number of rolls required so that one of the faces is observed twice. That is, X is the roll on which the first match occurs. Then the p.d.f. of X is

$$f(x) = \frac{M}{M} \; \frac{M-1}{M} \; \frac{M-2}{M} \; \cdots \; \frac{M-(x-2)}{M} \; \frac{x-1}{M}, \quad x = 2,3,4,\ldots,M+1$$

Find the value of $\mu = E(X)$ for several values of M.

Note that the mean of X is given by

$$\mu = E(X) = \sum_{x=2}^{M+1} x f(x)$$

so that it is not easy to solve for μ. However, by using a short computer program, the value of μ can be found easily for different values of M. For example, suppose that you have a large number of people in a room and you ask each person to give their birthday. On the average, how many people do you have to ask before a match occurs? It is interesting to show that $\mu = 24.6166$ when $M = 365$.

Because 6-sided dice are readily available, this problem could be simulated by letting $M = 6$ and rolling such a die. The students could compare probabilities and relative frequencies for $x = 2,3,4,5,6,7$. In addition, the sample mean could be compared with the theoretical mean $\mu = 3.775$. See [8] for a more complete solution.

4. In New Zealand a coin has a Kiwi on one side and Queen Elizabeth II on the other side. Flip such a coin successively.

a) Let X equal the number of flips that are required to observe the same face on consecutive flips. Show that the mean and variance of X are 3 and 2, respectively.

b) Let Y equal the number of flips that are required to observe Kiwis on consecutive flips. Show that the mean and variance of Y are 6 and 22, respectively.

This problem is perhaps too advanced theoretically for some students. However, a physical simulation is very easy. For part (a), simply count the number of flips of a standard coin that are required to observe the same face, heads-heads or tails-tails, on consecutive flips. For part (b), count the number of flips required to observe heads-heads, for example, on consecutive flips. For each experiment, calculate the sample means and sample variances.

For part (a), the p.d.f. of the random variable X is

$$f(x) = (\tfrac{1}{2})^{x-1} , \quad x = 2,3,4,\ldots$$

If you know that the mean and variance of a geometric random variable W with p.d.f.

$$h(w) = (\tfrac{1}{2})^{w-1} (\tfrac{1}{2}) , \quad w = 1,2,3,\ldots$$

are $\mu_W = 1/(\tfrac{1}{2}) = 2$ and $\sigma_W^2 = (\tfrac{1}{2})/(\tfrac{1}{2})^2 = 2$, it follows that $\mu^X = 3$ and $\sigma_X^2 = 2$.

For (b), let f_n equal the nth Fibonnaci number where $f_1 = 1$, $f_2 = 1$, and $f_n = f_{n-1} + f_{n-2}$ for $n = 3,4,5,...$ Then the p.d.f. of Y is

$$g(y) = \frac{f_{y-1}}{2^y}, \quad y = 2,3,4,...$$

It is possible to find the mean and variance of Y using sums of infinite series. However, a computer program can also be written to calculate the sums for you, and thus show that the mean and variance are 6 and 22, respectively. Extensions of this problem are given in [2].

5. Let X_1, X_2,... X_n be a random sample of size n from a normal distribution with mean μ and variance σ^2.

a) Show that an unbiased estimator of σ^2 is

$$S^2 = \frac{1}{n-1} \sum_{i=1}^{n} (X_i - \overline{X})^2 .$$

b) Show that an unbiased estimator of σ is cS where

$$c = \frac{\Gamma\left(\frac{n-1}{2}\right)\sqrt{n-1}}{\Gamma\left(\frac{n}{2}\right)\sqrt{2}} .$$

To illustrate this result, it is necessary for you to be able to simulate observations from a normal distribution. There are several ways to do this. We will look at three of them.

● This first method is perhaps the easiest and most intriguing. If U is a uniform random variable on the interval $(0,1)$, that is, the value of U is a "random number", then

$$Z = \frac{U^{0.135} - (1 - U)^{0.135}}{0.1975}$$

is approximately normal with mean 0 and variance 1, i.e, $N(0,1)$.

● If U has a uniform distribution on the interval $(0,1)$, then the mean and variance of U are 1/2 and 1/12, respectively. Applying the Central Limit Theorem, the sum of 12 observations of U has a distribution that is approximately normal with mean $12(\frac{1}{2}) = 6$ and variance $12(1/12) = 1$. Thus, the sum of 12 random numbers minus 6 gives the value of an approximate standard normal random variable. That is,

$$Z = \sum_{i=1}^{12} RND - 6 ,$$

where each of the 12 RND's represents a different "random number" selected from the interval $(0,1)$, has an approximate $N(0,1)$ distribution.

● The Box-Muller method gives an exact theoretical method for obtaining observations of standard normal random variables. Let U and V be independent uniform random variables on the interval $(0,1)$. Let

$$X = \sqrt{(-2)(\ln U)}\cos(2\pi V) ,$$

and

$$Y = \sqrt{(-2)(\ln U)}\sin(2\pi V) .$$

Then X and Y have independent $N(0,1)$ distributions.

To go from a standard normal random variable Z to a normal random variable X with mean μ and standard deviation σ, let

$$X = \sigma Z + \mu$$

It is now possible to write a computer program that would, for example, simulate $n = 5$ observations from a normal distribution with mean $\mu = 75$ and standard deviation $\sigma = 20$. If 100 samples of size 5 were simulated, and for each of them the sample mean $(\bar{x})$, sample variance (s^2), and cs were calculated, it should be true that the average of the $\bar{x}$'s is close to 75, the average of the s^2's is close to 400, and the average of the $c \cdot s$'s is close to 20 where

$$C = \frac{\Gamma\left(\frac{5-1}{2}\right)\sqrt{5-1}}{\Gamma\left(\frac{5}{2}\right)\sqrt{2}} = \frac{2}{\left(\frac{3}{2}\right)\left(\frac{1}{2}\right)\sqrt{\pi}\sqrt{2}} = \frac{8}{3\sqrt{2\pi}} .$$

The simulation also illustrates that the variance of the sample variance is very large. In fact, since we know that the distribution of $(n-1)S^2/\sigma^2$ is chi-square with $n-1$ degrees of freedom,

$$Var(S^2) = Var\left(\frac{\sigma^2}{n-1}\frac{(n-1)S^2}{\sigma^2}\right) = \left(\frac{\sigma^2}{n-1}\right)^2 2(n-1) .$$

So for our example,

$$Var(S^2) = \left(\frac{400}{4}\right)^2 (2)(4) = 80000 .$$

The following table shows the output for just 8 repetitions of this experiment.

	Simulation Results								Averages
$\bar{x}$	69.33	71.96	83.34	76.91	71.90	79.75	68.21	79.82	75.15
s^2	177.69	477.72	1062.98	249.19	679.15	188.23	815.41	570.11	527.56
cs	14.18	23.25	34.69	16.79	27.72	14.60	30.38	25.40	23.38

See [10] for a BASIC program and additional sample output.

For those of you who are not familiar with the gamma function, all that is needed for this example is that, when n is an integer,

$$\Gamma(n) = (n - 1)!$$

and when n is odd, say $n = 2k + 1$,

$$\Gamma\left(\frac{n}{2}\right) = \Gamma\left(\frac{2k+1}{2}\right) = \frac{2k-1}{2}\ \frac{2k-3}{2} \cdots \frac{1}{2}\sqrt{\pi}\ .$$

6. Let X1, X_2, ..., X_5 be a random sample of size $n = 5$ from a normal distribution with mean μ and variance σ^2. The endpoints for a 90% confidence interval for μ are

$$\bar{x} \pm 1.645\ \sigma/\sqrt{5}$$

when σ is known and

$$\bar{x} \pm 2.132\ s/\sqrt{5}$$

when σ is unknown. Note that the critical t value replaces the critical z value and s replaces σ. Which of these intervals is shorter?

First, it should be made clear that both intervals are 90% confidence intervals. And it is not too difficult to empirically find the answer to the question. The simulation depends on your ability to sample from a normal distribution. A method for doing this is given in the solution for the last example.

When the standard deviation is known, the length of the z confidence interval for μ is

$$\text{length} = 2(1.645)\ \sigma/\sqrt{5} = 1.471\ \sigma.$$

When the standard deviation is unknown, the length of the t confidence interval will vary from sample to sample, some being quite short and others very long. Simulation will illustrate this. So in a sense, the answer to the question is that sometimes the z interval is shorter and sometimes the t interval is shorter. However, on the average, the expected length of the t interval is longer. Using the result from the last example, we have that

$$E(length) = 2(2.132)\ \frac{\Gamma\left(\frac{5}{2}\right)\sqrt{2}}{\Gamma\left(\frac{5-1}{2}\right)\sqrt{5-1}}\ \frac{\sigma}{\sqrt{5}} = 1.792\sigma\ .$$

A graphical comparison of the two types of confidence intervals is extremely helpful. The graphical display in Figure 1 from [10] illustrates this with the t-intervals on the left and the corresponding z-intervals on the right.

Implementation

Examples like those that have been described can be used in a variety of ways. One of the most effective ways is to offer an optional computer based laboratory along with a probability and statistics class. In this laboratory, each of the students would be expected to write a computer

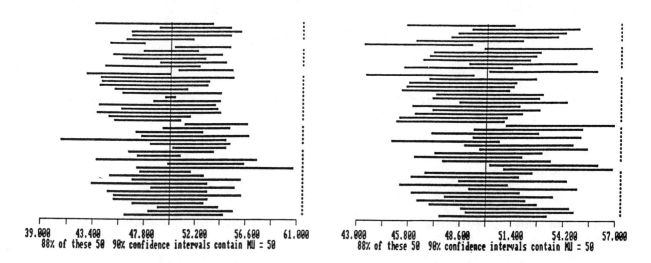

Figure 1

program to solve each problem. It is helpful in this setting if programs for the graphical output are provided for the students. The computer programs and the results of the students' simulations should be shared with all of the students in the class. Students often come up with ingenious ways for solving a problem. They should also be encouraged to modify the questions that are asked and to raise new questions.

If it is not possible to add a laboratory, a few students could solve the problems as an independent study project. It would then be possible for the student or the professor to present the solution to the class, getting the students in the class involved as much as possible in the solution of the problem, perhaps through physical simulation.

And, of course, the professor could solve a problem alone and present its solution to a class.

Much of the learning comes about by writing the computer program, rather than just watching the output from the program. Thus, the effectiveness of this approach increases as the student involvement increases.

Conclusions

Hopefully these examples give you some new ideas about ways in which the computer can be used. Additional suggestions and examples are given in other sections of this volume.

A computer disk for an IBM (compatible) computer is available that contains computer solutions for the listed problems as well as several others. Some of these solutions contain graphical output. If you would like to receive a copy of this disk for $10 to cover the cost, specify whether you would like a 5¼" or a 3½" disk, and send your request to the author.

References

1. Bush, L.E. (1961). "The William Lowell Putnam Mathematical Competition". *Amer. Math. Monthly*, **68**, pp. 18-33.

2. Chamberlain, M. (1979). "Coin-tossing problem revisited". *Two-Year College Math J.*, **10**, pp. 349-350.

3. Hogg, Robert V., and Elliot A. Tanis. (1987). *Probability and Statistical Inference, 3rd edition*. Macmillan Publishing Co., Inc., New York.

4. Jensen, U. (1984). "Some remarks on the renewal function of the uniform distribution". *Adv. Appl. Prob.*, **16**, pp. 214-215.

5. Russell, K.G. (1983). "On the number of uniform random variables which must be added to exceed a given level". *J. Appl. Prob.*, **20**, pp. 172-177.

6. Schultz, H.S. (1979). "An expected value problem". *Two-Year College Math J.*, **10**, pp. 277-278.

7. Schultz, Harris S. and Bill Leonard (1989). "Unexpected occurrences of the number e". *Mathematics Magazine*, **62**, No. 4, pp. 269-271.

8. Tanis, Elliot A. (1985). "Birthday problem and expected values". *Teaching Statistics*, pp. 21-25.

9. Tanis, Elliot A. (1987). "The answer is $1 - 1/e$. What is the question?". *Pi Mu Epsilon Journal*, **8**, No 6, pp. 387-389.

10. Tanis, Elliot A. (1987). "Computer simulations to motivate and/or confirm theoretical concepts". *American Statistical Association 1987 Proceedings of the Section on Statistical Education*, pp. 27-32.

11. Tanis, Elliot A. (1990). "Interplay between simulation and theory". *Proceedings of the Third International Conference on Teaching Statistics*, **Volume 2**, pp. 159-165.

Elliot Tanis is professor of mathematics at Hope College. He is the author of three textbooks in statistics, including *Probability and Statistical Inference* which he co-authored with Robert Hogg. Professor Tanis was recently awarded the first Distinguished College or University Teaching of Mathematics Award from the Michigan Section of the MAA. He is also interested in the mathematical properties of Escher's tessellations and has written computer programs to illustrate them.

Using Spreadsheets in Teaching Statistics

Deane E. Arganbright
Whitworth College

Introduction

The electronic spreadsheet, a popular microcomputer program initially created for business modeling, is a powerful and creative tool for the study of statistics. It provides a natural medium in which to examine statistical ideas interactively, implement algorithms, and generate insightful graphic displays. Use of spreadsheets in a class stimulates student interest and provides new perspectives. Acquiring spreadsheet experience also enhances a student's skills in using software which is highly valued in the business and academic worlds. Because the careers of many students will include spreadsheet usage, incorporating it in statistics can provide a foundation for a continuing study of that discipline.

There is another significant benefit to using spreadsheets in teaching statistics. Students are able to learn the fundamental operations, and to develop enough proficiency to use the program productively, in a very short time.

We present some ways in which spreadsheets can be used in teaching statistics, and provide some illustrative examples. A wide range of spreadsheets is available, each possessing different features and hardware requirements. Our examples are created using Quattro Pro, a powerful spreadsheet with excellent graphic capabilities. However, the basic ideas and examples are applicable to any spreadsheet, such as Lotus 1-2-3, SuperCalc, VP Planner, Excel, As-Easy-As, or MicroSoft Works.

Spreadsheet Fundamentals

We begin by describing some of the fundamental operations of using spreadsheets. The experienced reader may want to skip ahead to the next section.

The spreadsheet format consists of a large rectangular array, or table, a portion of which is displayed on the screen. Columns are labeled by letters and rows by positive integers. A location, or cell, in the array is identified by its column and row. Thus, C2 refers to the cell at Column C and Row 2.

One cell is highlighted by a cursor which can be moved throughout the spreadsheet by using the arrow keys. We can enter into that cell a descriptive label (or string), a number, or a formula which references other cells. The program computes the value of each formula, and displays labels and values on the screen. We create a spreadsheet model by entering the formulas and data of a mathematical model or algorithm into the spreadsheet's cells. If the value of any cell is changed, the entire spreadsheet is recalculated, and the screen display is updated. This enables a user to interrogate a spreadsheet model and investigate "What if...?"

questions simply by changing a model's parameters, data, and formulas and observing the results. Spreadsheet models can be developed, stored, and then recalled at a later time by a user, who has only to enter the desired input data.

Figure 1 provides the formulas and output of an elementary model. The model shows the number of touchdown passes by AFC Eastern Division teams during 1990. To construct the model, we first enter column headings in Row 1, then team names in Cells B2..B6, and output headings in Cells B7..B8. We next enter the numbers of touchdowns in Cells C2..C6.

	A	B	C	D
1		TEAM	TD	z
2	1	Pats	14	-1.09
3	2	Bills	28	1.55
4	3	Miami	21	0.23
5	4	Jets	14	-1.09
6	5	Colts	22	0.41
7		MEAN=	19.8	
8		SD=	5.31	
9	No:	3	TD=	21
10			Rank=	hi

Output

	A	B	C	D
1		TEAM	TD	z
2	1	Pats	14	(C2-C\$7)/C\$8
3	1+A2	Bills	28	(C3-C\$7)/C\$8
4	1+A3	Miami	21	(C4-C\$7)/C\$8
5	1+A4	Jets	14	(C5-C\$7)/C\$8
6	1+A5	Colts	22	(C6-C\$7)/C\$8
7		MEAN=	@AVG(C2..C6)	
8		SD=	@STD(C2..C6)	
9	No:	3	TD=	@VLOOKUP(B9,A2..D6,2)
10			Rank=	@IF(D9>C7,"hi","lo")

Formulas

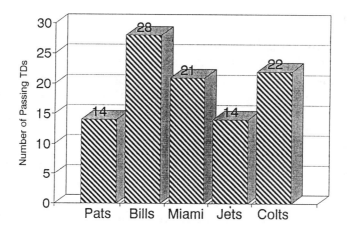

Touchdowns via Pass: 1990
AFC Eastern Division

Figure 1

Column A is used to number the teams. We enter 1 into Cell A2 and the formula $1+A2$ into Cell A3. The program evaluates the formula and displays 2. A copy command enables us to create large models efficiently. To copy the formula in Cell A3 down Column A, we first type /C. We then follow screen prompts and enter A3 as the source range, with A4..A6 as the destination. When the formula $1+A2$ in Cell A3 is copied, A2 is treated as a variable location, or "the cell above the present cell".

Spreadsheets possess many library functions which typically are indicated by @. We find the mean and standard deviation of the data by entering @AVG(C2..C6) and @STD(C2..C6) into Cells C7 and C8. We also compute a z-score for each team by entering the formula (C2-C\$7)/C\$8 into Cell D2 to find the z-score, -1.09, for the Pats. We then copy this into Cells D3..D6. A \$ in front of a row or column identifier makes that identifier an absolute reference in copying.

Row 9 contains a table-lookup function. @VLOOKUP(B9,A2..D6,2) vertically searches the first column of the range A2..D6 (i.e., Column A, which must be in ascending order) for the value B9. It is found at Cell A4. The function returns the value of the cell two rows to the right. We can name the range A2..D6 as TAB using the spreadsheet name command, and Cell D9 becomes @VLOOKUP(B9,TAB,2). The formula in Cell D10 shows a decision function useful for comparisons: If $D9 > C7$ Then "hi" Else "lo".

Most spreadsheets can also generate the types of graphs used in statistics. To create the graph shown in Figure 1, we choose a type (bar) and select the x-axis (B2..B6) and a y-range (C2..C6). We can also use internal labels (C2..C6) to put labels on the bars. As data is changed, we obtain an updated graph by pressing the graph key.

Spreadsheets possess many other useful commands that allow us to format and modify models. These can be introduced gradually in class lectures, and by using on-screen help facilities, or the many available books and manuals.

Teaching Features

A spreadsheet can be incorporated into statistics courses both as a demonstration tool and as a means for enabling students to explore and analyze topics on their own using a computer.

Classroom Demonstrations

Spreadsheets are excellent tools for interactive demonstrations. Using a computer with an overhead LCD display unit, an instructor can employ previously created models to add visual dimensions to classroom presentations. The graphs produced provide quick, dynamic, and accurate illustrations for topics ranging from probability distributions to confidence intervals. The screen display itself is effective in leading a class in "What if...?" investigations. Together, these features provide a good way to illustrate the consequences of changes in assumptions, parameters, and data.

In addition, spreadsheet models that read data from databases furnish interesting, concrete examples, while simulation models are valuable for generating data sets, illustrating distributions or sampling, and investigating topics such as the Central Limit Theorem. Creating models in class also provides a forum for discussing techniques in modeling and implementing statistical concepts.

Student Usage

Most statistics topics can be implemented naturally in spreadsheet models in a format that closely parallels text and class presentations. Thus, having students create these models in assignments not only engages them actively in the learning and experimenting process, but also reinforces the ideas being studied while promoting problem-solving skills.

As a part of assignments, students can create models for both text exercises and application sets specifically designed to exploit spreadsheet capabilities. Besides working on common problem sets, students can use databases or information from library research to illustrate current concepts. Work can be personalized by encouraging students to investigate topics chosen from such areas as sports, geography, or political science. Assignments can also utilize previously prepared models, or templates. This approach is useful in applications in which the spreadsheet format is convenient, but the actual construction is more complex.

Projects

A spreadsheet project can be a capstone for the spreadsheet component of a course. The project serves to unify statistics concepts, problem-solving skills, and spreadsheet techniques developed during the term. Various types of projects are possible, including a deeper study of a topic encountered in class, the analysis of a topic drawn from outside sources, or an application of statistics in another discipline. These projects should be more substantial than regular assignments. The models should include user instructions, documentation, and presentation-quality output. Projects can be assembled in a notebook for future reference in the laboratory or the library. This experience may also give students an indication of how individual study in statistics might continue.

Illustrative Models

Virtually any topic in a basic statistics course can be implemented via a spreadsheet model. This section provides examples that give a hint of the vast range of possible uses of spreadsheets in statistics. We provide characteristic spreadsheet and graphical output for each example.

Moving Average

Trends are often hard to determine from lists of raw data, especially if the data is influenced by periodic or erratic fluctuations. One way to detect trends, as well as to smooth out the fluctuations, is to compute and graph "moving" averages. Figure 2 determines moving 4-quarter averages for a firm's quarterly profits. Profits are entered in Column B. The first year's average is found in Cell C6 by @AVG(B3..B6). This is copied down Column C to generate successive averages.

This example is a good initial laboratory exercise for teaching spreadsheet usage and graph construction, and it illustrates one way in which data is smoothed to discover trends. It is easy for students to construct the model, save it, and then use it later as a template to investigate such topics as trends in mean July temperatures for successive years, SAT scores, sports performances, monthly new car sales, or weekly church offerings.

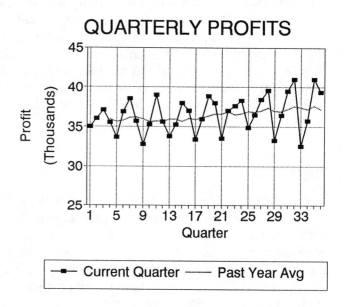

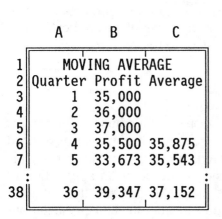

	A	B	C
1	MOVING AVERAGE		
2	Quarter	Profit	Average
3	1	35,000	
4	2	36,000	
5	3	37,000	
6	4	35,500	35,875
7	5	33,673	35,543
38	36	39,347	37,152

Figure 2

Correlation

A spreadsheet provides a very convenient way to construct, analyze, and display scatter diagrams. Figure 3 computes the linear correlation coefficient, $r = 0.609$, between strikeouts (y) and home runs (x) for American League teams in 1990. We enter data in Columns A..C. After x- and y- means and standard deviations are calculated in Rows 22..23, product moments $(x - m_x)(y - m_y)$ are found in Column D, and r is determined by

$$r = \frac{\sum (x - m_x)\,(y - m_y)}{(n-1)s_x\,s_y}$$

In our graph, we show (x,y) points by using team names, although a symbol such as * can be used instead. The solid vertical and horizontal lines represent the locations of the x- and y-means. Our model lets us easily investigate effects of changing data (for the 12 National League teams in 1990, $r = 0.071$).

This is an easy model for students to create for themselves. It can also be provided as a template in a lab, where students enter data sets and observe r for various data distributions. It is particularly effective to use with small data sets that are chosen to show how a change in only a few data points can greatly affect r. Student exercises can include modifying the model to produce a table of critical r values for various α to use in hypothesis testing, or designing a model for another method for computing r.

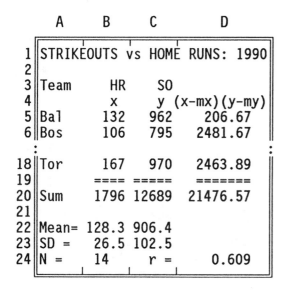

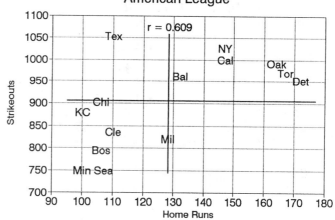

Figure 3

Linear Regression

Figure 4 determines the least-squares regression line for a set of data essentially as it would be done by hand. We enter data points (x,y) into Columns A and B of Rows 4..8. Columns C and D calculate x^2 and xy, with column sums found in Row 10. The slope and y-intercept of the least-squares line, $y = mx + b$, are determined in Cells D15 and F15, using intermediate computations in Rows 12..13 and the formulas

$$m = \frac{n \sum (xy) - \left(\sum x\right)\left(\sum y\right)}{n\left(\sum x^2\right) - \left(\sum x\right)^2}$$

$$b = \frac{\left(\sum y\right)\left(\sum x^2\right) - \left(\sum x\right)\left(\sum xy\right)}{n\left(\sum x^2\right) - \left(\sum x\right)^2}$$

Columns E and F give y-components for points on the regression line and the corresponding square-of-error terms. In Row 17, the model predicts a y value for a given x. Finally, spreadsheet graphs furnish a natural way to illuminate regression concepts. Our graph uses Columns B and E as y-ranges.

Lab exploratory activities can include investigating how the inclusion or exclusion of an outlier, or the adding of data points, affects the computation. Students can also use the SSE term (Cell F10) to compare the "fit" of other lines, $y = m_0 x + b_0$, by replacing Cells D15 and F15 by values m_0 and b_0, or by expanding the model to include separate computations and a graph for another trial line (see [4]). Two graphical projects for advanced students are to show vertical error lines, and to generate curves giving confidence regions for $E(y \mid x)$ about the regression line.

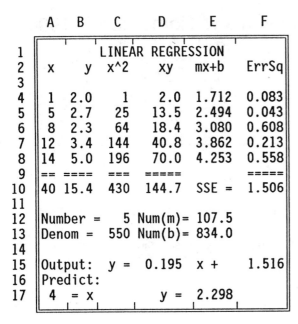

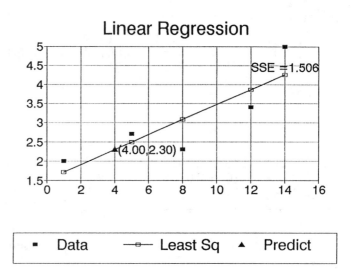

Figure 4

Discrete Probability

All too often, binomial probability problems are artificially limited because of a dependence on printed tables containing relatively few selected values of n and p. Spreadsheets can easily calculate such a probability for any n, p, and number of successes k. Virtually as easily, they can produce entire tables of binomial probabilities and the associated graphs for any p.

Figure 5 produces such a table of binomial probabilities using a recurrence relation. If p is the probability of success in one trial and $q = 1 - p$, we want the probability $P_{n,k}$ of k successes in n trials. There are only two ways in which, after $n+1$ trials, there are $k+1$ successes. After n trials there are either

(a) k successes (prob $P_{n,k}$) followed by a success on try $n+1$ (prob p), or

(b) $k+1$ successes (prob $P_{n,k+1}$) followed by a failure on try $n+1$ (prob q). Thus, the probability of exactly $k+1$ successes in $n+1$ tries is

$$P_{n+1,k+1} = pP_{n,k}+qP_{n,k+1}.$$

Since $P_{0,0} = 1$, we first enter 1 into Cell E6. We then place the basic formula +$F$3*D6+$H$3*E6 into Cell E7 and copy it throughout (blank cells are treated as 0). The graphical display in Figure 5 shows the distribution for several rows, and provides an ideal way to illustrate the normal approximation to the binomial distribution for sufficiently large n for any given p. Unfortunately, to display the different binomial distributions on the same set of axes, the graphs drawn may suggest that these curves are continuous. Consequently, it is necessary to point out to the students that the binomial distribution is discrete.

This model is useful for students to use in investigating a variety of binomial events, such as the probability of winning the majority of games in an n-game series for given p and varied n. Various modifications of the model are also possible for student assignments. Setting F3 =

H3 = 1 generates Pascal's triangle. We can also use Column A for variable probabilities, P_n, so that $P_{n+1,k+1} = p_n P_{n,k} + (1-p_n)P_{n,k+1}$. Students might also be challenged to design a model that generates only the nth line of the table to provide the probabilities of the outcomes for each k with one particular n.

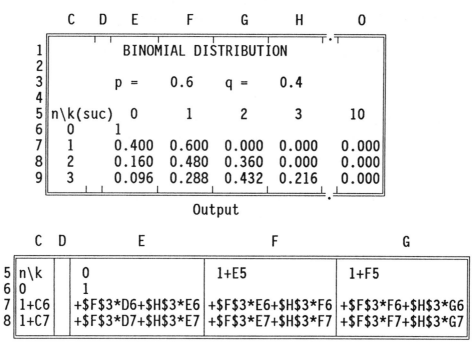

Output

C	D	E	F	G
5	n\k	0	1+E5	1+F5
6	0	1		
7	1+C6	+$F$3*D6+$H$3*E6	+$F$3*E6+$H$3*F6	+$F$3*F6+$H$3*G6
8	1+C7	+$F$3*D7+$H$3*E7	+$F$3*E7+$H$3*F7	+$F$3*F7+$H$3*G7

Formulas

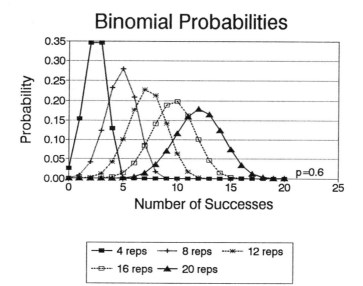

Figure 5

Many discrete probability calculations can be done in this way. One of them involves the popular lottery game, Lotto, and its casino cousin, Keno. Assume that balls are numbered $1,2,...,T$, with G "good" balls and $B = T-G$ "bad" balls. We wish to find the probability that, after n balls are drawn, exactly k are "good", or "hits". In one typical Lotto game, $T = 44$, $G = n = 6$, and the case $k = 6$ represents the jackpot.

Again $P_{0,0} = 1$. We want $k+1$ good balls after $n+1$ selections. Thus, after n selections either

(a) k good balls have been drawn (prob $P_{n,k}$) followed by a good ball (prob $(G-k)/(T-n)$),

or

(b) $k+1$ good balls have been drawn (prob $P_{n,k+1}$) followed by a bad one
(prob $(B - (n-1) + k)/(T-n)$).

The model of Figure 6 uses the recurrence relation

$$P_{n+1,k+1} = [x(G-k) + y(B-(n-1)+k)]/(T-n)$$

to generate the hypergeometric distribution. By changing n and k, students can compare how the designs of various state lotteries affect winning probabilities. These models can also provide students with the insight needed to construct tables for other discrete distributions, urn models, and the birthday problem.

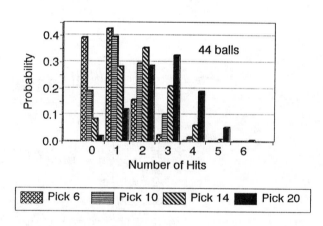

	C	D	E	F	G	H
1			KENO PROBABILITIES			
2						
3			Tot(T)=	44	Good(G)=	6
4					Bad(B) =	38
5	n\k(hits)		0	1	2	3
6	0		1			
7	1		0.864	0.136	0.000	0.000
8	2		0.743	0.241	0.016	0.000
9	3		0.637	0.318	0.043	0.002

Figure 6

Simulation

Simulation models are widely used in many disciplines that employ statistics. In teaching statistics, simulation is also useful for generating data sets, and for illustrating and investigating such topics as the central limit theorem, probability distributions, and correlation [5]. A spreadsheet provides a natural, easy way for students to design and implement many simulation models.

As an example, consider a baseball player with a lifetime batting average of $p = 0.300$ (Cell G3) (so we can assume that the probability of his getting a hit remains approximately constant on each at-bat). We consider the course of a full 162-game season in which he has 4 at-bats in each game. Figure 7 shows a simulation of this binomial event. The @RAND

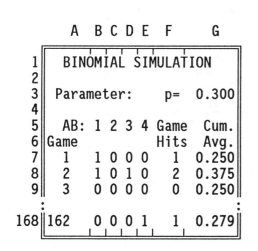

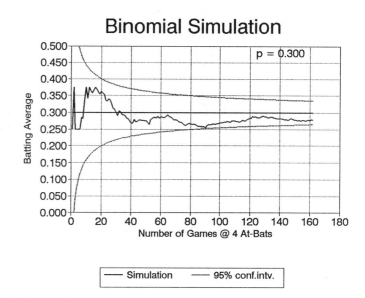

Binomial Simulation

Figure 7

function produces a random number in the range 0 to 1. Thus, entering the logical expression @RAND<$G$3 in Cell B7 gives 1 (true) with probability p, and 0 (false) otherwise. We copy Cell B7 into B7..E168, producing 1 (hit) or 0 (out) for each at-bat. The cumulative batting average is found in Column G. Our graph shows the average throughout the season, and the 95%-confidence interval for k games, $p \pm 1.96\sqrt{(p(1-p)/4k)}$.

A spreadsheet model gives us the ability to quickly examine repeated simulations. Once a model is constructed, we simulate another season by simply pressing the recalculation key to produce a new set of random numbers. From many sets of output, we see that a batting average often fluctuates considerably during a season, and frequently ends up quite distinct from 0.300. For 162 games, the 95%-confidence interval for a lifetime 0.300 hitter is 0.265-0.335. Such models can spark class discussions on the wisdom of a manager's decision to "bench" players who are in a hitting "slump", or an owner's decision to sack the manager of a team that wins fewer games than expected. Our simulation generates other data to investigate as well, such as multi-game hitting streaks.

Normal Distribution

Many topics in statistics involve the normal distribution. Figure 8 generates the standard normal table found in every statistics text, but in a column format. Column A of Rows 25..646 lists z values from -3.1 to 3.1 in steps of 0.01. Column B evaluates the density function $f(z) = \exp(-0.5z^2)/\sqrt{(2\pi)}$, while cumulative probabilities $P(u \leq z)$ are found in Column C using Simpson's rule.

```
            A        B        C       D        E        F    G
    1                        NORMAL  PROBABILITY
    2
    3   Parameters:
    4                       Mean=      50     S.D.=   12
    5
    6   Input/Output: P(   44.5   <=x<=    64      )= 0.5562
    7                      ====            ====
    8
    9   Computations:       z1=  -0.46   0.3228
   10                       z2=   1.17   0.8790
   11
   12   Conf.Int.:Pct =    0.05           26.48  <x< 73.40
```

```
            A         B         C         D        E
   21   Normal  Table:   con=     2.5066
   22
   23      z        f(z)     Pr(u<z)      x       g(x)
   24   -999                 0.0000
   25   -3.10     0.0033     0.0010     12.8     0.0003
   26   -3.09     0.0034     0.0010     12.9     0.0003
   27   -3.08     0.0035     0.0010     13.0     0.0003
        :                                                :
  645    3.10     0.0033     0.9990     87.2     0.0003
  646    3.11                1.0000     87.3
```

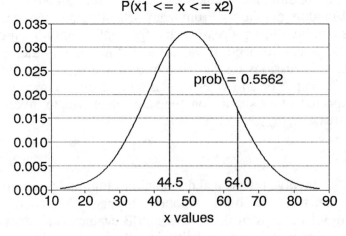

Normal Probability Distribution
P(x1 <= x <= x2)

prob = 0.5562

44.5   64.0

x values

Figure 8

This model provides us with both a convenient way to calculate normal probabilities and also an illustrative graph. After the mean and standard deviation of a normal variable, x, are entered in Cells D4 and F4, a list of x values corresponding to the z values is found in Column D by $x = \sigma z + \mu$. The density function for x is obtained in Column E by dividing $f(z)$ by σ.

To compute an x-interval's probability, we enter its endpoints into Cells C6, E6. Rounded z values are found in Cells D9, D10 and their table values in Cells E9, E10. The difference, +E10-E9, gives the probability in Cell G6. In Row 12 the model also finds a $(1-\alpha)100\%$ confidence interval for the mean.

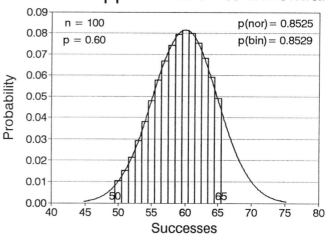

Figure 9

Students can use the model in the vast range of standard exercises involving normal probabilities. Another assignment can be to modify the model to compute normal approximations to binomial probabilities to use when binomial calculations are unwieldy. It is also possible to design a model that generates the graph of Figure 9. Students can use this model to experiment with various p and n to see when the best approximations are obtained.

Percentiles

Spreadsheets present us with superb opportunities to develop alternate models to illustrate a concept. For example, Figure 10 computes percentiles for grouped data. In Row 3 we enter the grouping parameters: lowest class limit (D3), class width (F3), and precision (H3). These generate class limits in Columns B, C, and F, and lower boundaries in Column A. We enter group counts in Cells D11..D19, and the percentile sought in Cell D5. Its value is found in Cell F5. The cumulative number of points in the prior intervals is found in Column E. In this model, our output locates the 55th percentile using standard formulas and interpolation. The ogive of Figure 10 is produced using Column E.

However, a spreadsheet also makes it easy for students to construct a simpler version of this model. This makes for a good exercise in estimation and conceptualization. This time, the calculation of percentiles is omitted. Instead, an additional column is used to find cumulative percentages. This column is then used as the y-range to produce an ogive. Students can use this simple graph to practice estimating percentiles. They can also design similar models for cumulative probability distributions.

Frequency Table

Grouped data concepts are encountered early in most statistics classes. Because grouping by hand is tedious, we often must be content with choosing one particular set of grouping

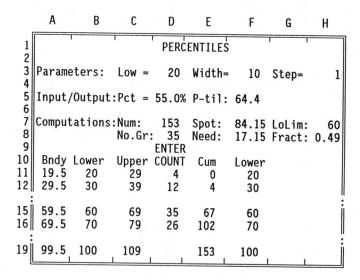

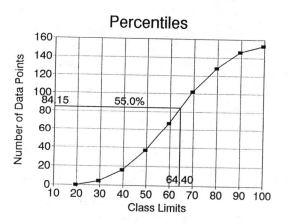

Figure 10

parameters for a given data set. Yet, the parameters we select affect the values of the grouped statistics. Figure 11 shows how a spreadsheet affords us with an excellent way to explore grouped data concepts.

In Row 3 we enter the grouping parameters that generate class limits in Columns A and B of Rows 6..13. Classes are numbered in Column C. We enter unsorted data in Column A of Rows 23..44. Column B determines an item's group number. Logical formulas in Columns C..J generate 0's and 1's, with 1 indicating that a row item is in the column class. Thus, since 31 is in Class 2, 0 appears in Cell C23 and 1 in Cell D23. Class totals are calculated in Row 47, and copied into Cells D6..D13. Columns E..G then use class marks to find the grouped mean and variance in Rows 18 and 19. Our bar graph presents a picture of the grouping.

Students can experiment with such a model to see readily how changes in the number of classes, the class width, or the starting point of the lowest group affect the values of the grouped mean and variance, and the resulting frequency chart. The model also allows them to compare grouped and actual parameters. Students can be assigned the task of gathering data, such as GPAs, faculty salaries, or physical measurements, and use the model to see how grouping can affect the presentation of a summary of the data.

Sorted Data

A spreadsheet's sort command can be used to order data, a very useful feature for many statistics applications. Figure 12 locates the median of a set of numbers, and displays the data in a stem-and-leaf graph. Unsorted data is first entered into Column C of Rows 13..72, and then copied into Column A using the copy command. Formulas in Column D copy the data, with 999 used to signify blanks. To sort the data, we issue the sort command, indicating the block to be sorted (C13..D72), the sort key (D13..D72), and the order (ascending). The mean is found in Cell D4. The median is determined after the data is sorted using standard statistics

```
      A     B     C     D      E      F       G
 1│               FREQUENCY TABLE
 2│ Parameters:
 3│      Low=  20 Width=    10  Step=      1
 4│
 5│ Low   Up Class Freq     x      nx      nx^2
 6│  20   29    1    3    24.5    73.5     1801
 7│  30   39    2    3    34.5   103.5     3571
 8│  40   49    3    2    44.5    89.0     3961
  ┊
13│  90   99    8    0    94.5     0.0        0
14│
15│ Count:      22         Sum   1179    69766
16│
17│ Output:        Actual Grouped
18│   Mean         53.73   53.59
19│   Variance     310.6   313.4
```

```
      A    B   C D E     I  J
21│ Data Num 1 2 3       7  8
22│
23│   31   2  0 1 0       0  0
24│   29   1  1 0 0       0  0
  ┊
44│   43   3  0 0 1       0  0
45│
46│ Summary:   1 2 3      7  8
47│ Count:     3 3 2      0  0
```

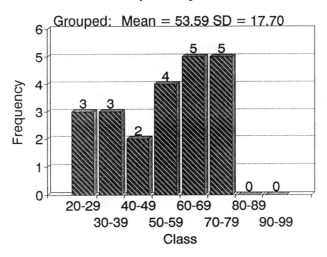

Frequency Chart

Grouped: Mean = 53.59 SD = 17.70

Figure 11

procedures. The rest of the model creates the "graph" by counting the numbers in each branch. Grouping parameters are entered in Row 1.

The stem-and-leaf model can be used in lab as a template, with students entering data specifically selected to illustrate, for instance, how skewed data affects the median and mean. The graph provides an excellent picture of the effect of skewing. Students can also investigate the effect that outliers, and changes of data, have on the mean, median, and the distribution's appearance.

	A	B	C	D	E	F
1	Parameters:	Low =		6	Width=	9
2						
3	Output:		Count:	58	Even:	1
4			Mean:	50.86	Median:	53.5
5						
6	Computations:			Mean	Median	Median
7			Median:		29	30
8			Values:	49	53	54
9			Offset:	11	2	3
10						
11	Orig	Count	ENTER	Sort	Group	GpCount
12						
13	55	1	16	16	2	1
14	16	2	18	18	2	2
15	83	3	23	23	2	3
⋮						
69	18	57	79	79	9	1
70	23	58	83	83	9	2
71		59		999		
72		60		999		

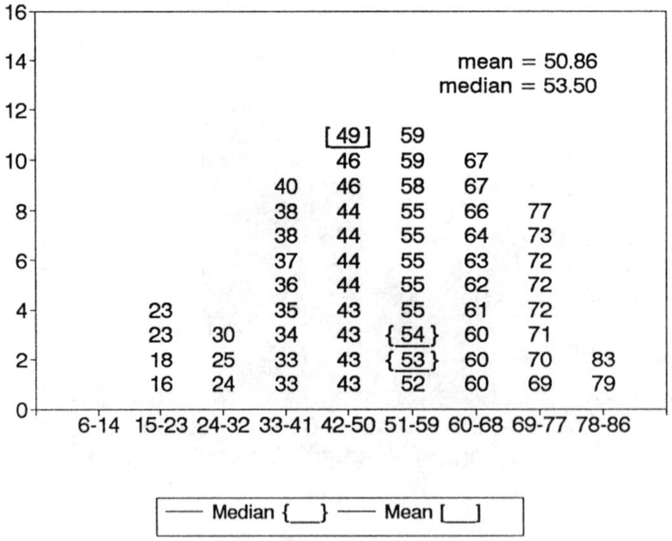

mean = 50.86
median = 53.50

```
16
14
12
                                [49]  59
10                                    46   59   67
                           40         46   58   67
 8                    38   44         55   66   77
                      38   44         55   64   73
 6                    37   44         55   63   72
                      36   44         55   62   72
 4         23         35   43         55   61   72
           23   30    34   43       { 54 } 60   71
 2         18   25    33   43       { 53 } 60   70   83
           16   24    33   43         52   60   69   79
 0
      6-14 15-23 24-32 33-41 42-50 51-59 60-68 69-77 78-86
```

— Median {__} — Mean [__]

Figure 12

Central Issues

Hardware and Software

For classroom work, it is advantageous to have a computer with a hard disk and an overhead LCD display unit. Students must either have good access to a computer laboratory with spreadsheets easily accessible, or have a spreadsheet on their own computer. A laboratory facilitates the use of template assignments by providing a central place to store and dispense templates. For example, an instructor can provide templates and database files in the lab (on either hard disks, a network, or floppy disks) for students to use to examine the correlation between, say, the GNP and infant mortality rates of third-world nations.

Any good spreadsheet with a full range of functions can be used. It should have good graphics and a straightforward menu tree. Larger programs that possess extensive options, such as Quattro Pro, require a hard disk and sufficient memory. However, As-Easy-As is an example of a smaller spreadsheet program that possesses the needed features as well as some additional ones (table interpolation, user-defined functions) not found on other programs.

Spreadsheet Instruction

Early in the term, an instructor can demonstrate spreadsheet use and graphic capabilities with statistical examples, and provide students with directions for constructing some basic models. If a laboratory is available, the instructor later can lead the class through a hands-on session. Generally, one or two class periods are adequate to provide enough basic spreadsheet instruction to enable students to get started.

Initially an instructor may need to prepare some materials to distribute to classes. However, books discussing the uses of spreadsheets in mathematics and statistics are appearing with increasing frequency [1,2,3,6,7,8].

Model Format

One shortcoming of spreadsheet models is their unstructured nature. Because of this, they can be somewhat hard to read and debug. Thus, it is important to adopt some standard programming conventions. Models should include comments, user instructions, and descriptions of any named objects. Separate well-defined areas should be used for listing parameters, output, and data.

Advanced Spreadsheet Features

Spreadsheets have many advanced features that can be used in statistics. User-defined macros allow a series of commands to be replaced by a single keystroke, and also enable the implementation of such programming operations as loops. Quattro Pro has built-in commands for matrix operations, linear regression, and finding data frequencies. As with the sort command, these are not interactive, but must be issued separately to update displays. Some spreadsheets support such database operations as searching, and most allow a user to protect all cells, except those used for input, against user changes. This is a valuable feature with template assignments.

Emphasis

Finally, we should keep in mind that the purpose of using spreadsheets is to strengthen statistics teaching, and not become overly concerned with ancillary matters. Also, while spreadsheets are excellent tools for teaching statistics, they are not designed to perform all tasks, some of which are more easily accomplished using other computer packages. However, spreadsheets do a vast range of statistical procedures exceptionally well, and represent vital, multi-purpose tools for teaching statistics. They provide a productive, readily accessible means for adding vitality and insight to our teaching of statistics.

Details of these and other examples will be included in a forthcoming book by the author on the use of spreadsheets in statistics. In the meantime, formula listings of selected examples from the present article may be obtained by writing to the author.

References

1. Arganbright, Deane E., *Mathematical Applications of Electronic Spreadsheets*, McGraw-Hill, 1985.
2. Bell, Peter C. and E. F. Newson, *Statistics for Business with Lotus 1-2-3*, Scientific Press, 1989.
3. Kilpatrick, Michael, *Business Statistics Using Lotus 1-2-3*, Wiley, 1987.
4. Mathews, John, "Finding Least Squares with Mathematica", *PRIMUS*, **1**, 1991, pp. 103-111.
5. Neuwirth, Erich, *Visualizing Correlation with Spreadsheets*, Technical Report, Institute of Statistics and Computer Science, Univ. of Vienna, Austria, 1990.
6. Piele, Don, *Introduction to Statistics with Spreadsheets*, Addison-Wesley/Benjamin Cummings, 1991.
7. Sope, Jean B. and Martin B. Lee, *Statistics with Lotus 1-2-3*, 2nd ed., Krieger, 1990.
8. Spero, Samuel W., *Graphing Functions: An Introduction to the Electronic Spreadsheet*, Scott, Foresman, 1990.

Deane Arganbright is professor of mathematics and computer science at Whitworth College. His primary interests are in mathematical modelling using spreadsheets, with an emphasis on graphical display. He has been a Fulbright professor at the University of Papua, New Guinea. He is the author of many publications on spreadsheets including *Mathematical Applications of Electronic Spreadsheets*.

Using HyperCard to Teach Statistics

Walter Chromiak and Allan Rossman
Dickinson College

The Problem

As instructors, we soon discover that no matter how well we present a lesson, some students fail to learn. Much of the problem concerns the level of presentation: the material is not challenging enough for some students, while we assume that others know more than they actually do. In the past thirty years, cognitive psychologists have amassed evidence which clearly shows that what a person already knows determines what that person can learn and understand [2].

Our task as instructors, then, is to discover what each of our students already knows and understands and to design our classes to match the students' current level of knowledge. Impossible, you say! Of course you're right, but we hope to convince you that it is possible to do better than most of us currently do. We certainly do *not* argue that instructors teach to the level of their weakest students or abandon all standards. We *do* contend that for good teaching to occur, instructors must take into consideration what their students currently know. And by what their students currently "know," we don't simply mean their level of mathematical proficiency. For example, explaining probability concepts in the context of a lottery does no good for students who don't know what a lottery is.

In this paper we describe an ideal solution to the problem of matching what we want to teach to the current state of a student's knowledge. We offer some ideas on how computers can help implement this solution, and we show how a specific computer software package, HyperCard™, can help to solve our problem of presenting material which is either not challenging enough or too challenging for students.

An Ideal Solution

Over the years, most of us have assembled a collection of examples and anecdotes which we believe helps students to grasp the central ideas that we try to teach. When a student fails to see our point we simply blame the student. After all, other students have told us how interesting and exciting our classes are. A cognitive psychologist would not be surprised, however, to find that some students think our explanations are clear and to the point while other students, in the same class, are wondering what we're talking about. In the best of all possible worlds, we would know exactly what each of our students understood before we began a lesson; we would also know exactly what interested each of our students. Unfortunately, this is not the case, so what can we do if we want to be sure to reach all-well, almost all-of our students?

Get the students involved

Someone once said that learning is something you **do**, not something which is **done to you**. Emphasizing students' active participation by helping them discover and explore fundamental concepts in statistics can help students to acquire a deeper and more lasting understanding of the material.

Give students immediate feedback

To help students learn, we typically give quizzes or homework assignments which take some time for us to grade and return to students with the constructive comments that we expect to further their understanding of the material. Psychologists have long known that the most effective way to learn is to receive immediate feedback on what you did right or wrong. This allows the learner to spend less time doing things incorrectly and more time performing correctly.

Present the same idea in many different ways

Presenting the same concept in as many interesting ways as possible can only increase your chances of having students understand the concept. For example, you can teach the idea of association between variables in many ways: (1) by describing a circumstance where we'd expect variables to be related to one another (e.g., height and shoe size), or (2) by producing a scatterplot (e.g., shoe size vs. height), or (3) by constructing a two-way contingency table and computing relevant proportions (e.g., the proportions of tall, of medium, and of short people who have large shoe sizes). Rather than searching for that one example or anecdote that rings true for you, collect as many examples or anecdotes as you can and see which ones seem to work.

Present ideas visually

Many of us benefit when we can see or imagine a concept we are trying to understand. Suppose that you're teaching a class about outliers. You could, of course, tell the students that an outlier is an observation which lies more than 1.5 times the inter-quartile range from its nearer quartile. Better yet, you could show them pictures which illustrate the same idea. Most students would probably be lost if you failed to use pictures.

Focus on ideas rather than on calculations

Each of us needs to guard against the temptation to focus on computation rather than the central ideas we're trying to teach. In many ways it is much easier to spend our time going over how to calculate a standard deviation or a correlation coefficient. It's much harder to try to convey the central idea of variability or association. When we spend our time focusing on computation, we turn our students into third-rate calculators rather than taking advantage of what they can do best which is deal with ideas.

Let's summarize what some of the parts of an ideal solution to our problem would be. First, we should engage our students so that they don't just passively listen to us lecture. Second, we should provide our students with immediate, constructive feedback on their performance. Third, we should present the same idea in a number of different ways through a number of different media including pictures, words, etc. Finally, we should spend our time focusing on key ideas rather than on tedious calculations.

How Computers Can Help

Ever since microcomputers have been in existence, experts have predicted that they would revolutionize education. Computers have a number of features which can make them, in theory at least, ideal teachers. A short list of these features follows:

■ Computers are patient (kind, reverent, cheerful, loyal...). One blessing of being a computer is that your teeth don't grind when you hear the same question for the thousandth time. Computers don't get too tired to explain what a standard score is at eleven o'clock at night or too bored to explain how to calculate a standard deviation for the millionth time. (We have been describing computers as if they were people and for many students they are just like people, or at least very special machines.)

■ Computers are also precise. They don't make errors in calculation and they don't fail to recall information. Because of this they can make computations trivially easy for our students–and we should allow them to.

■ Computers can provide students with immediate feedback on their performance. They can instantly tell students if what they have done is right or wrong; they can also guide students in the proper direction as needed.

■ Computers can present information using both pictures and text. We have already argued that we should present ideas in as many ways as possible. Even though most of us are not simultaneously great artists and raconteurs, the right computer and the right software can make each of us both a Picasso and a Bill Cosby.

■ Computers can simulate events. Many years ago Kenneth Craik, a British psychologist, argued that "[o]ne of the most fundamental properties of thought is its power of predicting events." [3] We are able to anticipate and to imagine what will happen rather than waiting for experience to present us with the results of events. In the same way, computers can present students with simulations of interesting phenomena.

■ Finally, many people simply think that it is fun to play with computers.

Why the Computer Revolution in Education Hasn't Occurred

Despite the predictions of how computers would change the way we teach, most instructors don't use computers very much. Oh, many students use computers to write papers; a few even use computers to do calculations, but most classroom instruction proceeds as if computers didn't exist. Why have computers failed to capture the hearts and minds of educators?

Putting aside for the moment the actual cost of a computer, one of the reasons that computers aren't used to their potential in education is that the programs we require are often very expensive–even after educational discounts. Expense has certainly not been the only barrier to the intelligent use of computers in education, though. Instructors often find that the programs they've purchased don't do exactly what they had hoped. Many of us would like to be able to "tweak" a program to get it to perform in the way we hoped it might. In most cases, programs can't easily be changed, and it's usually illegal to do so. Even if programs were easier to modify, most instructors simply don't want to become computer programmers. They became teachers not to program computers but rather to teach. Finally, writing a useful computer program is hard; it's hard to design useful programs and it's even harder to implement them.

HyperCard to the Rescue

Despite the problems we've described that have hampered the computer revolution in education, we suggest that you consider writing your own courseware. We make this suggestion because we think that the tools which instructors need to develop quality courseware are finally available in an attractive and easy-to-use package. HyperCard, which runs on a Macintosh computer, allows instructors to design and implement their own courseware quickly, easily, and effectively. (Toolbook™ is the name of a program by Asymetrix® (Bellevue, WA 98004) which mimics HyperCard in MicroSoft Windows® for IBM-type machines. We have not examined this product so we can't say how well it mimics HyperCard or whether it contains features which surpass HyperCard's.)

HyperCard has been described by its creator, Bill Atkinson, as "an 'erector set' for software ."[1] HyperCard is a computer program which comes **free** with every Macintosh computer. One of the reasons that the Macintosh is popular with students is that even new users very quickly find that they can use the computer with a minimum of fuss and bother. The Mac also makes it easy for instructors to present pictures, graphs, text, and even sound and animation. HyperCard takes advantage of these particular strengths of a Macintosh and places a variety of tools at the instructor's disposal.

To get a sense of how all this works, imagine that a student is sitting in front of a Macintosh, staring at the screen, reading material about correlations and comes across the term z-score. The student doesn't quite remember what a z-score is and normally would give a mental shrug and keep reading. However, this term has been marked with an asterisk, which indicates that more information is available.

The student uses the computer mouse to move an electronic pointer that looks like a hand over an icon which displays a question mark and presses the button on the mouse (see Figure 1). The computer screen changes to display a small box with a brief description of what a z-score is and a "button" that the student can "click" for more information, just as she did earlier with the term "z-score." This time the student decides she needs more information and is soon engaged in a brief demonstration of what z-scores are and how they are calculated. If need be, our student could do an exercise or two to reinforce the ideas she is reviewing; however, she decides she now understands and with a click of the mouse returns to exactly the same point where she was before the term z-score stumped her.

Clearly such a program allows the instructor to anticipate the students' needs. For the student who recalls what a z-score is, no side trip is necessary and that student can continue with the lesson. However, the student who has forgotten what z-scores are can quickly and easily refresh her memory and continue with the lesson. In addition, the help which is provided can vary in complexity, from the simplest explanation to a full-blown lesson, depending on the amount of help that is needed.

Obviously, a student can do much the same with a textbook, but doesn't. The advantage here is that the instructor can anticipate where the problems are, **highlight** the fact that a particular concept may be a problem and thus alert the student to it. We hope the student will respond to our suggestion and not just ignore it as happens with a book.

Let's consider how you might go about building a HyperCard application. First of all, it's convenient to think of HyperCard as an infinitely large deck of index cards. Imagine that you can place anything you would like on any individual index card. Also imagine that you can connect the picture(s) or words of one index card to some other picture or word on either the same card or on new cards. This, in a nutshell, is what HyperCard allows you to do: to make connections between ideas.

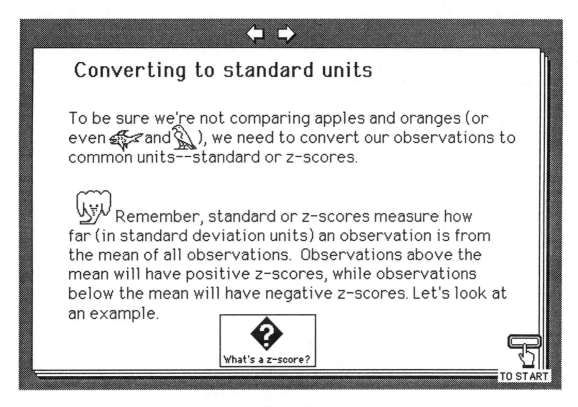

Figure 1

Since Bill Atkinson didn't expect that the people using his program were already expert artists or musicians, he made it easy for users to share resources. HyperCard comes with an extensive set of art, sounds, and graphics of various sorts. To use these materials in your application, you simply cut what you want to use from its source and paste it into your application. In most cases, the materials even have all of the necessary programming commands already in place, allowing you to not only cut and paste the pictures or sounds themselves, but also the attached commands which make them function.

A Brief Tour of HyperCard

Making a program easy to use often requires a tremendous amount of programming skill. Most computers require the user to remember commands in order to make the computer function. The Macintosh allows the user to use a computer mouse to point at items on the screen and then, by a click of a button on the mouse, initiate some action (Of course, the MS-DOS world of computers now has access to MicroSoft Windows® which mimics many of the functions of the Macintosh "user interface."). Letting a user simply point at an object on the computer screen and click to initiate some action is just one of the many "whistles and bells" which comes with HyperCard. In fact, let us describe some of the objects which are provided in HyperCard.

The most commonly used object is a "button." Buttons come in a variety of styles and they are completely under the control of the HyperCard programmer. Buttons can have names, buttons can have pictures, and buttons can be visible or invisible. Buttons allow you to initiate some event once the user has clicked the on-screen button. For example, clicking a button might send you to another card in a HyperCard stack. Alternately, clicking a button might start a process such as drawing a graph or performing a calculation. In Figure 2, the four icons on the right of the illustration are buttons. Three of these buttons take the user to a new section of the stack. There is also a "quit" button at the top of the illustration. Buttons are clearly powerful tools which make the computer very easy for the novice to use.

CORRELATION

Walter Chromiak and Allan Rossman
Dickinson College

> This is the first card of the CORRELATION HyperCard™ stack. The purpose of this stack is to help you learn the fundamental statistical concept of correlation. The stack contains a wealth of information for you to peruse and digest and also numerous opportunities for you to think, experiment and apply what you learn.
>
> Through this card, you can reach various parts of the stack or quit HyperCard. To quit, click on the QUIT! button at the upper left. To reach the applications or quiz part of the stack, click on the appropriate button on the right. If this is your first pass at this stack, simply click on the arrow in the lower right to reach the next card. To return to this card later, click on the TO START button from any card. If you see a confused penguin and feel a little lost yourself, click on him for moral support (and maybe some helpful information)!

Figure 2

Besides buttons, HyperCard stacks also contain fields of text. The main text of the card displayed in Figure 2 is in a text field. The size of such a field is completely determined by the HyperCard programmer. Fields can contain a limited but large amount of text. Technically, you're limited to about 32,000 characters in a field. Any particular card can have as many fields as you can fit on the card. Just as with buttons, fields can be visible or invisible; in fact, you can turn a field into a button (i.e., once the user has moved the cursor into a field, some event can occur). In fact, if the user clicks on the "question mark" button seen in Figure 2, a previously hidden field of text, shown in Figure 3, appears.

Finally, HyperCard stacks can contain pictures which the programmer has either created or copied from some other source. HyperCard provides all the facilities you might need to create new drawings and to modify existing graphics. You can even rotate parts of your drawing in the plane of the screen. Figure 4 contains graphics which were taken from stacks provided with HyperCard, as do Figures 1 through 3.

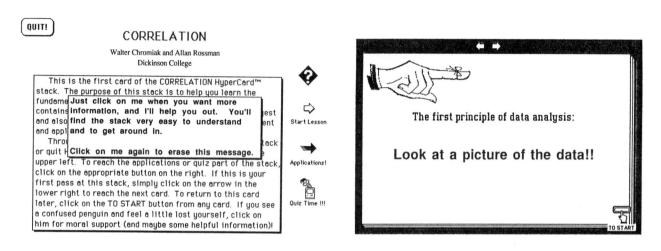

Figure 3   Figure 4

If you already have illustrations, you can have them digitized using inexpensive scanners and easily add them to your HyperCard stack. In fact, many people use HyperCard stacks as a repository for clip-art or illustrations. In that way they can easily cut and paste what they need from one stack into another stack.

While HyperCard has many features, it is also designed to be easily expanded. Often these new features become available either for free via bulletin boards or are sold commercially at very reasonable prices. All in all, HyperCard is designed to be easy to modify and easy to expand.

HyperCard for Introductory Statistics

While HyperCard can be a valuable teaching tool in any field, it is particularly well suited for use in statistics education for several reasons:

■ Students entering a beginning statistics course often do so with more trepidation than with other courses; they are often fulfilling some dreaded requirement. HyperCard's visual appeal and special effects can create an engaging means of allaying students' apprehensions.

■ Students in elementary statistics courses often display a very wide disparity of abilities and of interests, so the capacity of the instructor to meet each student at different levels using HyperCard is especially valuable.

■ Numerical calculations are perhaps the most daunting impediment to students' acquiring an understanding of statistical ideas. HyperCard can allow students to perform the calculations at the touch of a button, freeing them to concentrate on the principles rather than on rote memorization of computational formulas.

■ Many statistical principles are best presented graphically; HyperCard can convey these ideas with very attractive and informative pictures and graphical displays.

■ Students can discover statistical ideas for themselves by doing specially constructed interactive exercises with HyperCard. The immediate feedback provided allows the students to hone their statistical skills.

■ The instructor's ability to adapt HyperCard materials easily to suit his or her own purposes is especially useful for statistics instruction, since courses in statistics are taught in many different departments with varying emphases for different groups of students.

■ HyperCard can provide an efficient information storage device for newspaper and magazine articles, data sets and descriptions, and reference listings for instructors to share and for students to peruse. With the current emphasis on teaching data-driven courses and on viewing statistics as a liberal art, the availability of shared resources such as these would be quite welcome.

Illustration: A HyperCard Stack for Teaching Correlation

We have developed a stack to teach the concept of correlation that illustrates many of these uses of HyperCard. From the very first card of the stack (shown in Figure 2), the students make choices (depending on their interests and levels of understanding of and comfort with the material) that lead them on a path through the stack. For example, a student may proceed to the start of the lesson, to an applications section which presents real-world examples of uses of correlation, or to a quiz section which tests her knowledge of the material. The student can return to this card from any card in the stack (and therefore move on to any section of the stack) at any time. "Detours" are also available at several points. For instance, the correlation coefficient is derived in terms of standardized or z-scores; if the student wants a quick refresher on z-scores when she reaches this point, the touch of a button produces that refresher (see Figure 1); the student who is familiar and comfortable with z-scores need not take this detour. Obviously, the student can move through the cards at her own speed, further attesting to the stack's being virtually custom-tailored to each student.

The stack makes extensive use of HyperCard's ability to create visual displays. "Cutesy" pictures, such as an elephant or a string tied around a finger to urge the student to remember a point, make the stack more entertaining. More importantly, plots and displays of data abound to emphasize a visual approach to understanding the ideas. For example, while the properties of the correlation coefficient are being examined, the student is presented with a plethora of scatterplots which illustrate that the data points fall closer to a straight line as the correlation coefficient approaches ± 1. Figure 5 shows the choices students have in deciding what type of correlation they want to view. These choices appear after the students "click" the "Analyze Data" button. Figure 6 shows the result of choosing "strong +."

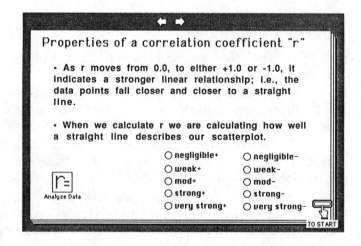

Figure 5

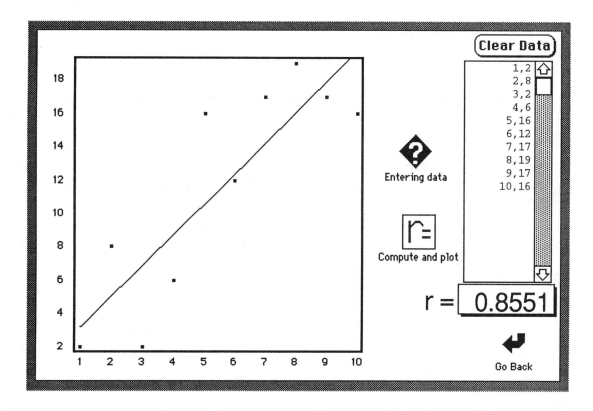

Figure 6

HyperCard also affords the students the opportunity to explore the properties of correlation for themselves, to experiment interactively and to receive immediate feedback. For example, a student can change one number in a data set and immediately see the resulting change in the scatterplot and in the value of the correlation coefficient. (Figure 6 shows the scrolling field of data which the user can edit. "Clicking" the "Compute and plot" button after editing the data produces a new scatterplot and calculates the new "*r*" value.) By experimenting and receiving immediate feedback, the student can discover for herself that the correlation coefficient is not resistant to outliers and that it may not detect a nonlinear relationship between two variables.

The quiz section of the stack allows the student to test her understanding of the material and to prepare for exams on the material; it also can let the student know whether she is ready to move on to the next lesson. The types of quiz questions asked include true/false, multiple choice (see Figure 7), and experimentation questions; for instance, the student might be asked to change one data point in a set to make the correlation drop to near zero (see Figure 8). These questions let the student know immediately whether her answer is correct.

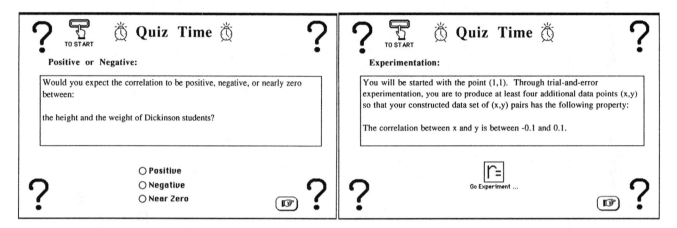

Figure 7   **Figure 8**

Further Illustrations: AIDS Testing and Normal Curves

A much simpler stack that we have designed consists of an exercise intended for use in a quantitative reasoning course among liberal arts students with no statistical knowledge and little mathematical preparation. The exercise guides the student to discover for herself the important idea that when testing for a rare disease, most positive test results are "false positives," even with fairly reliable tests. To motivate the student's interest in the principle, the exercise is phrased in terms of AIDS testing; real data concerning testing reliability are used and the student is asked to assess a claim made in an actual Red Cross brochure.

The exercise asks the student to imagine the AIDS test being given to a hypothetical population of 1,000,000 people. The student uses percentages given to determine how many of those people would carry the AIDS virus and how many of them would test positive (see Figure 9). The student then calculates how many people from the population would not carry the AIDS virus and how many of them would test positive.

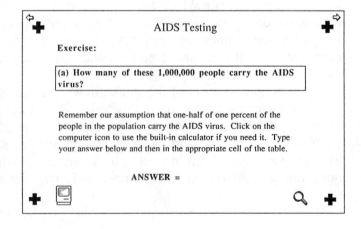

Figure 9

All these calculations are done using a 2×2 contingency table (see Figures 10 & 11). This set-up makes plain, with calculations no more involved than adding, subtracting, and taking percentages, the idea in question. HyperCard can even supply the student with a calculator so that she can do the arithmetic on the computer; it also allows the student to move back and forth from the statement of the exercise questions to the two-way table in which the student can enter the relevant numbers and see the idea developing. The exercise concludes by asking the student to articulate the principle that she has discovered, to criticize the brochure's claim, and to think of other settings in which the principle would apply.

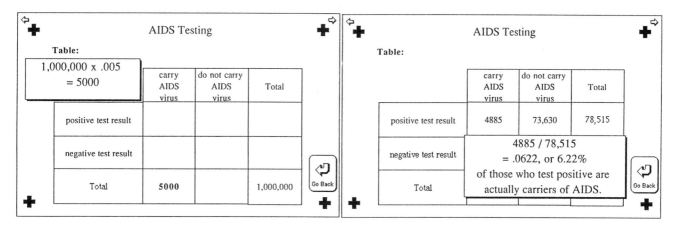

Figure 10   **Figure 11**

A more mundane example, but one in which HyperCard's visual capabilities shine, calculates areas (and therefore probabilities) under normal curves. With this card, the student enters the name of the (normally distributed) variable and the mean and standard deviation of its distribution. The card then displays the variable's normal curve and asks the student for the interval whose probability is desired. This region is then shaded in on the graph and the z-scores are calculated and shown on the axis. Finally, the probability of the interval is calculated and displayed (Figure 12). One might object that this card does too much for the student, that learning to read a table of probabilities is a useful skill to acquire. Perhaps, but we would argue that this demonstration can help students understand what they are finding when they do read a table.

These graphics and calculations are performed by programs called scripts that are written in HyperTalk, the language which comes with HyperCard. While HyperTalk's capabilities are limited, external commands (XCMDs) and functions (XFCNs), which can be written in Pascal or C, greatly enhance HyperCard's potential.

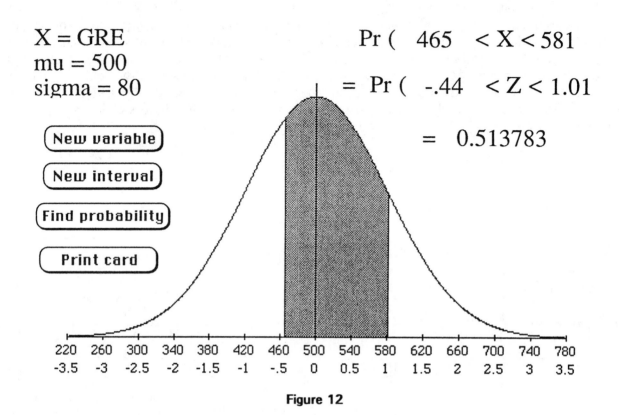

Figure 12

Use of HyperCard in the Classroom

How one uses HyperCard in the statistics classroom depends, of course, on the extent of the computer resources available. Another obvious consideration is the willingness of the instructor to abandon the traditional norms of lecturing in class and assigning readings and problems from a text outside of class. In the best of all possible worlds, we envision that the classroom would be equipped with microcomputers enabling each student (or perhaps a pair of students) to work through HyperCard stacks, exploring ideas, drawing connections, and experimenting with data for themselves. In such an environment, HyperCard could naturally become the primary teaching tool for the course, replacing lectures and perhaps even the textbook. Although one would certainly devote portions of class periods to lecture, discussion, and other class activities, HyperCard is an ideal medium for providing students with an interactive learning environment. Once instructors prepare enough material for HyperCard stacks, the need for a course textbook could also disappear, although we discuss below some reasons for textbooks never becoming obsolete. Of course, we expect that students will have access to computers outside of class to continue working on their assignments.

Even if classrooms could not be equipped with computers, HyperCard could be used very effectively if students had access to computers outside of the classroom. With this environment, students could spend their time out of class by working through HyperCard stacks and exercises, such as the correlation and AIDS testing stacks, rather than reading the text and practicing with traditional homework exercises. Alternatively stacks could serve to supplement and expand on material from a textbook or lecture.

Instructors with very limited computer facilities and those who want to retain the use of lectures and textbooks can still use HyperCard as a teaching tool. For instance, an overhead projector and appropriate hardware will allow an instructor to display a computer screen for an entire class to see. The instructor could then take advantage of HyperCard's visual and graphics capabilities which probably exceed his or her own blackboard artistic ability, as with the normal curve calculation stack described above.

At this point, a warning about an inappropriate use of HyperCard is in order. Instructors must avoid the temptation to fill a HyperCard stack with nothing but text, making it in effect a textbook reprinted on a computer screen. We should use to their fullest both HyperCard's visual capabilities and its capacity to make connections between disparate ideas, as these are virtues not readily possessed by traditional textbooks.

Speculations About the Future

Will HyperCard survive?

There is reason to believe that it will. For example, one major software developer, Microsoft, includes a HyperCard stack with its release of the spreadsheet program Excel for the Macintosh. The stack is designed to introduce the novice to the features of Excel and provides an easy way for both new and old users to refresh their memories of how a particular feature is used. In addition, a number of commerical stacks are available. For example, one of us uses commercial stackware to introduce students to various concepts in visual perception in a cognitive psychology course. These stacks include animation as a primary feature and also allow students to explore perceptual phenomena in a convenient and interesting way. A site license and a computer network allow the students access to the stacks. Finally, a number of programs such as Toolbox are making available all or most of the features of HyperCard to users of IBM machines and their clones.

Will HyperCard replace textbooks?

On a good day, when we are filled with optimism, we are certain that we can and should replace textbooks in our courses with HyperCard stacks. Realistically, textbook publishers will not give up books for HyperCard stacks. Books offer at least the advantage of true portability, which even portable computers do not. Books also don't require batteries, are affordable, don't require accessories, and can be borrowed from a library.

HyperCard, however, does offer some advantages to publishers which books do not. First, it would be easy to put together a very useful instructor's guide using HyperCard. For example, an instructor could have easier access to publisher-supplied quiz and exam questions if they were available in HyperCard. Publishers wouldn't need to worry about instructors having the right software package to run their quiz or exam generating programs. Also, instructors could easily modify or add their own questions to the pool provided by the publisher. Publishers could also include useful demonstration programs, allowing instructors to not only read about, but actually use good demonstrations. Finally, it would be easy for publishers to revise their instructor's manuals frequently and accept contributions from a variety of sources since HyperCard would be the glue which holds the various contributions together. It wouldn't be necessary to wait until the next edition of a text before new materials, demonstrations, or questions were available. Instructors could download these materials from a publisher's computer bulletin board.

A truly interactive student workbook for a course would be a boon for both students and instructors. As we've mentioned, a well designed stack would allow students to progress at their own pace, explore avenues of interest, and provide glossary and tutorial material at the student's finger tips. In our imaginations we see the day when we can design an assignment which includes a computerized pretest to help students determine if they have the skills to tackle the next assignment, a tutorial which provides students with as little or as much practice as they need, and a posttest which helps students determine that they've actually mastered the material before they come to class. What would we do in class? We could explore the application of the concepts we are trying to teach and examine their limitations. Analyzing data and deciding how to interpret results is more fun and useful, for both students and instructors, than going over the mechanics of how to calculate some statistic such as a correlation coefficient or z-score.

What can we realistically expect?

We've discussed what we hope might happen, but what might we realistically expect in the immediate future? We suspect that both software developers and various individuals will produce HyperCard stacks to serve a variety of purposes. For example, Heizer Software of Pleasant Hill, California provides a catalog of a variety of HyperCard stacks and add-ons. In addition, Clear Lake Research of Houston, Texas has developed a series of HyperCard stacks to cover most of the topics of a basic statistics course. These stacks might even be commercially available by the time you read this. It appears that at least some of our daydreams might come true.

Traditionally, software has always served as a supplement to a textbook and series of lectures. You might expect to see the same happening with HyperCard stacks initially. We believe that it will take some time for publishers and instructors to explore the possibilities of interactive software such as HyperCard. As both instructors and publishers become familiar with what can and can't be done, we expect to see a greater emphasis on the use of software in teaching. If we're right, the computer will actually become a useful classroom tool rather than a novelty or glorified typewriter/calculator.

Summary

We believe that HyperCard provides instructors with the tools they need to produce effective presentations which allow students to explore statistical concepts in engaging and interesting ways. In a properly designed HyperCard stack, students can choose how they wish to proceed through a lesson. Students can be given reminders or more complete help at appropriate times, without intruding on the main ideas being developed. This capability, in a sense, allows the material to be tailored to the needs of each student. In addition, it is easy to produce a number of different examples and explanations for a concept and to provide them as needed to match not only students' ability levels, but also their academic intestests. Finally, by including a variety of quizzes and providing the students with immediate feedback on their performance, the students learn more quickly and efficiently.

We have just scratched the surface for what is possible with HyperCard. Rather than reading about it, we urge the reader to play with a HyperCard stack. As we mentioned, they are provided at no cost with every Macintosh computer, so a colleague with a Mac can show you HyperCard or a brief visit to a computer store will do the trick. If you are as impressed as we think you'll be, you might want to consult some books on HyperCard which may be found in most bookstores. (We've found HyperCard books by Danny Goodman to be particularly useful.) We welcome any comments or questions.

References

1. Atkinson, W. (1988) In Kaehler, C. *HyperCard™ Power: Techniques and Scripts*. Reading, MA: Addison-Wesley Publishing Company, Inc., p. xv.

2. Bransford, J. D. (1979). *Human Cognition: Learning, Understanding and Remembering*. Belmont, CA: Wadworth Publishing Co.

3. Craik K. (1967). *The Nature of Explanation*. Cambridge University Press, p. 50.

Walter Chromiak is associate professor of psychology at Dickinson College where he teaches courses in research methods, statistical analysis, and human cognitive processes. He and his colleagues in other disciplines are also interested in the use of computers in teaching statistics.

Allan Rossman is assistant professor of mathematics at Dickinson College whose primary teacing interests are probability and statistics, and their application to the social sciences. His research area is Bayesian statistical inference and decision theory.

Graphing Calculator Enhanced Introductory Statistics

Iris Brann Fetta
Clemson University

The Clemson Experience

Careers which require mathematical ability and technical competence, including those which require a basic understanding of fundamental statistical processes, are ever-increasing. The concepts of probability and statistics are now presented to a mass of students from a variety of different disciplines. While students majoring in mathematical sciences, engineering, and computer science have traditionally taken such courses, students at Clemson University majoring in areas such as financial or industrial management, visual arts studies, architecture, biology, accounting, economics, secondary education, journalism, microbiology, or medical technology are also now required to take at least one course in statistics.

In an effort to increase student interest, involvement, comprehension, and retention of the subject material, Clemson University integrated the use of graphing calculators into the undergraduate curriculum in the fall of 1989. One calculator-enhanced section of the statistics course servicing the above-mentioned majors has been taught for each of the last six semesters with funding provided by a three-year grant from the Fund for the Improvement of Postsecondary Education (FIPSE). The following topics are in the multisection course syllabus (numbers indicate number of 50 minute class periods): Graphical and Numerical Descriptions of Data (3), Probability (5), Discrete and Continuous Random Variables (10), Sampling Distributions (3), One and Two Sample Estimation and Tests of Hypothesis (13), and Simple Linear Regression (5). The prerequisite for this course is one semester of "regular" calculus or business calculus. The majority of the sections are typically taught in lecture format with daily homework assignments, between 3 and 5 major tests and a final examination. Students in the regular sections are encouraged to use scientific calculators, but tests are normally designed so that a student with a calculator with statistical capabilities does not have an advantage. Calculator instruction is only offered in the graphing calculator sections.

During the first two years of the project, even though the calculator-enhanced section had a special designation in the master schedule, only a few students knew they were registering for a calculator section of the course. However, in the fall and spring semesters of 1991, students who had used Hewlett-Packard calculators in their calculus courses and those who had received recommendations from fellow students registered for the graphing calculator sections. After an initial explanation of the philosophy of the special section, students are given the opportunity to change to a "regular" section of the course. To date, only two students have selected this option. Sharp EL-5200 calculators and operational manuals are loaned to students each semester for the duration of the course (under a signed Calculator Loan Agreement).

The Sharp EL-5200 was initially chosen for the Clemson course because of its relatively low cost, its simple operation, its graphics capabilities, its ease of programming and its relatively large memory for program and data storage. Such a calculator eliminates hand calculations and allows more time for concepts and analysis of results. The programming capability allows commonly used equations to be entered and stored in the student's own notation. The programs are provided in the instructional materials [10], so that novices (instructors as well as students) need not know how to program the calculator to use it effectively. The subject matter and enhanced content can be taught using any calculator with similar capabilities.

For the first two years of the FIPSE project, fundamental calculator operations were worked into the course as needed. Students were given specific programs to be entered outside of class. Many of the students in this course suffer from math anxiety. As the semester progressed, the students built on small successes with increasing confidence. Some attempted to enter the statistical formulas on their own or modify the programs they were given to enter outside of class. This often raised interesting questions on statistical and mathematical principles for class discussion. Many of the students found it tedious to enter the longer programs in the calculator. As a result, for the third year of the project, the Sharp EL-5200 calculators were "loaded" with the programs necessary for the course via the Sharp CE-50P Printer and Cassette Interface. Students with HP48 calculators obtain simulation programs via the calculator's infrared communication. As the quality of calculator specific instructional material improved and the student time spent on keying in programs was eliminated, the time spent in class on the manual operation of the calculators dramatically decreased.

This instructor has taught another section of this course each semester along with the calculator-enhanced section. Students in the control group used a variety of calculators, but none had graphics calculators and very few had programmable models. Both sections used the same syllabus and text and similar exams were given. Even though the number of points missed due to conceptual misunderstanding has been roughly the same in the two groups, higher scores have generally resulted in the calculator-enhanced section where significantly fewer points were lost due to computational errors. Students in the calculator-enhanced statistics sections generally comment that the calculator makes the course much less time consuming, and therefore, it seems easier and more interesting. Many students agree that calculator use reduces anxiety over testing because careless errors are not readily made. Quite a few of the students in these enhanced sections have found applications for their calculator in other courses, research, and off-campus jobs, and many indicated that they would purchase their own calculator in the near future.

Student reactions were obtained at the end of each semester using an extensive evaluation form prepared by the project's external evaluator. Below is a summary of several key questions from the first three semesters (size = 63):

Question	Disagree or Strongly Disagree	Neutral	Agree or Strongly Agree
1. The graphics calculator helped me understand the course material.	25%	32%	43%
2. The graphics calculator allowed me to do more exploration and investigation.	23%	27%	50%
3. The graphics calculator was useful in solving problems.	1%	3%	97%
4. I could have learned more if I had not used a graphing calculator.	54%	27%	19%
5. Learning the graphics calculator was so difficult that it detracted from learning the material in the course.	72%	24%	4%

6.	Time devoted to instruction in the use of the graphics calculator meant less material was covered in the course.	59%	16%	21%

Responses to the question "What did you like best about using a graphics calculator in this course?" range from "For once in my life, I have seen something in a math course that made sense. The use of this calculator was an incredible idea!" to "If you knew what you were doing, it was great. Unfortunately, this was not the case for me in most instances." One student wrote "At first I was so confused, and I was afraid I would fail the course just because I couldn't use the calculator. However, bit by bit I got the hang of it and really came to appreciate how it helped me understand the material."

The underlying concern of many students at both the beginning and end of the course seems to be that they rely on the calculator too much and will not be able to perform well in subsequent courses without it. They may be denied the use of the calculator by instructors in subsequent courses, or if they do not have their own calculator, they may not feel financially able to purchase one. It is helpful to convince students that they can, if necessary, perform the mathematics involved using pencil and paper.

One tactic I employ is to ask the students to find a hypergeometric probability where the number of elements in the population is 70 or more. Since the Sharp EL-5200 results in a memory overflow when the combinations key is employed, students are initially confused when their calculator program for hypergeometric probabilities results in an error message. A few students will answer "0", some will indicate "this problem cannot be worked", but the majority will revert to the formula and work it out by hand. One student even came up with the binomial approximation to the hypergeometric distribution before it was discussed in class. I also have students pretend their calculator is broken and have them work a problem (with pencil and paper) employing the independent sample t-test. Students then are asked to work the problem using the program in the calculator and realize that the procedure is the same but the calculator enables them to perform the numerical calculations with accuracy in a fraction of the time. It is a good idea to discuss frankly with students their concerns about calculator dependency several times during the course.

Implementing a Calculator-Enhanced Statistics Course

One difficulty encountered with the course is that there is no text that effectively integrates this new technology. Many problems in current texts become exercises in the use of the calculator rather than problems that involve mathematical reasoning. To fully take advantage of the experimentation possible using technology, some of the material in the current syllabus must be deleted and replaced with content and exercises that encourage students to use their knowledge and explore. Since statistics is often a prerequisite for other courses, the revisions must also be carefully examined in light of the goals of the course and the sequence of subsequent courses in many different disciplines.

From an instructor's point of view, the calculuator-enhanced course is a delight to teach for many reasons. Attendance is higher and the percentage of students who withdraw is much smaller than in the regular sections. A "closeness" with the class is common due to the joint exploration and discussion by instructor and students learning to employ this new technology. Variation in techniques more easily arise and there is strong interaction between instructor and students. Exams are much easier to design and grade because problems do not have to be "fixed" to come out with "nice" answers, but assigning partial credit may be difficult since the work is done in the calculator and not on the student's paper.

For example, consider the problem: "A clothing store manager has noticed that 63% of all purchases are made with credit cards. He randomly selects 58 purchases during a given day and notes the method of payment. What is the probability that at least 39 but less than 43 purchases were made using a credit card?" Most texts do not include binomial tables for $n = 58$ and $p = .63$, so this becomes an exercise in time-consuming arithmetic of four applications of the binomial formula for the student with a nonprogrammable calculator. With the computations done by a programmed binomial formula, one may test for recognition of the binomial variable and the knowledge that the individual probabilities should be added. To avoid loss of credit due to an erroneous keystroke, students are asked to identify the variables involved and the calculator program they use if they wish partial credit on the problem.

Class projects and group work are a natural extension of the use of the technology. For instance, consider the following exploration which uses programs in the student's calculator:

a) Construct a graph of the binomial distribution for $n = 10$ and $p = 0.8$. If your calculator will not store graphs, record the graph on paper while you notice its skewness, mean and standard deviation.

b) Construct a graph of the Poisson distribution for $\lambda = 8$. Does it appear that the graph of the Poisson distribution would "fit" over the graph of the binomial distribution you constructed in a)? Why or why not? What is the relation of the mean of the binomial distribution to the mean of the Poisson distribution in this problem?

c) Repeat parts a) and b) of this problem with $n = 50$, $p = .1$ and $\lambda = 5$. Discuss your results with other members of the group. Under what circumstances do you think Poisson probabilities could be used to approximate binomial probabilities?

d) Overlay the normal distribution on the graph you constructed in part a) of this problem. Repeat the procedure for the binomial distribution with $n = 25$, $p = .8$ and for $n = 50$ and $p = .8$. Formulate a conjecture as to when you feel the normal distribution could be used to approximate binomial probabilities.

Since constructing the graphs of the distributions is easy with the calculator programs, most students will explore further using values other than those given in the problem. Many "what if's" are heard from the groups and an understanding of concepts is apparent.

What does calculator use cost in terms of time and distraction from standard material? Benefits to students will not be realized unless the students feel comfortable using their calculators. The first few class meetings need to be spent with this idea in mind. After that, instruction in the use of the calculator should not occupy class time. It may be necessary, however, to have a graduate assistant or upper-level student familiar with the machine hold help sessions for the slower or less motivated students.

Rationale for Calculator Use

The problems that many students encounter with the mathematics inherent in a probability and statistics course may be more emotional than intellectual. This often results in inattention, poorly preparation, and avoidance -- not attending class, not doing the assigned homework, and, consequently, not learning the material. The focus on detail and accuracy in the course can be very unpleasant, frustrating and annoying to the non-technical student, thereby increasing math anxiety. However, this anxiety is quite often a question of attitude rather than aptitude.

Teachers should realize that drill and practice usually leads to an attempt at short-term memorization of the material. One of the views on how students learn mathematics is constructively, a process of making mental constructions to handle mathematical concepts. "The implications for mathematics teaching is that the emphasis should shift from ensuring that a student can correctly replicate what he or she has been shown to concentrating on helping her or him organize and modify her or his 'mental schemas'." [19] To overcome their lack of confidence and be successful, students must be given opportunities to form generalizations and develop concepts about the subject matter. Interesting and realistic problems should be used to motivate and teach, not just to reinforce concepts at the conclusion of a lecture.

Graphic interpretation aids in conceptual understanding. "Mathematics teachers have long known that a good picture can be worth much more than a thousand words. A graph not only effectively captures a wealth of information about a function or relation, it also provides a visual heuristic for interpreting that information. Calculators with interactive capabilities to display graphic images are exciting new tools for teaching and learning mathematics." [8] The *NCTM Curriculum and Evaluation Standards for School Mathematics* assume that "Scientific calculators with graphing capabilities will be available to all students at all times" [17] in grades 9 through 12.

Statements such as these reflect the changing attitude of the mathematics community toward this new technology. Graphing calculators add a significant dimension to mathematics instruction when effectively used and should help alleviate math anxiety and increase the level of student interest in the subject matter. "Technology can help students think more deeply about mathematics, facilitate generalization, empower students to solve difficult problems, and furnish concrete links between geometry and algebra, algebra and statistics, and real problem situations and associated mathematical models." [6]

Calculator-enhanced probability and statistics places more emphasis on graphing, less emphasis on computation, and offers hands-on experience with simulation. The student is actively involved while in the classroom. The lecture format of a class can become a more informal exploration of concepts by both instructor and students. Nothing is gained in the total learning process by teaching students to press keys to obtain final answers. Pedagogical programming should be used rather than programming that offers shortcuts to arrive at "quick" solutions. While this new technology can free the student in an introductory probability and statistics course from tedious and time-consuming calculations, major advantages are that more detailed explanation of concepts can be offered due to the relative ease of simulation and that drill and practice exercises can be effectively supplemented with realistic problems which encourage exploration and experimentation.

One question that naturally arises is: "Isn't a computer better in a statistics course?" When implementing the use of any technology into a course, one must consider the intent and goals of the course as well as which technology is available and affordable. Use of technology should be convenient and students should be able to use it on a regular basis. Unfortunately, most college and university classrooms are not permanently equipped with computer workstations. Consequently, students must be sent to use computers outside the classroom and cannot use the computer technology on an in-class basis. Testing is the focus of the course for the great majority of students. If we wish to convince them that it is appropriate to use technology in mathematics courses, why deny them the use of technology on that which they consider most important?

Therefore, while a graphing calculator will not replace a computer in higher-level statistical analysis, it may be more appropriate for use in an introductory course. Its portability, personalized nature and relatively inexpensive purchase price make it attractive. Students are able to do serious exploratory work on their own, witness the dynamic nature of mathematical processes, and engage in more realistic applications. Their learning becomes more active and hence more effective [22].

Another reason why the graphing calculator can be a powerful tool to assist students in learning is the bond that develops between the calculator and the student. "The calculator seems to have become an important part of the student, due to its portability and utility. The relationship is a dynamic one because the machine invites growth, discovery and modification." [1].

Where is the use of a graphing programmable calculator appropriate or inappropriate? When any of the calculator's capabilities can be utilized in achieving the objective of the lesson more simply or with enchanced understanding, then its use is appropriate. It is important that the instructor and students realize that the use of a calculator in an enhanced course should be designed for interactive learning. Calculator use should parallel hand calculation for student understanding. If the objective is to develop other skills, the calculator should play a minor role. Appropriate uses are routine computation, estimation, simulation, exploration and enhancement in problem solving. Programs can be developed to reinforce mental computation and estimation skills in a game activity. Histograms can be used to recognize patterns and predict distribution shapes using those patterns. (Programming examples involving these ideas will be given later in this paper.) Quick feedback is available from data in the form of summary statistics that can help the student develop awareness in using numbers to understand quantitative relationships.

It is most important that the student realize early in the course that the calculator is not a replacement for all pencil-and-paper skills. There should be a proper balance between calculator use and hand performance of traditional methods. Have students initially construct a histogram by hand, compute means, variances, medians, confidence intervals, statistics used in hypothesis testing, etc. at least once using pencil-and-paper. Have students help construct the basic programming steps necessary for the calculator to perform these same tasks. They soon realize that, with enough time and patience, all the computations involved could be done by hand. They also see that they tire easily and that objectives can become lost in the drudgery of the tedious pencil-and-paper computations. Time saved from the task of arithmetic computation can be spent in the development of new concepts, extensions of familiar concepts, or more detailed interpretation of familiar topics.

Which model calculator should be used? The graphing and programming capabilities of Casio's fx7000G, fx7500G, fx7700G, Hewlett-Packard's HP-28S, HP48SX, HP48S, Sharp's EL-5200, EL9200, EL9300 and Texas Instruments' TI-81 and TI-85 calculators offer many pedagogical advantages over standard scientific calculators in a course in introductory probability and statistics. In addition to the visual representations, expressions can be entered, evaluated, edited, and reevaluated while the information is available to the student with a multiple-line text display. All have good graphic displays and allow the user to overlay one graph on top of another, and all have very easy to use editing features. Programs and graphic screen contents are automatically saved when each calculator is turned off, and each of these have a "power off" feature when not in use.

Examples of Graphing Calculator Usage

Which topics can be more efficiently studied with programmable graphing calculators? The use of the calculator is more appropriate than pencil and paper in any situation where the student is asked to interpret the numerical results that are obtained, when a detailed record of intermediate work is not needed, and when a significant amount of time is involved in developing or testing some mathematical idea or pattern.

Box plots provide a useful technique for describing data, and the graphing calculator can be programmed to construct and graph the box plot, identify and graph outliers and return the five-number summary to the student [10]. (Sharp's EL9200 and EL9300 calculators will construct multiple box plots with a single keystroke). Interpretation and understanding of data characteristics are natural results when time that would otherwise be spent in calculation of numerical results is available.

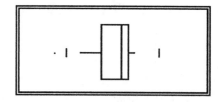

Figure 1

Another application for the calculator is histogram construction. Each model except the HP-28S has built-in graphic interfaces associated with the statistics routines. [10] To draw histograms on the Casio fx7000G, program storage must be expanded in accordance with the number of bars before data is input. Modes must manually be changed and data cannot be retrieved once the display is cleared. On the Sharp EL-5200, random data may be entered directly into the statistical matrix, but one must manually switch modes before using the histogram routine. The other Casio and Sharp models, the TI-81 and TI-85, as well as the HP-28S and HP48, can be programmed to generate random data, store it, and draw the histogram of results with a single keystroke. The Sharp models have the unique feature of the trace function giving the frequencies of the data as the y-coordinates and the class boundaries for the x-coordinates. All the other models only give pixel coordinates when the trace function is activated with the histogram on the display screen. The autoscaled barplot command on the HP48 is unique in that it plots bars of both positive and negative heights, and the autoscaled histogram plotting command plots a histogram using relative frequencies. On the HP calculators, control over the number of intervals plotted is obtained by drawing the histogram of the matrix of frequencies or the matrix of original data values. Since statistical data memory is shared with program memory on all these calculators, the amount of data that can be generated is a function of the calculator memory size.

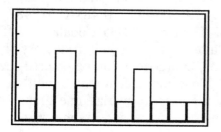 or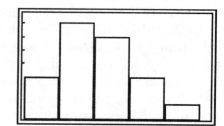

Figure 2

Students must understand the definition of class width of a frequency distribution in order to obtain a meaningful picture on the screen when using the calculator. Once the viewing window is set according to the students' specifications, the histogram can be drawn, as shown in Figure 2 [10]. Using a two-keystroke sequence on the Sharp models, the student can overlay a broken-line graph or a normal distribution graph, as shown in Figure 3.

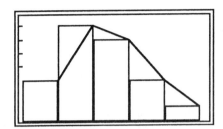   and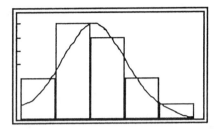

Figure 3

Another enhancement is in-class simulation. Simulation programs are both interesting and very effective when used to foreshadow many topics in probability and statistics. Simulation techniques have the advantage over actual experiments in that many repetitions can be performed quite easily with the aid of a programmable calculator. Each student's answer is unique when simulations involve the calculator's random number generator. This alone encourages discussion and finding "the answer" loses some of its importance to the math-anxious student.

Each of the calculators discussed in this paper has a built-in program for generating pseudo-random numbers on the interval [0,1]. The outcomes for the experiment of tossing one fair coin may be simulated with a keystroke sequence which will compute the integer portion of twice a random number. (Letting INT represent the keystroke that will return the integer portion of a number and RAND be the random number generator, the keystroke sequence would be INT 2RAND.) Students may calculate the percentages of zeros (heads) and ones (tails) obtained as the number of tosses increase and should collectively find the relative frequencies of each of the numbers of zeros and ones approaching the theoretical probabilities of the simple events in the sample space {TAIL, HEAD}. To simulate the experiment of the toss of one fair die, have students randomly generate the six distinct integers {1, 2, 3, 4, 5, 6} with the keystroke sequence INT 6RAND + 1. Relative frequencies of occurrences may then be computed and compared to the theoretical probabilities. In addition to observing the numerical frequencies for increasingly larger numbers of trials, a graph of the outcomes will help the student visually compare the experimental outcome to the theoretical results. [10]

An interesting simulation that can be used to aid student understanding of the Central Limit Theorem is the experiment of tossing four fair coins. [10] Suppose that four identical fair coins are tossed and the number of heads (or tails) is counted using a program written for the calculator. The function

INT 2RAND + INT 2RAND + INT 2RAND + INT 2RAND

produces the values 0, 1, 2, 3, and 4 and simulates tossing four fair coins n times with the number of heads being recorded for each of the n tosses. A pedagogical advantage is that students must utilize theoretical binomial probabilities and the concept of expected value when setting the calculator range parameter for the maximum frequency. As the number of times the coins are tossed increases and the results are graphed in the form of a histogram, students visually see the shape of the histogram approaching the shape of the histogram for the theoretical binomial probability distribution of the number of heads obtained in tossing the four coins.

Recall that the function RAND generates random numbers x such that $0 \leq x < 1$. Thus, the command INT 5RAND will generate random numbers between and including 0 and 4. Students will very likely ask "Can't the expression

INT 2RAND + INT 2RAND + INT 2RAND + INT 2RAND

be replaced by INT 5RAND and achieve the same results?" Instructors, instead of answering this question, may ask students to explore and use graphical representation of the results of both simulations to see that histograms of similar shape are *not* obtained using INT 5RAND in place of INT 2RAND + INT 2RAND + INT 2RAND + INT 2RAND when performing the experiment of tossing four fair coins. The Central Limit Theorem is a natural explanation of these results. This simulation can be performed on any of the graphics calculators.

By partitioning the set of random numbers at a value other than 0.500, one may use a general probability simulation program with user entry of the probability of success. [22, 11] The program may be used to generate the results of the trials, the success and failure ratio and, if desired, to draw a histogram of the results. These ideas may also be used to motivate the theoretical results based on simulated results.

Another application that is appropriate in an introductory statistics course is the generation of random numbers from a Poisson distribution. The programming capabilities of the calculator may be used for the generation and accumulation of values from a Poisson distribution with a user-specified mean. [24] Summary statistics for the generated data may be obtained from the statistics menu of the calculator and a histogram drawn of the results.

When the normal distribution is studied, students can experiment and explore by drawing normal distribution graphs to see what effect changing μ and σ has on the graph of the density function. Students should be asked to notice the calculator screen range settings and will discover that even though the normal density function has domain of all real numbers, practically "all" of the graph appears between $\mu - 3\sigma$ and $\mu + 3\sigma$. The program which draws the normal density function for varying values of μ and σ could easily be adapted to a "guessing game" to find μ and one of several preset values of σ.

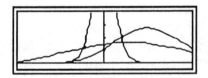

Figure 4

A natural extension would involve an investigation of the shape of the distribution of sample means. The graph of the sampling distribution of the sample mean can be overdrawn on the graph of the normal density function for a population illustrating the effect of increasing the sample size.

Programs can be written to generate probabilities for and to construct probability histograms of discrete distributions. The graph in Figure 5 shows the binomial distribution for $n = 12$ and $p = 0.45$ with a normal distribution overlay; it was obtained from such a program written for the HP48.

The formulas for single and two-sample confidence intervals and hypothesis tests can be entered into the calculator with prompts for values using, in most cases, the same symbols as in the text. When using the Hewlett-Packard calculators, formula entry is not even necessary due to the built-in upper tail probabilities for the normal, t, chi-square and F distributions.

Students may find confidence intervals and perform hypothesis tests without using any of the textbook tables!

Scatter diagrams and regression models are definite application areas for the graphing calculator. The ease of obtaining the best regression model varies slightly on the calculators discussed in this paper, but all can display the model superimposed on the scatter diagram. Many realistic problems can be done in the classroom and can be assigned for homework due to the ease of regression analy-

Figure 5

sis when graphing calculators are used. The figure below shows the scatter diagram and linear regression model for the records for the mile run from 1911 to 1985 [10]. Also shown are two "zoomed" graphs obtained from the HP48SX using linear, logarithmic, exponential, and power regression analysis on the mile-run data. The variables of interest are the year in which the record was set and the time, in seconds, of that record. Notice that students can easily see the dangers of extrapolation.

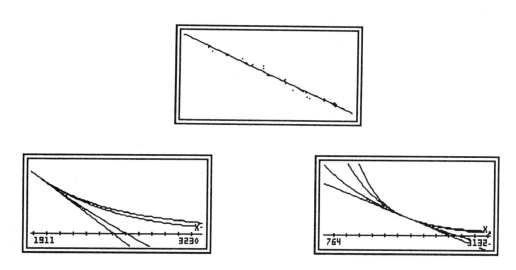

Figure 6

Conclusion

The preparation time for a "first time" calculator instructor is significant if the course is to truly enhance the teaching and learning of the subject matter. Many questions arise from student discussions and the exploratory nature of the classroom. Instructors must learn to feel comfortable saying "I'm not sure - let's try this and see what happens". The rewards in the advancement of the total learning process definitely make the time well-spent.

Can programmable graphing calculators genuinely complement conceptual understanding? Students are more enthusiastic about the course. Preliminary studies show that mathematical anxiety diminishes as students build more confidence. Graphic displays provide strong visual images representing the numeric calculations resulting from calculator programs. Definitions arise more naturally and students are more apt to read the text to explain what they see in the graph. Verbal interaction and active involvement are encouraged when the students are asked to communicate what they see. If students are actively involved, they will remember what they do much better than if they only hear an instructor talk about what they should do. Graphing calculators do add a significant dimension to mathematics instruction.

Bibliography

1. Bertram, B. and W. Francis, Student + HP-28S: A Symbiotic Relationship, *Proceedings, Second Conference on Technology in Collegiate Mathematics: Teaching and Learning with Technology*, Addison-Wesley, Reading, MA, 1990, 125-128.

2. Casio fx-7000G Owner's Manual, Casio, Inc., Dover, NJ.

3. Cheney, W. and D. Kincaid, *Numerical Mathematics and Computing*, 2nd Ed, Brooks/Cole, Monterey, CA, 1985.

4. The College Board, The Use of Calculators in the Standardized Testing of Mathematics, College Entrance Examination Board, NY, 1989.

5. Demana, F. and J. Harvey (ed.), *Proceedings, Conference on Technology in Collegiate Mathematics: The Twilight of Pencil and Paper*, Addison-Wesley, Reading, MA, 1990.

6. Demana, F. and B. Waits, The Role of Technology in Teaching Mathematics, *The Mathematics Teacher* (January, 1990), 27-31.

7. Demana, F., J. Harvey, L. Taylor and B. Waits (ed.), *Proceedings, Second Conference on Technology in Collegiate Mathematics: Teaching and Learning with Technology*, Addison-Wesley, Reading, MA, 1990.

8. Demana, F., B. Waits, T. Dick, G. Musser, J. Harvey and J. Kenelly, Graphing Calculators - Comparisons and Recommendations, *The Computing Teacher* (April 1990), **10**.

9. Demana, F., Waits, B., Foley, G. and Osborne, A. (1990), *Graphing Calculator and Computer Graphing Laboratory Manual* to accompany *Precalculus Mathematics, A Graphing Approach*, Addison-Wesley, Reading, MA, 1990.

10. Fetta, I., *Calculator Enhancement for Introductory Statistics*, Harcourt Brace Jovanovich, San Diego, CA, 1991.

11. Fetta, I., *Explorations in Algebra, Precalculus, Statistics: A Manual of Applications for the TI-81 Graphing Calculator*, Saunders College Publ., Philadelphia, PA, 1992.

12. Fetta, I. and J. Harvey, Technology is Changing Tests and Testing, *UME Trends*, **1**, (January-February 1990).

13. Fetta, I. and J. Kenelly, The Informed Consumer's Instructional Guide to Graphing Calculators, MAA Minicourse, AMS/MAA joint meeting, Louisville, January 1990.

14. Hewlett-Packard HP-28S Advanced Scientific Calculator Owner's Manual, 4th Ed, Corvallis, OR, 1988.

15. Hewlett-Packard HP-48SX Scientific Expandable Calculator Owner's Manual, Volumes 1 and 2, 3rd Ed, Corvallis, OR, 1990.

16. Mucinio, M. Buyer's Guide to Graphic Calculators, *The Mathematics Teacher*, **81** (December 1988), 705-708.

17. National Council of Teachers of Mathematics, Commission on Standards for School Mathematics, Curriculum and Evaluation Standards for School Mathematics, NCTM, Reston, VA, 1989.

18. Schultz, J. and R. Rubenstein, Integrating Statistics into a Course on Functions, *The Mathematics Teacher*, **83** (November 1990), 612-616.

19. Selden, A. and J. Selden, Constructivism in Mathematics Education: A View of How People Learn, *UME Trends*, **1** (March 1990).

20. Sharp EL-5200 Graphic Scientific Calculator Owner's Manual, Sharp Electronics Corp., Mahwah, NJ.

21. Texas Instruments TI-81 Advanced Scientific Graphics Calculator Guidebook, Texas Instruments Inc, Lubbock, TX, 1990.

22. Tucker, T. (ed.), *Priming the Calculus Pump: Innovations and Resources*, CUPM Subcommittee on Calculus Reform and the First Two Years, MAA Notes #17, Clemson University Project, D. LaTorre, Project Director, November, 1990.

23. Vonder Embse, C., Point Plotting and Probability Simulation Programs for the Casio fx-7000G Graphing Calculator, *Proceedings, Second Conference on Technology in Collegiate Mathematics: Teaching and Learning with Technology*, Addison-Wesley, Reading, MA, 1990, 67-73.

24. Wickes, W. C., HP-28 Insights, Principles and Programming of the HP-28C/S, Larken Publ., Corvallis, OR, 1988.

25. Wickes, W. C., HP48 Insights, Larken Publ., Corvallis, OR, 1991.

Iris Fetta is assistant professor of mathematical sciences at Clemson University. She has had substantial experience in undergraduate instruction and coursework development, especially calculator-enhanced algebra, precalculus, calculus and introductory statistics. She is also the author of a variety of texts incorporating the use of graphing calculators into many portions of the undergraduate curriculum.

Part IV

Resources

for

Statistical Education

Bibliography of Materials for Teaching Statistics

Thomas L. Moore and **Jeffrey A. Witmer**
Grinnell College   Oberlin College

Introduction

The following short bibliography is a somewhat personal list of valued readings in statistics and the teaching of statistics.

1. The Short List

If you can read but three articles, we suggest these:

Cobb, G. W. "Introductory Textbooks: A Framework for Evaluation."

Moore, D. S. "Should Mathematicians Teach Statistics?"

Garfield, J. and A. Ahlgren "Difficulties in Learning Basic Concepts in Probability and Statistics."

You will find the complete citations below.

2. Background readings for the mathematician new to the teaching of statistics

Box, G. E. P. (1984), "The Importance of Practice in the Development of Statistics," *Technometrics*, **26**, 1-8.

Box uses examples from the history of statistics to argue his thesis.

Cobb, G. W. (1992), "Teaching Statistics: More Data, Less Lecturing," in *Heeding the Call for Change*, Lynn Steen, editor, MAA Notes Number **22**, 3-43.

This is a report on the teaching of statistics from the MAA's Curriculum Action Project Focus Group. A preliminary version was published in *UME Trends*, Vol. **3**, No. 4.

Cobb, G. W. (1987), "Introductory Textbooks: A Framework for Evaluation," *Journal of the American Statistical Association*, **82**, 321-339.

This article gives an excellent structure for reviewing textbooks and also gives specific criticisms of 16 books.

CUPM, *Recommendation for a General Mathematical Science Program* (1981), MAA.

> This latest CUPM report calls for a data-driven statistics course for mathematics majors.

Garfield, J. and A. Ahlgren (1988), "Difficulties in Learning Basic Concepts in Probability and Statistics," *Journal for Research in Mathematics Education*, **19**, 44-63.

> The authors review the research on how students learn and give some recommendations on how we should teach.

Hoaglin, D. C. and D. S. Moore (editors) (1992), *Perspectives on Contemporary Statistics*, The Mathematical Association of America: MAA Notes Number **21**.

> Eminent statisticians provide overview articles for beginning teachers on topics central to the teaching of statistics. Topics include: data analysis, computers and modern statistics, samples and surveys, statistical design of experiments, probability, inference, diagnostics, and resistant and robust procedures.

Iversen, G. R. (1985), "Statistics in Liberal Arts Education," *The American Statistician*, **39**, 17-19.

> This provides a philosophical argument for the fundamental place of statistics within the liberal arts.

Kempthorne, O. (1980), "The Teaching of Statistics: Content Versus Form," *The American Statistician*, 34, 17--21.

> Kempthorne gives philosophical underpinnings on how to teach statistics.

Misra, S. C., H. Sahai, A. P. Gore, and J. K. Garrett (1987), "A Bibliography on the Teaching of Probability and Statistics," *The American Statistician*, **41**, 284-310.

> This is a very comprehensive bibligraphy.

Moore, D. S. (1988), "Should Mathematicians Teach Statistics?" (with discussion), *The College Mathematics Journal*, **19**, 3-35.

> Moore's first sentence is "No!" He goes on to discuss how statistics differs from mathematics and how the difference should affect the teaching of statistics.

Moore, T. L. and J. A. Witmer (1991), "Statistics Within Departments of Mathematics at Liberal Arts Colleges," *The American Mathematical Monthly*, **98**, 431-436.

> This article describes how statistics and mathematics differ in fundamental ways and discusses the practical implications these differences have for departments of mathematics at liberal arts colleges.

Moore, T. L. and R. A. Roberts (1989), "Statistics At Liberal Arts Colleges," *The American Statistician*, **43**, 80-85.

This article summarizes the role both statistics and a statistician can play at a liberal arts college. It includes results of a survey on liberal arts college curriculum and staffing in statistics.

Peters, W. S. (1987), *Counting for Something: Statistical Principles and Personalities*, Springer-Verlag, New York.

This charming and casual introduction to elementary statistical concepts gives the reader an excellent sense of the history of the discipline.

Singer, J. D., and J. B. Willett (1990), "Improving the Teaching of Applied Statistics: Putting the Data Back Into Data Analysis," *The American Statistician*, **44**, 223-230.

The authors argue for the use of real data sets for teaching statistics, describe the practicalities of doing so, and include a long bibliography of sources of usable, published data sets. [Note: An updated and expanded version of this article appears elsewhere in this volume as well as an expanded list of the sources.]

3. Statistical Misuses

Each of the following books contains many interesting and entertaining examples of statistical misuse in everyday life.

Campbell, S. (1974), *Flaws and Fallacies in Statistical Reasoning*, Englewood Cliffs, NJ: Prentice Hall.

Hooke, R. E. (1983), *How to Tell the Liars From the Statisticians*, New York: Marcel Dekker.

Huff, D. (1954), *How to Lie With Statistics*, New York: W. W. Norton.

4. Statistics in the World

The following present statistics in "real-world" settings.

Gould, S. J. (1981), *The Mismeasure of Man*, New York: W. W. Norton.

In this book, written for a general audience, Gould discusses the social impact of statistical abuse, devoting much of the work to IQ tests. Summary data are included, as well as an elementary presentation of factor analysis.

Mosteller, F., and W. Fairly (1977), *Statistics and Public Policy*, Reading, MA: Addison-Wesley.

This is an anthology of articles dealing with real applications of statistical reasoning.

Tanur, J. M., F. Mosteller, W. H. Kruskal, E. L. Lehman, R. F. Link, R. S. Pieters, and G. R. Rising (1989), *Statistics: A Guide to the Unknown* (3rd ed.), San Francisco: Holden-Day.

> This volume is a collection of accessible short articles on applications of probability and statistics in a wide variety of settings. You might look at previous editions of this book for good articles no longer present in the recent edition.

5. Classroom Capsules

The following is a list of sources of short articles that contain ideas for use in the classroom.

McKean, K. (1985), "Decisions, Decisions," *Discover*, 6, 22--31.

> This article, which is full of fascinating examples, highlights the work of Daniel Kahneman and Amos Tversky, who have studied common (mis)perceptions of probability and human judgement.

Proceedings of the Section on Statistical Education, American Statistical Association.

> This yearly volume contains papers presented at the annual ASA meeting, many of which are directed at undergraduate education.

Chance Magazine, New York: Springer-Verlag.

> This quarterly contains many wonderful applications of probability and statistics written in a lively style.

Stats, American Statistical Association.

> This is ASA's quarterly "magazine for students of statistics." It contains short expository articles of applied statistics as well as articles on career opportunities and graduate programs.

Teaching Statistics

> This journal is full of ideas for teaching statistics to students aged 9-19. It is edited at the University of Sheffield in England.

6. Textbooks

The following are textbooks with a difference. They tend to contain excellent examples, many of which include real data sets.

Bishop, Y. M. M., S. E. Fienberg, and P. W. Holland (1975), *Discrete Multivariate Analysis: Theory and Practice*, Cambridge, MA: MIT Press.

> The authors analyze several real (and interesting) sets of data, which are summarized in an index to data sets. (We wish more books had such indexes.)

Box, G. E. P., W. G. Hunter, and J. S. Hunter (1978), *Statistics for Experimenters*, New York: John Wiley.

This book is slanted toward engineers, chemists, etcetera.

Cox, D. R., and E. J. Snell (1981), *Applied Statistics: Principles and Examples*, London: Chapman & Hall.

Devore, J., and R. Peck (1986), *Statistics: The Exploration and Analysis of Data*, St. Paul: West.

This introductory textbook is loaded with real data from all areas of application.

Ehrenberg, A. S. C. (1978), *Data Reduction*, New York: John Wiley.

Ehrenberg, A. S. C. (1982), *A Primer in Data Reduction*, New York: John Wiley.

The two books by Ehrenberg explore methods for handling and interpreting data. Warning: Ehrenberg downplays the use of graphics.

Freedman, D., R. Pisani, R. Purves, and A. Adhikari (1991), *Statistics* (2nd ed.), New York: W. W. Norton.

This highly nonstandard elementary textbook is devoid of equations but loaded with examples. Many instructors have told us that they learned a great deal when teaching from this book.

Koopmans, L. (1987), *Introduction to Contemporary Statistical Methods* (2nd ed.), Boston: Duxbury.

Koopmans was among the first authors to thoroughly integrate the modern topics of exploratory data analysis, graphics, and robust methods into a coherent textbook.

Landwehr, J. M. and A. E. Watkins (1986), *Exploring Data*, Palo Alto: Dale Seymour Publications.

Even though it is aimed at a secondary school audience, this book contains many useful, concrete, and real examples of data analysis. There is also a *Teacher's Edition*.

Larsen, R. J., and M. L. Marx (1986), *An Introduction to Mathematical Statistics and Its Applications* (2nd ed.), Englewood Cliffs, NJ: Prentice Hall.

This textbook provides background information for each of the many data sets included. Caution: Many of the data sets in this book may not satisfy the assumptions of the methods they purport to illustrate.

Millekin, G. A., and D. E. Johnson (1984), *Analysis of Messy Data*, Volume 1: *Designed Experiments*, Belmont, CA: Lifetime Learning Publications.

Moore, D. (1991), *Statistics: Concepts and Controversies* (3rd ed.), San Francisco: W. H. Freeman.

This book is written for humanities and social science students.

Moore, David S. and George P. McCabe (1989), *Introduction to the Practice of Statistics*, New York: W. H. Freeman.

This textbook features real data, statistical graphics, and a concern for model assumptions and robustness.

Olkin, I., L. J. Gleser, and C. Derman (1978), *Probability Models and Applications*, New York: Macmillan.

This is an excellent source of real data to use in a probability course.

Rice, J. A., (1988), *Mathematical Statistics and Data Analysis*, Belmont, CA: Wadsworth.

This introductory textbook aimed at advanced undergraduates provides that audience with a true appreciation of the range of statistics and features the use of real data, exploratory data analysis and graphics, and a concern for model assumptions.

B. F. Ryan, B. L. Joiner, and T. A. Ryan, Jr. (1985), *Minitab Handbook* (2nd ed.), Boston: PWS/Kent.

The Handbook is an excellent companion textbook to the highly popular statistical package Minitab. The book discusses the several larger data sets that come with the package and includes many good exercises about them.

Tufte, E. R. (1970), *Quantitative Analysis of Social Problems*, Reading, MA: Addison-Wesley.

Tufte gives an excellent introductory exposition in applying statistics (especially regression) in the social sciences.

Weisberg, S. (1985), *Applied Linear Regression* (2nd ed.), New York: John Wiley.

This book has nice data sets for regression.

Witmer, J. A. (1992), *Data Analysis*, Englewood Cliffs: Prentice Hall.

7. Statistical Graphics

The following discuss statistical graphics.

Cleveland, W. S. (1985), *The Elements of Graphing Data*, Monterey, CA: Wadsworth.

This is a practical guide to graphical presentation with many references to the scientific literature on perception.

Tufte, E. R. (1983), *The Visual Display of Quantitative Information*, Cheshire, CT: Graphics Press.

Examples of what to do and what not to do.

Wainer, H. (1984), "How to Display Data Badly," *The American Statistician*, **38**, 137-147.

The title says it all. Wainer corrects many of the mistakes that he presents.

8. Student Projects

The following authors discuss their experiences with having students carry out and analyze their own experiments.

Hunter, W. (1977), "Some Ideas About Teaching Design of Experiments, With 2^5 Examples of Experiments Conducted by Students," *The American Statistician*, **31**, 12-17.

Tanner, M. (1985), "The Use of Investigations in the Introductory Statistics Course," *The American Statistician*, **39**, 306-310.

9. Miscellaneous Articles and Books

"Statistical Computing Section" of *The American Statistician*.

This section appears in each issue of *The American Statistician*. Besides general articles related to the teaching or practice of statistical computing it includes the two subsections: "New Developments in Statistical Computing" and "Statistical Computing Software Reviews".

Andrews, D. F., and A. M. Herzberg (1985), *Data*, New York: Springer-Verlag.

A collection of interesting sets of data taken from a variety of fields.

Chatillon, G. (1984), "The Balloon Rule for a Rough Estimate of the Correlation Coefficient," *The American Statistician*, **38**, 58-60.

This article helps the reader develop a feeling for correlation.

Colton, T. (1974), *Statistics in Medicine*, Boston: Little, Brown.

An introductory book based on real data. Colton's final chapter is an excellent discussion of critical reading of the scientific literature.

Stigler, S.M. (1986), *The History of Statistics: The Measurement of Uncertainty Before 1990*, Cambridge, MA: Harvard.

Stigler provides serious background in the history of statistics. It makes for a good seminar book for the statistics enthusiasts at an institution.

Walton, M. (1987), *The Deming Management Method*, New York: Putnam.

Walton gives a non-technical exposition on the life and methods of W. Edwards Deming, the statistician who became a chief architect of the post-war Japanese economic revolution.

Thomas Moore teaches statistics in the mathematics department at Grinnell College. He is co-founder of the Statistics in the Liberal Arts Workshop (SLAW) and has been active in the statistics education committees of MAA and ASA. He has also spent a year's leave as an applied statistician at the Eastman Kodak Company.

Jeff Witmer is associate professor of mathematics at Oberlin College. He has written research papers in the area of Bayesian decision theory and is the author of a book, *Data Analysis: An Introduction*. He has been active in the ASA's Quantitative Literacy Project for several years.

CHANCE: Case Studies of Current Chance Issues

J. Laurie Snell
Dartmouth College

Introduction

CHANCE is a new course being developed by faculty from six colleges -- Dartmouth, Grinnell, King, Middlebury, Princeton, and Spelman. The CHANCE course treats issues that are currently reported in the news: statistical problems related to AIDS, the effects of lowering serum cholesterol on heart attacks, the use of DNA fingerprinting in the courts, maintaining quality of manufactured goods in the face of variation, informed patient decision making, reliability of political polls, and so forth. Students study the newspaper accounts of these topics, articles in general science journals, such as *Science, Nature, New England Journal of Medicine* and *Scientific American*, and finally original research papers. The necessary probability and statistics are discussed in the context of the issue under consideration. Students use statistical packages to analyze relevant data.

Under an NSF grant, the CHANCE group will develop materials -- summary modules, student projects, classroom experiences, data sets, and bibliographies based on the various CHANCE courses, and will disseminate these materials to both college and pre-college teachers. To this end, a CHANCE electronic data base will be established and maintained at Dartmouth College. We provide here a bibliography of some of the material used in CHANCE courses during the year 1991-2 that may be useful to teachers of introductory statistics courses.

For further information about the project, interested readers should contact the author.

CHANCE Bibliography of Newspaper and Popular Science Articles

Maintaining quality of manufactured goods in the face of variation

W. Edwards Deming, "Making things right", in *Statistics: A guide to the unknown*, (second edition), Judith Tanur et al, Holden-Day, 1978.

Mary Walton, *The Deming management method*, Putnam, 1987.

Alan Goldstein, "Guru who preached quality finally gets to say 'I told you so.'", *The Los Angeles Times*, 31 March 1987.

Nancy R. Mann, "Why it happened in Japan and not in the U.S.", *Chance*, **1** No 3, 1988.

Lloyd Dobyns, "Ed Deming wants big changes, and he wants them fast", *The Smithsonian,* 21, August 1990, p 74.

David Mehegan, "Quality management to rival Japan's", *The Boston Globe*, 1 November, 1990.

Michael Schrage, "If statistics are the key to quality, our students need some chance encounters", *The Washington Post*, 15 March 1991.

Daniel J. Boorstin and Gerald Parshall, "History's hidden turning points; the true watersheds in human affairs are seldom spotted quickly amid the tumult of headlines broadcast on the hour", *U.S. News & World Report*, **110**, 22 April 1991, p 52.

Dyan Machan, "Eager pupils", *Forbes*, 16 September 1991.

K. O. Browman, T. H. Hopp, R. N. Kacker and R. J. Lundegard, "Statistical quality control technology in Japan", *Chance*, **4** No 3, 1991.

Scoring streaks and records in sports

Julian L. Simon, "'Batter's slump and other illusions", *The Washington Post*, 9 August 1987.

James Gleick, "The cold facts about the 'hot hands' phenomenon: a myth?" *The New York Times*, 19 April, 1988.

Amos Tversky and Thomas Gilovich, "'Hot hands' in basketball", *Chance*, **2** No 1, 1989.

Stephen Jay Gould, "The streak of streaks", *Chance*, **2** No 2, 1989.

"Basketball, baseball, and the null hypothesis", *Chance*, **2** No 4, 1989.

Patrick D. Larkey, Richard A. Smith and Joseph B. Kadane, "It's okay to believe in the 'hot hand'", *Chance*, **2** No 4, 1989.

Amos Tversky and Thomas Gilovich, "The 'hot hand': statistical reality or cognitive illusion", *Chance*, **2** No 4, 1989.

Michael Specter, "Record hot readings in 1980's boost global-warming theory", *The Washington Post*, 13 January 1990.

Barbara McGarry Peters, "How to balance reason and intuition", *The Washington Post*, 23 October, 1990.

Richard A. Kerr, "Global temperature hits record again", *Science*, **251**, 18 January 1991.

Ronald N. Forthofer, "Streak shooter -- the sequel", *Chance*, **4** No 2, 1991.

Kristine Mary Zaleskas, *Where have you gone, Joe DiMaggio?*, Senior Thesis, Harvard University, 1991.

"What do economists know about the stock market", *The Journal of Portfolio Management*, Winter, 1991.

Alfie Kohn, "Folk Wisdom is all wet", *The San Francisco Chronicle*, 15 March 1992.

Cholesterol and heart disease

Joel B. Greenhouse and Samuel W. Greenhouse, "An Aspirin a Day ... ?", *Chance*, **1** No 4, 1988.

Gina Kolata, "Study offers first evidence that cutting cholesterol levels saves lives", *The New York Times*, 4 October 1990.

Bea Lewis, "No magic bullet against cholesterol", *The Los Angeles Times*, 21 February 1991.

Jane E. Brody, "The two faces of niacin: a substance that lowers cholesterol levels, but not without side effects", *The New York Times*, 7 March 1991.

Jane E. Brody, "In cholesterol struggle, a giant among winners", *The New York Times*, 28 March 1991.

"Cholesterol debate flares anew", *The Los Angeles Times*, 1 April 1991.

Malcolm Gladwell, "New studies: oat bran can cut cholesterol; findings counter earlier report", *The Washington Post*, 10 April 1991.

"Serum cholesterol concentration and coronary heart disease in population with low cholesterol concentrations", *The Journal of the American Medical Association*, **267**, 1 January 1992, p 34.

George Davey Smith and Juha Pekkanen, "Should there be a moratorium on the use of cholesterol lowering drugs?", *British Medical Journal*, **304**, 15 February 1992, p 431.

Gail Vines, "Trials 'a must' for disputed cholesterol drugs", *New Scientist*, **22** February 1992, p 11.

Padus Emrika, "The cholesterol plunge!", *Prevention*, **44**, February 1992, p 34.

C. G. Isles et al, "Relation between coronary risk and coronary mortality in women of the Renfrew and Paisley survey: comparison with men", *Lancet*, **339**, 21 March 1992, p 702.

H. Engelberg, "Low serum cholesterol and suicide", *Lancet*, **339**, 21 March 1992, p 727.

G. R. Cooper et al, "Blood lipid measurements. Variations and practical utility.", *The Journal of the American Medical Association*, **267**, 25 March 1992, p 1652.

Greg Gutfeld et al, "Leapin' lipids: ratio may predict heart attacks best", *Prevention*, **44**, March 1992, p 14.

Debra Wise, "A guide to smart drinking", *Mademoiselle*, **98**, April 1992, p 112.

Statistics and the law

M. Bar-Hillel, Probablistic analysis in legal factfinding, *Acta Psycologica*, **55**, 1984, p 91.

Stephen E. Fienberg, Samuel Krislov and Miron L. Straf, "Statistics, expert witnesses, and the courts", *Chance*, **1** No 4, 1988.

The Evolving Role of Statistical Assessments as Evidence in Courts, Steven E. Fienberg editor, New York: Spinger-Verlag, 1989.

Michael O. Finkelstein and Bruce Levin, *Statistics for Lawyers*, Spinger-Verlag, 1990.

Air safety

The New York Times, every day.

"It's safer, better to fly". *Consumers' Research Magazine*, **72**, March 1989, p 25.

Arnold Barnett, "Air Safety: End of the gold age?" *Chance*, **3** No 2, 1990.

"Is flying safe?", *The Washington Times*, 8 March 1992.

Michael Moss and Karen Freifeld, "Where to sit? When to fly?", *Newsday*, 5 April 1992.

Environmental risks

"Study finds link between cancer and power lines", *New York Times*, 30 November 1989.

M. Granger Morgan, "Electric and magnetic fields from 60 Hz electric power: possible health risks?" *Chance*, **2** No 4, 1989.

Peter H. Lewis, "Of magnetism and monitors", *New York Times*, 7 August 1990.

"EPA stands firm on electrical report". *Los Angeles Times*, 14 January 1991.

John C. Bailar III, "Scientific inferences and environmental health problems", *Chance*, **4** No 2, 1991.

Andy Coghlan, "Are power lines bad for you?", *New Scientist*, 11 April 1992, p 22.

The undercount problem in the 1990 U.S. Census

Peter A. Bownpane and Thomas A. Jones, "Automation of the 1990 U.S". *Chance*, **1** No 1, 1988.

Edwin D. Goldfied, "What we expect to learn from the 1990 decennial census", *Chance*, **3** No 2, 1990.

James Gleick," Why we can't count", *New York Times Magazine*, 15 July 1990.

Felicity Barringer, "Head of census defends urban tallies", *New York Times*, 12 September 1990.

David A. Freedman, "Adjusting the 1990 census", *Science*, **252**, 31 May 1991.

Stephen E. Fienberg, "An adjusted census in 1990? Commerce says 'no'", *Chance*, **4** No 3, 1991.

Felicity Barringer, "Commerce Dept. declines to revise '90 census counts; 5 million people missed; critics say administration's ruling will hurt minority groups and big cities." *New York Times*, 16 July 1991.

Felicia R Lee, "Already struggling, big cities are hit hard", *New York Times*, 16 July 1991.

Peter Skerry, "Sense on the census". *New York Times*, 18 July 1991.

"Bush officials to give census tapes to Congress", *The New York Times*, 10 January 1992.

Dick Dahl, "In the realm of the census", *Massachusetts Lawyers Weekly*, 9 March 1992.

William Grady, "Court says it can't fix census problems", *Chicago Tribune*, 17 March 1992.

"Census", *United States Law Week*, 31 March 1992.

Rhonda Richards, "Study: many women in business uncounted", *USA Today*, 31 March 1992.

Extraterrestrial communication

William J. Broad, "Hunt for aliens in space:the next generation", *New York Times*, 6 February 1990.

Arnold Barnett, "Chatting with extraterrestrials -- what are the odds?" *Chance* 1, **3** No 2, 1990.

John Gribbin, "Is anyone out there?", *New Scientist*, 25 May 1991, p 29.

The use of DNA fingerprinting in the courts

Alec J. Jeffreys et al. "Individual-specific 'Fingerprints' of Human DNA", *Nature*, **316**, 4 July 1985.

Nakamura Yusuke et al, "Variable Number of Tandem Repeat (VNTR) Markers for Human Gene Mapping", *Science*, **235**, 27 March 1987.

James D. Watson, N. Hopkins, J. Roberts, J. Steitz and A. Weiner, *Molecular Biology of the Gene,* Menlo Park, California: The Benjamin/Cummings Publishing Company, Inc., 1987.

Eric S. Lander, "DNA fingerprinting on trial", *Nature*, **339**, 15 June 1989.

Gerald J. Sheindlin, (Judge), *New York vs. Castro*, 143 Misc. 2d 276; 540 N.Y.S. 2d 143; 1989 N.Y. Misc. LEXIS 214.

Gerald J. Sheindlin, (Judge), *New York vs. Castro*, 144 Misc. 2d 956; 545 N.Y.S. 2d 985; 1989 N.Y. Misc. LEXIS 548.

Jack Ballantyne, George Sensabaugh and Jan Witkowski, eds., Banbury Report 32: *DNA Technology and Forensic Science*, United States: Cold Spring Harbor Laboratory Press, 1989.

Franklin S. Billings, Jr. (Chief US District Judge), *USA vs. Jakobetz*, Crim. No. 89-65, 747 F. Supp. 250; 1990 U.S. Dist. LEXIS 12714.

Gina Kolata, "Some scientists doubt the value of 'genetic fingerprint' evidence", *New York Times*, 29 January 1990.

Thompson Ford, W. C., "Is DNA Fingerprinting Ready for the Courts?", *New Scientist*, 31 March 1990, p 38.

U.S. Congress, Office of Technology Assessment, *Genetic Witness: Forensic Uses of DNA Tests*, OTA-BA-438 Washington D.C: U.S. Government Printing Office, July 1990.

Christopher Joyce, "High profile: DNA in court again", *New Science*, 21 July 1990, p 24.

James G. Carr (US Magistrate), *USA vs. Yee*, Docket No. 3:89 CR0720, 1990 U.S. Dist. LEXIS 15908.

Donald A. Berry, "DNA fingerprinting: what does it prove?", *Chance*, **3** No 3, 1990.

Joel E. Cohen, "DNA fingerprinting: what really are the odds?" *Chance*, **3** No 3, 1990.

Donald A. Berry, "More on DNA fingerprinting", *Chance*, **3** No 4, 1990.

Lorne T. Kirby, *DNA Fingerprinting: An Introduction*, New York: Stockton Press, 1990.

Gina Kolata, "Gene test barred as proof in court; DNA fingerprinting requires more scientific scrutiny, Arizona judge rules", *The New York Times*, 14 February 1991.

B. Devlin, Neil Risch and Kathryn Roeder, "Estimation of Allele Frequencies for VNTR Loci", *American Journal of Human Genetics*, 1991, 48.

Donald A. Berry, "Inferences using DNA profiling in forensic identification and paternity cases", *Statistical Science*, **6** No 2, 1991.

Ranajit Chakraborty and Kenneth K. Kidd, "The utility of DNA typing in forensic work", *Science*, **254**, 20 December 1991.

R. C. Lewontin and Daniel L. Hartl, "Population genetics in forensic DNA typing", *Science*, **254**, 20 December 1991.

Terry Burke, Dolf Gaudenz, Alec J. Jeffreys and Roger Wolff, eds., "DNA Fingerprinting: Approaches and Applications", Berlin: Birkhäuser Verlag, 1991.

Warren E. Leary, "Genetic record to be kept on members of military", *The New York Times*, 12 January 1992.

Rorie Sherman, "DNA evidence dispute escalates", *The National Law Journal*, 20 January 1992.

Joseph Palca, "The case of the Florida dentist", *Science*, **255**, 24 January 1992.

Lewis Cope, "Genetic clues; (DNA fingerprinting use, potential for abuse, raise issues)", *Star Tribune*, 8 February 1992.

Rorie Sherman, "DNA is on trial yet again", *The National Law Journal*, 16 March 1992.

DNA Technology in forensic science, report of Committee on DNA Technology in Forensic Science prepared for the National Research Council, National Academy Press, Washington, D. C. 1992.

American Association for the Advancement of Science, *The Genome, Ethics and the Law: Issues in Genetic Testing*, Berkeley Springs, WV: AAAS, 1992.

John T. Elfvin (Judge), *Jenkins vs. Scully*, Docket N. CIV-91-298E, 1992 U.S. Dist. LEXIS 1907.

Gina Kolata, "US panel seeking restriction on use of DNA in courts", *The New York Times*, 14 April 1992.

Gina Kolata, "Chief Says Panel Backs Courts' Use of A Genetic Test (Correction Appended)", *The New York Times*, 15 April 1992.

"FBI, Critics Laud Points in Study of DNA Tests", *The Atlanta Journal and Constitution*, 15 April 1992.

Gina Kolata, "DNA Fingerprinting: Built-In Conflict", *The New York Times*, 17 April 1992.

"A Science Panel Backs DNA Tests", *Chicago Tribune*, 10 May 1992.

Charles Seabrook, "DNA links dentist to AIDS in 5 patients, study shows", *The Atlanta Journal and Constitution*, 22 May 1992.

B. Devlin, Neil Risch and Kathryn Roeder, "Forensic Inference From DNA Fingerprints", *Journal of the American Statistical Association*, June 1992, **87** No 418.

Earthquake prediction

Richard L. Williams, "Science tries to break new ground in predicting great earthquakes, *Smithsonian*, **14**, July 1983, p 41.

Robert L. Ketter, "Eastquake: countdown to catastrophe", *Washington Post*, 4 December 1988.

Steve Fishman, "Questions for the cosmos", *New York Times Magazine*, 26 November 1989.

Paul A. Rydelek, "California aftershock models uncertainities", *Science*, 19 January 1990.

Sandra Blakeslee, "Quake theory attacks prevailing wisdom on how faults slip and slide", *The New York Times*, 14 April 1992.

Randomized clinical trials in assessing risk

Steven Budiansky, "Playing roulette with experimental drugs", *U.S. News & World Report*, **103**, 13 July 1987, p 58.

John A. Swets, "Measuring the accuracy of diagnostic systems", *Science*, **240**, 3 June 1988.

Paul Meier, "The experimental evaluation of relative risk", *Chance*, **3** No 4, 1990.

"Cholesterol drugs may cause other problems, study finds", *The Toronto Star*, 20 February 1992.

The role of statistics in the study of AIDS

Joseph L. Gastwirth, "The statistical precision of medical screening procedures: application to polygraph and AIDS antibodies test data", *Statistical Science*, **2**, 1987.

"Random Testing for AIDS?" Opinion column in *Chance*, **1** No 1, 1988.

H. D. Gayle et al, "Prevalence of the human immunodeficiency virus among university students", *The New England Journal of Medicine*, 1990.

N. E. MacDonald et al, "High-risk STD/HIV behavior among college students", *The Journal of the American Medical Association*, 1990.

Ron Bookmeyer and Mitchell H.Gail, "A statisical history of the AIDS Epidemic", *Chance*, **3** No 4, 1990.

Joseph Palca, "The sobering geography of AIDS". *Science*, **252**, 19 April 1991.

Janet Elder, "Many favor wider testing for AIDS virus, poll finds". *New York Times*, 19 May 1991.

David Bloom and Sherry Glied, "Benefits and costs of HIV testing". *Science*, **252**, 28 June 1991.

Ron Bookmeyer, "Reconstruction and future trends of the AIDS epidemic in the United States". *Science*, **253**, 5 July 1991.

David E. Rogers and Bruce G. Gellin, "AIDS and doctors: the real dangers". *New York Times*, 16 July 1991.

Lawrence K. Altman, "An AIDS puzzle: what went wrong in dentist's office?" *New York Times*, 30 July 1991.

Mireya Navarro, "Yale study reports clean needle project reduces AIDS cases". *New York Times*, 1 August 1991.

P. Yam, "Has AIDS peaked? Quastions persist in modeling the deadly, a typical infection", *Scientific American*, **256**, September 1991, p 30.

P. Aldhous, "AIDS treatment. Confusion over therapy", *Nature*, **355**, 9 January 1992, p 102.

The use of graphics in statistics

Gina Kolata, "Computer graphics comes to statistics", *Science*, **217**, 3 September, 1982.

Edward R. Tufte, *The Visual Display of Quantitative Information*, Graphics Press, Cheshire, Connecticut, 1983.

Howard Wainer and David Thissen, "Plotting in the modern world: statistical packages and good graphics", *Chance*, **1** No 1, 1988.

Forrest W. Young, "Visualizing six-dimensional structure with dynamic statistical graphics", *Chance*, **2** No 1, 1989.

Edward R. Tufte, *Envisioning Information*, Graphics Press, Cheshire, Connecticut, 1990.

Paradoxes in probability and statistics

Gabor Szekely and J. D. Reidel, *Paradoxes in probability theory and mathematical statistics*, Dordrecht-Boston, Mass., 1986.

John Allen Paulos, *Innumeracy: Mathematical illiteracy and its consequences*, Hill and Wang, New York, 1988.

Julie Baumgold, "In the kingdom of the brain", *New York Magazine*, 6 February 1989.

Ask Marilyn, Marilyn Vos Savant, *Boston Globe*, Parade magazine, 9 September, 199;, 2 December, 1990; 17 February, 1990.

Ask Marilyn, Marilyn Vos Savant, *Boston Globe*, Parade magazine, 26 January 1991.

John Tierney, "Behind Monty Hall's doors: puzzle, debate and answer?" *New York Times*, 21 July 1991.

J. P. Morgan, N. R. Chaganty, R. C. Dahiya, and M. J. Doviak, "Let's make a deal: The players dilemma", *The American Statistician*, **45** No 4, November 1991, p 284.

Marilyn Vos Savant, Letters To The Editor, *The American Statistician*, **45** No 4, November 1991, p 347.

Eduardo Engel and Achilles Venetoulias, "Monty Hall's probability puzzle", *Chance*, **4** No 2, 1991.

Ask Marilyn, Marilyn Vos Savant, *Boston Globe*, Parade magazine, 5 January 1992.

Work of Kahneman and Tversky on Fallacies in human statistical reasoning

Kevin McKean, Decisions, *Discover*, **6**, June 1985, p 22.

Don Colburn, "You bet your life; weighing the risks in an age of uncertainly", *Washington Post*, 21 May 1986.

John J. Curran, "Why investors make the wrong choices". (Special Issue, The 1987 Investor's Guide), *Fortune*, **14**, 1986, p 63.

John Rubin, "Weighing anchors", *Omni*, **12**, June 1990, p 20.

"Probability blindness, neither rational nor capricious", Massimo Piattelli-Palmarini, *Bostonia*, April 1991.

The stock market and the random walk hypothesis

"The irrational stock market", *Scientific American*, **256**, March 1987, p 60.

Dennis Kneale, "Market chaos: scientists seek pattern in stock prices", *Wall Street Journal*, 17 November 1987.

John R. Dorfman, "Pros gamely go up against 'dartboard'", *Wall Street Journal*, 4 October 1988.

Robert Savit, "Chaos on the trading floor", *New Scientist*, 11 August 1990, p 48.

Burton G. Malkiel, *A random walk down Wall Street*, 5th edition, Norton N.Y. 1990.

Michael Schrage, "Trying to probably predict what might maybe happen", *The Washington Post*, 3 January 1992.

Gary Weiss and David Greising, "Poof! Wall Street's sorcerers lose their magic", *Business Week*, 27 January 1992.

Cynthia Mitchell, "Morehouse prof takes 3rd place in stock contest", *The Atlanta Journal and Constitution*, 6 March 1992.

Herb Greenberg, "Business insider", *The San Francisco Chronicle*, 31 March 1992.

Roger Lewin, "Making maths make money", *New Scientist*, 11 April 1992, p 31.

Demographic variations in recommended medical treatments

Ronald Kotulak, "Surgical roulette", *Reader's Digest*, **126**, March 1985, p 19.

Joe Davidson, "Research mystery: use of surgery, hospitals varies greatly by region", *Wall Street Journal*, 5 March 1986.

James Barron, "Unnecessary surgery", *New York Times Magazine*, 16 April 1989.

Robert S. Boyd, "Much of medicine remains guesswork", *San Jose Mercury News*, 18 February 1990.

Informed patient decision making

Stephen Pauker and Jerome P. Kassirer, "Decision analysis", *The New England Journal of Medicine*, **316**, 29 January 1987.

Michael J. Barry et al, "Watchful waiting vs. immediate transurethral resection for symnptoma- ticprostatism; the importance of patients' preferences", *The Journal of the American Medical Association*, **259**, 27 May 1988.

Edward Faltermayer, "Medical care's next revolution", *Fortune*, 10 October 1988, p 126.

Elisabeth Rosenthal, "2d Opinions: new world for patients", *New York Times*, 5 June 1991.

Nicholas Regush, "MD's video lets patient choose treatment; computerized system outlines risks and possible side-effects", *The Gazette (Montreal)*, 27 February 1992.

Coincidences

Persi Diaconis, "Statistical problems in ESP research", *Science*, **201**, 14 July 1978.

Michael Guillen, "Life as a lottery: wherein we demonstrate that baffling coincidences areas common as cornflakes", *Psychology Today*, **17**, 1983.

James S. Trefil, "Odds are against your breaking that law of averages", *Smithsonian*, **15**, September 1984, p 66.

Rudy Rucker and David Povilaitus, "Puzzles in thoughtland: the powers of coincidence; do hidden forces link seemingly synchronous events?", *Science* 85, **6**. Jan-Feb 1985.

Persi Diaconis and Frederick Mosteller, "Methods for studying coincidences", *Journal of the American Statistical Association*, **84** No 408, 1989.

Gina Kolata, "1-in-a-trillion coincidence, you say? not really", *New York Times*, 27 February 1990.

David Holzman, "Coincidence isn't such a long shot after all", *The Washington Times*, 4 April, 1991.

Random and pseudo-random sequences

Gina Kolata, "What does it mean to be random?" *Science*, **1068**, 7 March 1986.

Kevin McKean, "The orderly pursuit of pure disorder", *Discover*, **8**, January 1987, p 72.

James Gleick, "The quest for true randomness finally appears successful: advance could meet important needs in science and industry.", *New York Times*, 19 April 1988.

Richard Preston, "The mountains of Pi", *The New Yorker*, 2 March 1992, p 36.

The reliability of political polls

Philip E. Converse and Michael W. Traugott, "Assessing the accuracy of polls and surveys", *Science*, **234**, 28 November 1986.

Everett C. Ladd, "Polls: they're useful tool but take with a grain of salt", *Christian Science Monitor*, 1 November 1988.

E. J. Dionne, Jr, "In the confusing land of electoral polls, which numbers count the most?", *The New York Times*, 17 August 1988.

Erik Larson, "Watching americans watch T.V.", *The Atlantic Monthly*, March 1991, p 66.

David Holmstrom, "What political polls show can be more shadow than substance", *The Christian Science Monitor*, 9 April 1992.

David W. Moore, *The Superpollsters: How they Measure and Manipulate Public Opinion*, Four Walls Eight Windows, 1992.

Card shuffling, lotteries, and other gambling issues

Darrell Huff, "Mathematics of sex, gambling, and insurance", *Harper's Magazine*, **219**, September 1959.

Martin Gardner, "Mathematical games: on the theory of probability and the practice of gambling", *Scientific American*, **205**, December 1961, p 150.

Gina Kolata, "Perfect shuffles and their relation to math", *Science*, **216**, 30 April 1982.

Gina Kolata, "In shuffling cards, 7 is winning number; computers confirm gamblers' intuition on how decks mix.", *New York Times*, 9 January 1990.

John Rennie, "Who's the dealer? What controls gene shuffling in the immune system?", *Scientific American*, **262**, March 1990, p 30.

Dan Kadell and Donald Ylvisaker, "Lotto play: the good the fair, and the truly awful", *Chance*, **4** No 3, 1991.

"Virginia tries to deter bulk lottery buying", *Chicago Tribune*, 25 February 1992.

Edmund L. Andrews, "Patents; for haters of jangling pocket cash", *The New York Times*, 29 February 1992.

The bootstrap method

Persi Diaconis and Bradley Efron, "Computer-intensive methods in statistics", *Scientific American*, **248**, May 1983, p 116.

Gina Kolata, "The art of learning from experience", *Science*, **225**, 13 July 1984.

Bradley Efron and Robert Tibshirani, " Statistical data analysis in the computer age", *Science*, **253**, 26 July 1991.

Testing

Stephanie A. Shields, "The variability hypothesis: the history of a biological model of sex differences in intelligence", in *Sex and Scientific Inquiry*, edited by Sandra Harding and Jean F. O'Barr, University of Chicago Press, Chicago 1975.

Daniel Goleman and Andrew L. Yarrow, "Girls and math: is biology really destiny?" *New York Times*, 2 August 1987.

Nancy W. Burton, Charles Lewis and Nancy Robertson, *Sex differences in SAT scores*, College Board Report no. 88-9 ETS RR no. 88-58, College Entrance Examination Board, New York, 1988.

Steven Goldberg, "Numbers don't lie: men do better than women", *New York Times*, 5 July 1989.

Joseph Berger, "All in the game", *New York Times*, August 6 1989.

Phyllis Rosser, *The SAT gender gap*, Center For Women Policy Studies, Washington D.C. 1989.

Don Oldenburg, "Nailing down the boards the SAT: isn't once enough?", *The Washington Post*, 6 November 1990.

Frederick L. Smyth, "SAT coaching, what really happens to scores and how we are led to expect more", *The Journal of College Admissions*, 1990, p 7.

Jennifer Toth, "Women's seeming ineptitude with figures doesn't add up", *The Los Angeles Times*, 14 August 1991.

Karen De Witt, "U.S. study shows pupil achievement at level of 1970", *The New York Times*, 1 October 1991.

Edmund L. Andrews, "Patents; giving tests with fewer questions", *The New York Times*, 26 October 1991.

Loenard Ramist, Charles Lewis and Laura McCamley, "Student differences in predicting college grades", College Board Report. October 1991.

Anthony DePalma, "SAT coaching raises scores, report says", *The New York Times*, 18 Decmeber 1991.

Gina Kolata, "Which students are worst at science", *The New York Times*, 24 December 1991.

Michael Olinick, "Probabilistic evidence of sexism?", Newletter from Department of Technology and Society, State University of New York at Stony Brook Center, Stony Brook NY.

Carol Innerst, "PC comes to the SAT; Tests aim to reflect diversity", *The Washington Times*, 23 February 1992.

Carol Innerst, "Tilt to minorities upsets some takers", *The Washington Times*, 23 February 1992.

Left-Handedness

Diane F. Halpern, "Do right-handers live longer?", *Nature*, **333**, 1988, p 213.

E. K. Wood, "Less sinister statistics from baseball records", *Nature*, **335** No 6187, 1988, p 212.

Max G. Anderson, "Lateral preference and longevity", *Nature*, **341**, 14 September 1988, p 112.

Wayne P. London, "Left-handedness and life expectancy", *Perceptual and Motor Skills*, **68** No 3, June 1989, p 1040.

Stanley Coren and Diane F. Halpern, "Left-handedness: A marker for decreased survival fitness", *Psychological Bulletin*, **109** No 1, 1991, p 90.

Diane F. Halpern, "Handedness and life span", *The New England Journal of Medicine*, 4 April 1991, p 998.

Michael Peters, "Sex, handedness, mathematical ability, and biological causation", *Canadian Journal of Psychology*, **45** No 3, September 1991, p 415.

Letters In Response to Halpern and Cohen, "Left-Handedness and Life Expectancy", *The New England Journal of Medicine*, **325** No 14, 3 October 1991, p 1041.

Browlee Shannon, "The southpaw's secret semantics. (left-handedness)", *U.S. News & World Report*, **112** No 7, February 1992, p 66.

Harold L. Klawans, "Don't wait for lefty, he's dead", *The New York Times*, 1 March 1992.

Joel Achenbach, "The spice of life? variety, pal", *The Washington Post*, 27 March 1992.

Nona Yates, "Culture watch; left or right? give 'em a hand", *Los Angeles Times*, 20 April 1992.

Gender Issues

Nicholas D. Kristof, "Stark data on women: 100 million are missing", *The New York Times*, 5 November 1991.

Stephen G. Brush, "Women in science and engineering", *American Scientist*, **79**, 1991.

"How schools shortchange girls", A Study of major findings on girls and education. *American Association of University Women*, The Wellesley College Center For Research on Women. 1992.

Irene Sege, "The grim mystery of the world's missing women; In some cultures in Asia, the ratio of boys to girls is way out of line", *The Boston Globe*, 3 February 1992.

Susan Chira, "Bias against girls is found rife in schools, with lasting damage", *The New York Times*, 12 February 1992.

Marcia Barinaga, "Profile of a field: neuroscience the pipeline is leaking", *Science*, **255**, 13 March 1992.

Ann Gibbons, "Key issue: Mentoring", *Science*, **255**, 13 March 1992.

Ivan Amato, "Profile of a field: Chemistry. Women have extra hoops to jump through", *Science*, **255**, 13 March 1992.

Data Points, *Science*, **255**, 13 March 1992.

Ann Gibbons, "Key issue: Two-career science marriage", *Science*, **255**, 13 March 1992.

Paul Selvin, "Heroism is still the norm", *Science*, **255**, 13 March 1992.

Ann Gibbons, "Key issue: Tenure", *Science*, **255**, 13 March 1992.

Ansley J. Coale, "Excess female mortality and balance of the sexes in the population: An estimate of the number of 'missing females'", (to be published).

Cancer

Tim Beardsley, "Nuclear numbers: living near power reactors may not add cancer risk", *Scientific American*, **264**, February 1991, p 30.

Karen Wright, "Going by the numbers: armed with facts, figures and a sharp tongue", *New York Times Magazine*, **41**, 15 December 1991, p 58.

Elaine Blume and Clayton T. Cowl, "A switch in time saves a statistical bind", *The Journal of The American Medical Association*, **266**, 18 December 1991, p 3284.

Jane E. Brody, "Personal health", *The New York Times*, 12 February 1992.

Sandra Blakeslee, "Better odds; faulty math heightens fears of breast cancer", *The New York Times*, 15 March 1992.

Associated Press, "Breast radiation called small risk", *The New York Times*, 19 March 1992.

Implants

Philip J. Hilts, "Panel to consider what sort of rules should control Gel implants", *The New York Times*, 18 February 1992.

Philip J. Hilts, "Studies see greater implant danger", *The New York Times*, 19 February 1992.

Philip J. Hilts, "Experts suggest U.S. sharply limit breast implants", *The New York Times*, 21 February 1992.

Felicity Barringer, "Breast-implant plan upsets surgeons", *The New York Times*, 22 February 1992.

Philip J. Hilts, "Biggest maker of breast implants is said to be abandoning market", *The New York Times*, 19 March 1992.

Caffeine

"Is coffee safe?", *Consumer Reports*, **54**, September 1987, p 529.

Deiderick E. Grobbee et al, "Coffee, caffeine and cardiovascular disease in men", *New England Journal of Medicine*, **323**, 11 October 1991, p 1026.

"The buzz on caffeine", *USA Weekend*, 5 January 1992.

Prentice Thomson, "Beware", *The Times*, 27 January 1992.

Herman Robin, "Coffee and cholesterol; no link to heart disease has yet been found", *The Washington Post*, 18 February 1992.

Howard Rothman, "You love the java jive, but does it love you? (coffee and health)", *Nation's Business*, **80**, February 1992, p 55.

"Tea gunpowder tea", *The Atlanta Journal and Constitution*, 9 April 1992.

Laurie Snell is Benjamin Cheney Professor of Mathematics at Dartmouth College. He worked with John Kemeny and Gerald Thompson to develop the Finite Mathematics course and the textbook based on it. He has published numerous articles and books; the most recent is a Carus Monograph, *Random Walks and Electronic Networks*, with Peter Doyle. He and Doyle are currently editing a forthcoming volume in the MAA Notes series, *Topics in Contemporary Probability*. Professor Snell is also serving as Project Director of the CHANCE project.

Annotated Bibliography of Sources of Real-World Datasets Useful for Teaching Applied Statistics

Judith D. Singer and John B. Willett[1]
Harvard University

Adkins, P. G. (1969). Deficiency in comprehension in non-native speakers. *TESOL Quarterly*, **3**, 197-201.

Ninth grade student achievement (40 subjects) on tests of seventh grade material, with and without idioms and figures of speech. Useful for bivariate analysis.

Afifi, A. A., and Azen, S. P. (1979). *Statistical Analysis: A Computer Oriented Approach*, 2nd edition. New York: Academic Press.

Several extensive data sets describing the blood chemistry (cholesterol, blood pressure, etc.), cardiovascular state, socioeconomic status, and year of death. Some censored cases, could be used in the teaching of survival analysis. Other datasets include body flexibility, diet, testosterone levels in right and left testes of mice (!), weaning of rats, and infant cognitive development.

Afifi, A. A., and Clark, V. (1984). *Computer Aided Multivariate Analysis*. Belmont, CA: Lifetime Learning, 30-39.

Depression scores and selected covariates for 294 participants in the Los Angeles Depression Study. Data set includes individual item responses for a 20 question depression scale, person background characteristics and selected health variables.

Aickin, M. (1983). *Linear Statistical Analysis of Discrete Data*. New York: Wiley.

A large variety of categorical data sets including: tenure in American universities, dolphin sightings, transitions between Piagetian stages, college expectations and participation in high school athletics, political preferences, religion and marajuana, sudden infant death.

---

[1]An earlier version of this bibliography was published in *The American Statistician*, 1990, **44**(3), 223-230. The order of the authors has been determined by randomization.

Aldrich, J. H., and Nelson, F. D. (1984). *Linear Probability, Logit and Probit Models.* Sage University Paper Number 45. Beverly Hills, CA: Sage.

Data concerning the effect of the "Personalized System of Instruction" on course grades in an intermediate macroeconomics course, useful for logit analysis and log-linear modeling.

Allison, T. and Cicchetti, D. V. (1976). Sleep in mammals: Ecological and constitutional correlates. *Science*, **194**, 732-734.

Average brain weights and body weights for 62 species of mammals. Both variables are very skewed, but logarithmic transformations alleviate the skewness and improve the linearity of the scatterplot.

Andrews, D. F. and Herzberg, A. M. (1985). *Data: A Collection of Problems from Many Fields for the Student and Research Worker.* New York: Springer-Verlag.

71 sets of raw data. Many substantive areas are included, but the emphasis is generally on the physical and natural sciences. Several interesting social science examples are given, including: unemployment statistics, insurance rate information, literary data sets (the Federalist papers data set and another on Platonic prose rhythm) and the birth-day/deathday problem.

Angell, R. C. (1951). The moral integration of American cities. *American Journal of Sociology*, **53** 1-140.

Measures of the moral integration, ethnic heterogeneity, crime, welfare effort, integration and mobility of residents in 43 American cities.

American Association of University Professors (1987). The annual report on the economic status of the profession, 1986-1987. *Academe*, **73**, 1-88.

Salary data by rank, sex, and tenure status for faculty at 1,901 colleges and universities. Institutions are categorized according to Carnegie classifications.

Aylward, G. P., Harcher, R. P., Leavitt, L. A., Rao, V., Bauer, C. R., Brennan, M. J., and Gustafson, N. F. (1984). Factors affecting neobehavioral responses of preterm infants at term conceptual age. *Child Development*, **55**, 1155-1165.

Contingency table of the relationship between gestational age and neurological status for 505 babies. Also see detailed log-linear analysis of these data in: Green, J. A. (1988), Loglinear analysis of cross-classified ordinal data: Applications in developmental research, *Child Development*, **59**, 1-25.

Bain, R. (1939). Verbal stereotypes and social control. *Sociology and Social Research*, **23**, 431-446.

Multiple multiway contingency datasets on the use of proverbs and mechanical cliches used by 133 college freshmen in 1937. Data are broken down by urban/rural, gender, etc.

Barrons' Publications (1990). *Profiles of American Colleges*, Sixteenth Edition. New York: Barron's Publications.

One of many sources describing the more than 1,500 four-year colleges in this country. Relevant data include: number of applicants, number of students accepted, number of students enrolling, mean SAT scores of incoming freshman, mean class rank of incoming freshmen, faculty/student ratios, financial aid available, number of part-time students and faculty, percent of faculty with doctorates, sex composition of student body. Can be supplemented with information from American Association of University Professors salary survey and endowment data given in the *Digest of Education Statistics*.

Bell, J. C. (1914). A class experiment in arithmetic. *Journal of Educational Psychology*, **5**, 467-170.

Individual data for 25 college sophomores at the University of Texas on the speed and accuracy with which they solved four types of arithmetic problems (addition, subtraction, multiplication and division).

Bell, J. C. (1916). Mental tests and college freshmen. *Journal of Educational Psychology*, **7**, 381-399.

Scores on nine tests for 37 of the "best" students and 37 of the "worst" students, with notations of class rank, designed "to be of assistance to college authorities in aiding freshmen to adjust themselves to their environment" (p. 381).

Berenson, M. L., Levine, D. M., and Goldstein, M. (1983). *Intermediate Statistical Methods and Applications*. Englewood Cliffs, NJ: Prentice Hall.

A large variety of non-social science data sets (lawn service, real estate market, professional sports, foreign food), with some social science data sets scattered here and there: categorical data on health issues in children by graduating class of pediatrician, starting salaries of MBA graduates, etc.

Berndt, E. R. (1991). *The Practice of Econometrics: Classic and Contemporary*. Reading, MA: Addison-Wesley.

Dozens of data sets relevant for economics on topics such as prices of color television sets, common stocks, gold, and other commodities. An added benefit is that all the data sets are entered onto a floppy disk included with every book.

Bishop, D. W., & Ikeda, M. (1970). Status and role factors in the leisure behavior of different occupations. *Sociology and Social Research*, **54** 190-208.

Multidimensional data on leisure activities by occupation. Useful for multidimensional scaling, cluster analysis, factor analysis, principal components analaysis, discriminant analysis, etc.

Bock, R. D. (1975). *Multivariate Statistical Methods in Behavioral Research*. New York: McGraw Hill.

A variety of educational growth data sets suitable for repeated measures/ MANOVA analysis, including data on responses to inkblot plates by grade and IQ over time, longitudinal (4 grades) data on scaled vocabulary scores for boys and girls, and so on. (Data repeated in Finn, J. D., & Mattsson, I. (1978). *Multivariate Analysis in Educational Research*. Chicago, IL: National Educational Resources.

Boli, J., Allen, M. L., & Payne, A. (1985). High ability women and men in undergraduate mathematics and chemistry courses. *American Educational Research Journal*, **22**, 605-626.

Multi-dimensional categorical data describing perceptions of course performance among men and women in physics and chemistry courses at Stanford University.

Bullen, A. K. (1945). A cross-cultural approach to the problem of stuttering. *Child Development*, **16**, 1-88.

Raw data for 46 children divided into four groups -- stutterers, well-adjusted, medium adjusted and poorly adjusted. Measures include age, achievement, receptivity to education, physical condition, social-personality traits, insightfulness, family background, somotype and anthropometrics.

Chambers, J. M., Cleveland, W. S., Kleiner, B. & Tukey, P. A. (1983). *Graphical Methods for Data Analysis*. Belmont, CA: Wadsworth.

33 raw datasets. Many substantive areas are included, and many of these data sets are just plain interesting, such as the ages of signers of the Declaration of Independence, murder/suicides by crashing private airplanes, heights of singers in the New York Choral Society.

Chapman, J. C. (1914). *Individual Differences in Ability and Improvement and their correlations*. Teacher's College, Columbia University Contributions to Education Number 63. New York: Teacher's College.

Ten longitudinal data sets (10 waves) on measures of computation, color-naming and opposites-naming for 22 college-age males in New York at the turn of the century. Suitable for growth curve analysis.

Cobb, M. C. (1917). A preliminary study of the inheritance of arithmetic abilities. *Journal of Educational Psychology*, **8**, 1-20.

Data for on the mother, father and children in eight families, with the age of each family member and their scores on five tests (addition, subtraction, multiplication, division and copying figures). The author concludes that "it is difficult to avoid the conclusion that ... likeness is due to heredity" (p.16).

Cooley, W. W., & Lohnes, P. R. (1985). *Multivariate Data Analysis*. Malabar, FL: Robert E. Kreiger.

Two large data sets: (1) a 20 variable subset of the PROJECT TALENT data (234 males, 271 females); and (2) the RECTANGLES data set on the dimensions of 100 rectangles, useful for factor analysis and principal components analysis.

Cox, D. R. & Snell, E. J. (1981). *Applied Statistics: Principles and Examples*. London: Chapman and Hall.

39 raw datasets. Relevant examples include: educational plans of Wisconsin school boys, statistical aspects of literary style, satisfaction with housing conditions.

Council of Great City Schools (1983). *Statistical Profiles of the Great City Schools*. Philadelphia, PA: Author.

Educational and demographic descriptors for 32 large urban school districts, including data on how these characteristics have changed over time.

Crymes, R. (1971). The relation of study about language to language performance with special reference to nominalization. *TESOL Quarterly*, **5**, 217-230.

Categorical and continuous data on the number of nominals produced by native and non-native speakers. Data for experimental and control groups on pre- and post-test measures.

Devore, J. and Peck, R. (1986). *Statistics: The Exploration and Analysis of Data*. St. Paul, MN: West Publishing.

The data sets tend to be small, and many are from the sciences, but there are dozens of them. One interesting example is the movie production and promotion costs for "Dumb Movies," such as Revenge of the Nerds and Police Academy.

Draper, N. R., and Smith, H. (1981). *Applied Regression Analysis*, 2nd edition. New York: Wiley.

Some data sets with broad interest are submerged among many others, including: sex differentials in teacher pay, aptitude and age of first word, nutrition of preschoolers, ailments of university alumni, and the relationship between length of lifeline and length of life.

Dunshee, M. E. (1931). A study of factors affecting the amount and kind of food eaten by nursery school children. *Child Development*, **2**, 163-183.

Data on the eating habits of 37 children, including age, sex, means (and standard deviation) for total amount of calories eaten and minutes spent at table.

Dunteman, G. H. (1984). *Introduction to Linear Models*. Beverly Hills, CA: Sage.

Data for 300 participants in the National Longitudinal Study on reading, math, gender, race, college status, SES, High school program, High school grades, creativity, stress avoidance, etc.

Eash, M. (1983). Educational research productivity of institutions of higher education. *American Educational Research Journal*, **20**, 5-12.

Research productivity rankings for 25 schools of education. Measures are based on presentations at AERA and articles published in AERA journals. The paper also presents data comparing AERA program contributions in 1975 and 1976 with those in 1980 and 1981. Compare with smaller data sets presented in West, C. K. (1978). Productivity ratings of institutions based on publication in the journals of the American Educational Research Association, *Educational Researcher*, **7**, 13-14 and Schubert, W. H. (1979). Contributions to AERA annual programs as an indicator of institutional productivity. *Educational Researcher*, **8**, 13-17.

Educational Research Service (1987). *Scheduled Salaries for Professional Personnel in Public Schools 1986-1987.* Arlington, VA: Author.

Raw data for 1,031 school districts on enrollment, per pupil expenditures, and salaries for superintendents, central office administrators, principals, teachers, staff and support personnel.

Erickson, B. H., & Nosanchuck, T. A. (1977). *Understanding Data.* Toronto, Canada: McGraw Hill Ryerson.

An introductory textbook that melds together Tukey's exploratory data analysis and the more traditional confirmatory approaches. Many interesting data sets, including frequency of teacher criticism by student IQ, sex differences in reactions to hostile treatment by an experimenter, experimenter artifacts in social psychology research, characteristics of social networks.

Fales, E. (1933). A comparison of the vigorousness of play activities of preschool boys and girls. *Child Development*, **4**, 144-157.

Age, IQ scores and activity ratings for 16 boys and 16 girls. Two activity ratings are available for each child.

Ferdinand, T. N., & Luchterhand, E. G. (1970). Inner-city youth, the police, the juvenile court, and justice. *Social Problems*, **17**, 510-527.

Many multiway categorical datasets on the crimes and police disposition of inner-city youth, by race, gender, type of crime, age, etc.

Finn, J. D. (1974). *A General Model for Multivariate Analysis.* New York: Holt, Rinehart and Winston.

Raw data for four studies of relevance to the social sciences: Creativity and achievement, memory for words, essay grading practices and effects of programmed instruction.

Fleming, J. T. (1967). The measurement of children's perception of difficulty in reading materials. *Research in the Teaching of English*, **1**, 136-156.

Individual data on 60 subjects. Data include background characteristics (gender, age, IQ, etc.) and achievement (comprehension, reading, word knowledge, spelling, language, study skills) for school-age children. Suitable for multiple regression, multivariate analysis, etc.

Fox, J. (1984). *Linear Statistical Models and Related Methods with Applications to Social Research*. New York: John Wiley.

Several interesting data sets including: relationship between status, authoritarianism and conformity, methods to enhance recall of words, causes of the 1907 Romanian Peasant rebellion.

Fraumeni, J. F. Jr. (1968). Cigarette smoking and cancers of the urinary tract: Geographic variation in the United States. *Journal of the National Cancer Institute*, **41**, 1205-1211.

Aggregate cigarette smoking and cancer death rates, by type of cancer and state.

Garwood, A. N. (Ed.) (1986). *Massachusetts Municipal Profiles*. Wellesley Hills, MA: Information Publications.

Sociodemographic characteristics for 353 Massachusetts towns and cities, including data on age, race, sex, income, labor force participation, voter registration, police, fire, crime, taxation, libraries and schools. The company publishes similar books for other states; write to them at Box 356, Wellesley Hills, MA 02181.

Gastwirth, J. L. (1984). Statistical methods for analyzing claims of employment discrimination. *Industrial and Labor Relations Review*, **38**, 75-86.

Several data sets used in employee discrimination cases against organizations including the University of Texas at Dallas, the city of Wichita Falls, and the Federal Reserve Bank.

Gelb, S. A., & Mizokawa, D. T. (1986). Special education and social structure: The commonality of "exceptionality." *American Educational Research Journal*, **23**, 543-557.

State-level data on percentage of children in each category of special education and sociodemographic composition of the states. Washington, DC is a high-leverage outlier for the relationship between percent of students classified as educably mentally retarded and percent of population that is black.

Gerlach, M. (1939). A study of the relationship between psychometric patterns and personality types. *Child Development*, **10**, 269-278.

Raw data for 61 maladjusted children on two IQ tests, as well as information on their sex, age, parentage (both Foreign, both American, Mixed) and maladjustment type (agressive or asocial). Authors explore relationship between IQ and all these predictors.

Gnanadesikan, R. (1977). *Methods for Statistical Data Analysis of Multivariate Observations*. New York: John Wiley.

Several multivariate data sets from a variety of disciplines including engineering, manufacturing, biology and mining. The volume includes several well-known data sets such as Fisher's iris data (1936) and Rothkopf's Morse-code confusion data (1957) which have utility for the teaching of principal components analysis, factor analysis, multidimensional scaling and cluster analysis.

Gordon, N. J., Nucci, L. P., West, C. K., Hoerr, W. A., Vguroglu, M., Vukosavich, P., & Tsai, S. L. (1984). Productivity and citations of educational research: Using educational psychology as the data base. *Educational Researcher*, **13**, 14-20.

Citation frequencies and dates of birth for 187 prominent educational researchers.

Gulliksen, H. (1934). A rationale equation of the learning curve, based on Thorndike's law of effect. *Journal of General Psychology*, **11**, 395-434.

Longitudinal data on rats -- some brighter than others -- trained to discriminate 18 cm and 12 cm circles.

Haberman, S. J. (1978). *Analysis of Qualitative Data*. New York: Academic Press.

Several categorical data sets of wide interest: Suicides by day of the week, homicides by month, stressful events, etc.

Hahn, H. H., & Thorndike, E. L. (1914). Some results of practice in addition under school conditions. *Journal of Educational Psychology*, **5**, 65-83.

Individual data from an experiment on the effects of time lapsed between pre-tests and post-tests for 167 students in grades 4, 5, 6, and 7.

Hamilton, L. C. (1990). *Modern Data Analysis: A First Course in Applied Statistics*. Pacific Grove, CA: Brooks-Cole.

Dozens of interesting data sets on topics such as the unemployment of graduates from 44 British universities, the drinking habits of 120 mothers and their adolescent children, average salaries and overall performance of 26 baseball teams, and marriage rates in the 50 states (Nevada is a very high outside value).

Hand, D. J. & Taylor, C. C. (1987). *Multivariate Analysis of Variance and Repeated Measures: A Practical Approach for Behavioral Scientists*. London: Chapman and Hall.

Raw data from 8 studies in psychology and psychiatry, on topics as diverse as headaches, smoking and Alzheimer's disease.

Harris, J. A., Jackson, C. M., Paterson, D. G., & Scammon, R. E. (1930). *The Measurement of Man*. Minneapolis, MN: University of Minnesota Press.

Many unique and interesting data sets on the link between physical and psychological characteristics. Do blondes have more fun? Do lunatics eyebrows join together in the middle? Are manic depressives thin or fat?

Harris, S., & Harris, L. B. (1986). *The Teacher's Almanac*. New York: Facts on File.

Assorted education data by state and school district, including teacher salaries, high school graduation rates, functional illiteracy rates, and presence of computers in schools.

Heyneman, S. P. (1976). Influences on academic achievement: A comparison of results from Uganda and more industrialized societies. *Sociology of Education*, **49**, 200-211.

Aggregate data on school achievement and economic development for 18 nations. Includes measures of preschool influence, GNP, percent enrollment in primary and secondary school.

Hirsch, N. D. M. (1928). An experimental study of the East Kentucky mountaineers: A study in heredity and environment. *Genetic Psychology Monographs*, **3**, 183-244.

Ages and IQ scores of siblings in 44 families.

Hirsch, N. D. M. (1930). An experimental study upon three hundred school children over a six-year period. *Genetic Psychology Monographs*, **7**, 487-546.

IQ scores over a six year period for 343 children.

Hodges, J. L., Krech, D., & Crutchfield, R. S. (1975). *Statlab: An Empirical Introduction to Statistics*. New York: McGraw-Hill.

Data on two parents and one child from each of 1296 families. Variables include information on height, weight, and psychological tests.

Hollander, M., & Proschan, F. (1984). *The Statistical Exorcist: Dispelling Statistics Anxiety*. New York: Marcel Dekker.

A host of data sets of different sizes on many different topics including: blood pressure and obesity of Mexican-Americans, baseball data, 1970/1 draft lottery, promotion rates among male and female pharmacists, leisure time companions of black women, ranking of rum brands by different nationalities, preference for Charlie's Angel's actors, longevity and environment, color of canned tuna, etc.

Howard, G. S., Cole, D. A., & Maxwell, S. E. (1987). Research productivity in psychology based on publication in the journals of the American Psychological Association. *American Psychologist*, **42**, 975-986.

Productivity, reputation and size of psychology departments at 75 universities.

Izenman, A. J. (1972). Reduced rank regression for the multivariate linear model. Doctoral dissertation, Department of Statistics, University of California at Berkeley.

Body length of crickets as a function of geographical location and weather throughout the USA.

Jensen, A. R. (1974). Kinship correlations reported by Sir Cyril Burt. *Behavioral Genetics*, **4**, 1-28.

"Burt's final assessments" of IQs of monozygotic twins reared apart, with "social class" ratings of the homes. For information on Burt's falsification of the data, see Dorfman, D. D. (1978). The Cyril Burt Question: New Findings. *Science*, **201**, 1177-1186; other sources include correspondence related to Dorfman's article: Stigler, S. M. (1979).

Letter to the editor. *Science*, **204**, 242-245; Rubin, D. B. (1979). Letter to the editor. *Science*, **204**, 245-246; Dorfman, D. D. (1979). Letter to the editor. *Science*, **204**, 246-254; Hearnshaw, L. S. (1979). *Cyril Burt: Psychologist*. London: Hodder and Stoughton, Ch. 12.

Jensen, A. R. (1970). IQ's of identical twins reared apart. *Behavioral Genetics*, **1**, 133-148.

Original data from four studies of IQs of identical twins reared apart: Newman, Freeman & Holzinger (1937), Shields (1962), Juel-Neilsen (1965) and Burt (1955).

Johnson, B., & Courtney, D. M. (1931). Tower building. *Child Development*, **3**, 161-162.

Twenty-five children were asked to build towers on each of two occasions. Each time they were given: (a) a set of cubes; and (b) a set of cylinders. Raw data are given on the number of blocks of each type used each time, and how many minutes it took to construct the tower.

Johnson, W. D., & Koch, G. G. (1971). A note on the weighted least squares analysis of the Ries-Smith contingency table data. *Technometrics*, **13**, 438-447.

Large four-way contingency table derived from a randomized experiment in which several brands of detergent were compared under a variety of user-related conditions.

Jones, L. V., Lindzey, G., & Coggeshall, P. E. (1982). *An assessment of research-doctorate programs in the United States: Social Sciences*. Washington, DC: National Academy Press.

"Quality" rankings and characteristics of university departments in the social sciences, by discipline. Data include number of faculty, number of students, productivity of faculty, number of grants awarded, follow-up placement of doctoral students.

Karelitz, S., Fisichelli, V. R., Costa, J., Karelitz, R., & Rosenfeld, L. (1964). Relation of crying in early infancy to speech and intellectual development at age three years. *Child Development*, **35**, 769-777.

Data for 38 infants on crying in early infancy and later measures of IQ.

Koch, H. L. (1933). Popularity in preschool children: Some related factors and a technique for its measurement. *Child Development*, **4**, 164-175.

Popularity scores for 17 children: percent each child was named first, percent each child name last, effects of ordering and the effects of sex.

Kroc, R. J. (1984). Using citation analysis to assess scholarly productivity. *Educational Researcher*, **13**, 17-22.

Citation frequency data for 51 schools of education. Measures include mean citation rate, percentage of faculty with 10-100 citations and percentage of faculty with no citations. Kroc correlates these measures with five rankings of schools of education; these analyses could be recreated by abstracting data from the five sets of rankings cited in his bibliography.

Leinhardt, G., & Leinhardt, S. (1980). Exploratory data analysis: New Tools for the analysis of empirical data. *Review of Research in Education*, **8**, 85-157.

Three measures of reading instruction for a sample of 53 learning disabled students, by curricular approach and school.

Leinhardt, S. and Wasserman, S. S. (1979). Teaching regression: An exploratory approach. *The American Statistician*, **33**(4), 196-203.

Life expectancy and per capita income for 105 nations divided into five wealth classifications (industrialized, petroleum exporting, higher, middle and lower).

Louano-Kerr, J., Semles, V., & Zimmerman, E. (1977). A profile of art educators in higher education: Male/female comparative data. *Studies in Art Education*, **18**, 21-37.

Multiway contingency table data on the qualifications (Masters vs. doctorate) of art educators, by year and gender.

Maresh, M. M., & Deming, J. (1939). The growth of the leg bones in 80 infants: Roentgenograms versus anthropometry. *Child Development*, **10**, 91-106.

Individual data for 80 children on the sizes of 10 bones, measured by both x-rays and anthropometry at each of 3, 4, or 5 occasions, by sex. The authors construct lots of individual growth curves.

Mason, T. J., & McKay, F. W. (1974). US Cancer Mortality by County: 1950-1969. Washington, DC: US Government Printing Office.

Lung cancer mortality by degree of urbanization and gender, in Louisiana.

Mickey, M. R., Dunn, O. J., & Clark, V. (1967). Note on the use of stepwise regression in detecting outliers. *Computers and Biomedical Research*, **1**, 105-109.

Gesell adaptive scores and age at first word (in months) for 21 children with cyanotic heart disease. The data set contains some interesting outliers and high leverage cases.

Mosteller, F., Hyman, H., McCarthy, P. J., Marks, E. S., & Truman, D. B. (1949). *The Pre-Election Polls of 1948*. New York: Social Science Research Council.

"Dewey beats Truman" is one of the classic headlines of the twentieth century. The data contained in this book provide an opportunity to examine the pre-election polls to ask why this headline seemed plausible. Data are provided by the respondents' social class, as well as the month in which the pre-election poll was conducted. Also see analyses in Baker, S. G., & Laird, N. M. (1988), Regression analysis for categorical variables with outcome subject to nonignorable nonresponse, *Journal of the American Statistical Association*, **83**, 62-69.

Mosteller, F. & Tukey, J. W. (1977). *Data Analysis and Regression: A Second Course in Statistics*. Reading, MA: Addison-Wesley.

Raw data for 13 data sets across several disciplines. Relevant examples include a subset of 20 from the Coleman Report, educational expenditures for Massachusetts school districts, municipal bond data for 20 US cities.

National Center for Education Statistics (1987). *The Condition of Education.* Washington, DC: US Department of Education, and National Center for Education Statistics (1987). *Digest of Education Statistics.* Washington, DC: US Department of Education.

Annual reports issued by the Department of Education providing descriptive information on education, often over time, sometimes by state, occasionally by school district. The data on university endowments can be used in conjunction with other university level data, such as that given in Barron's (1987).

National Education Association (1965). Annual report on public-school financing. *NEA Research Bulletin*, **43**, 90-92.

State-by-state listing of per-pupil expenditures (in 1955-56 and 1964-65) and percent of school revenues from state and local sources. The entire journal series contains a wealth of information on class size, salaries, public expenditures for education, etc.

National Education Association (1926). The ability of the states to support education. *Research Bulletin of the National Educaton Association*, **4**.

Several chapters of state-level data documenting wealth, per capita income, educational expenditure, population size, number of adults per child, and so on.

National School Boards Association (1986). *A Survey of Public Education in the Nation's Urban School Districts.* Alexandria, VA: author.

Data for 61 school districts on educational policies and practices, as well as selected education and economic descriptors.

Nelson, W. (1982). *Applied Life Data Analysis.* New York: John Wiley.

Several data sets presenting time-to-failure of products, useful for examining industrial product reliability.

Neuman, S., & Ziderman, A. (1985). Do universities maintain common standards in awarding first degrees with distinction?: The case of Israel. *Higher Education*, **14**, 447-459.

Number of first degree graduates with and without distinction, by school and faculty within school. Suitable for log-linear modeling.

Opening lines (1985, September). *Harper's*, pp. 29-30.

Selected results from a study of opening lines used in singles bars in the St. Louis area. Two-way contingency table describing the relationship between type of opening line (compliments, propositions, etc.) and time of evening.

Phillips, D. P. (1978). Deathday and birthday: an unexpected connection. In Tanur, J.M., Mosteller, F., Kruskal, W. H., Link, R. F., Pieters, R. S., Rising, G. R., and

Lehmann, E. L. (eds.), *Statistics: A Guide to the Unknown*, 2nd ed. San Francisco, CA: Holden-Day, 71-85. See, also, Phillips, D. P. (1977), Motor vehicle fatalities increase just after publicized suicide stories, *Science*, **196**, 1464-1465; Phillips, D. P. (1978), Airplane accident fatalities increase just after newspaper stories about murder and suicide, *Science*, **201**, 748-750; Phillips, D. P., and Carstensen (1986), Clustering of teenage suicides after television news stories about suicide, *New England Journal of Medicine*, **315**, 685-689 (and related articles in this issue); Schultz, R., Bazerman, M. (1980), Ceremonial occasions and mortality: A second look, *American Psychologist*, **35**, 253-261.

David Phillips has made a cottage industry of looking at what many might term coincidences -- birthdays and deathdays and copycat suicides after popularized accounts in the media. These are but a handful of articles, each listing the detailed raw data on deaths following these events that led him to his conclusions.

Plackett, R. L. (1981). *The Analysis of Categorical Data*. New York: MacMillan).

A large variety of categorical data sets including: fingerprints, family size, work conditions and work quality, behavioral problems and birth order, high school rank by gender and socioeconomic status.

Powell, B. & Steelman, L. C. (1984). Variations in state SAT performance: Meaningful or misleading? *Harvard Educational Review*, **54**, 389-412.

Mean SAT scores and percent of high school seniors taking the SAT, by state for 1982. For additional data, and a critique of their analyses, see: Wainer, H, Holland, P. W., Swinton, S., & Wang, M. H. (1985), On "State Education Statistics", *Journal of Educational Statistics*, **10**, 293-325. Also see: Rosenbaum, P. R. & Rubin, D. B. (1985), Discussion of "On State Education Statistics": A difficulty with regression analyses of regional test score averages, *Journal of Educational Statistics*, **10**, 326-333. and Wainer, H. (1986), Five pitfalls encountered while trying to compare states on their SAT scores, *Journal of Educational Measurement*, **23**, 69-81.

Rubin, E. (1972). Statistical exploration of a medieval household book, *The American Statistician*, **26**, 37-39.

Number of meals served, breads baked and ale brewed at the de Bryene household from October 1412-September 1413, by month. That's right, the fifteenth century.

Ryan, B. F., Joiner, B. L., & Ryan, T. A. Jr. (1985). *Minitab Manual*, 2nd edition. Boston, MA: Duxbury.

Thirty data sets of small to moderate sizes, on topics ranging from education to cartoons. The educational data sets include information on school strikes and freshman SAT verbal and math scores.

Sandvern, J. (1971). Causes of lacking sense of well-being in school. *Scandinavian Journal of Educational Research*, **15**, 21-60.

Many multiway contingency tables dealing with students' liking of school in Norway, broken down by type of school, SES, etc.

Scarcella, R. C. (1984). How writers orient their readers in expository essays: A comparative study of native and non-native english writers. *TESOL Quarterly*, 671-688.

Categorical data on the language background and language proficiency of native and non-native speakers and how this influences their choice of writing device.

Shearer, L. (1987). How will history rate Nancy Reagan? *Parade Magazine*, 14 June 1987, p. 8.

Rankings for 17 first ladies from Florence Harding through Nancy Reagan on 10 dimensions ranging from integrity, leadership and accomplishments.

Skodak, M., & Skeels, H. M. (1949). A final follow-up study of one hundred adopted children. *Journal of Genetic Psychology*, **75**, 85-125.

Raw data for 100 children who were adopted at birth. Measures include: natural mother's IQ and education level, foster mother's IQ and education level, foster father's occupation and child's IQ on each of 5 occasions, from infancy through pre-adolescence.

Stevens, J. (1986). *Applied Multivariate Statistics for the Social Sciences*. Hillsdale, NJ: Lawrence Erlbaum.

Many artificial data sets as well as approximately 20 interesting small to moderately sized real data sets, including: pre/post data on the influence of Sesame Street, risk of reading problems among kindergartners, behavior reversal, programmed music instruction of elementary school children, IQ testing, etc.

Stewart, L. H. (1955). The expression of personality in drawings and paintings. *Genetic Psychology Monographs*, **51**, 45-103.

Data for 28 boys given including: 2 IQ scores (Terman and Stanford-Binet), 2 socioeconomic status measures (parents education and father's occupation), somotypes (endomorph, mesomorph, ectomorph), and drawing type.

Supreme court ruling on death penalty (1988). *Chance*, **1**, 7-8.

Three-way contingency table on the relationship between race of victim, race of defendant and use of the death penalty, showing that the death penalty is not uniformly applied.

Taagepera, R. (1972). The size of national assemblies. *Social Science Research*, **1**, 385-401.

Aggregate data on the size of the national assemblies in more than 100 countries, with data on population, literacy, working age, etc.

Tanner, M. A. (1990). *Investigations for a Course in Statistics*. New York: MacMillan.

Dozens of exercises for generating raw data from students in the class. on variables such as the distance between eyebrows, and the ability to judge the lengths of lines and distances between points.

Timm, N. H. (1975). *Multivariate Analysis with Applications in Education and Psychology.* Belmont, CA: Wadsworth.

Raw data for a handful of data sets gathered in educational settings, including: effects of delay in oral practice on second language learning (pp. 228-229), relationship between recall and sentence structure (p. 233), predictors of student performance on the Peabody Picture Vocabulary Test (p. 281).

Tufte, E. R. (1974). *Data Analysis for Politics and Policy.* Englewood Cliffs, NJ: Prentice-Hall.

Several real data sets on topics of political and social relevance, thoroughly analyzed using fairly elementary statistical methods. Topics include the relationship between bureaucracy size and population, the number of radio receiver licenses issued and the proportion of mental defectives in Britain (per 100,000 population), and the relationship between dietery fat consumption and death rates in selected countries around the world.

Tufte, E. R. (1978). Registration and Voting. In Tanur, J. M. et al., *Statistics: A Guide to the Unknown*, 2nd edition. San Francisco: Holden-Day), 195-204.

Data on percent on population registered and percent of population voting in the 1960 election for 104 cities. Data are analyzed in greater detail in: Kelly, S. Jr., Ayres, R. E. and Bowen, W. G. (1967). Registration and Voting: Puting first things first. *American Political Science Review.* **61**, 359-379.

United Nation's Children's Fund (1987). *The State of the World's Children.* New York: Oxford University Press.

Sociodemographic, education, health and economic indicators for 130 countries.

Walberg, H. J., & Rasher, S. P. (1974). Public school effectiveness and equality: new evidence and its implications. *Phi Delta Kappan*, **66**, 3-9, and Walberg, H. J., & Rasher, S. P. (1976). Improving regression models. *Journal of Educational Statistics*, **1**, 253-277.

The authors analyze data for the 50 states on the relationship between failure on the selective service exam administered during 1969-1970 and contextual and education descriptors of the states. The selection bias inherent in analyses of state level SAT scores is present here, but the data set is interesting.

Weisberg, S. (1980). *Applied Linear Regression.* New York: Wiley.

Raw data for several interesting data sets including Cyril Burt's IQ data, Allison and Cicchetti's brain weight and body weight data, three time points for 26 boys and 32 girls who participated in the Berkeley Guidance study (anthropometric information only, however.)

Whiting, J. M. & Child, I. L. (1962). *Child Training and Personality*. New Haven, CN: Yale University Press.

> Many characteristics in dozens of societies around the world, including age at weaning, toilet training, fear of ghosts, rituals, etc.

Wilson, M. E., & Mather, L. E. (1974). Life expectancy [Letter to the editor]. *Journal of the American Medical Association*, **229**(11) 1421-1422.

> Age of person at death (in years) and the length of the person's lifeline (in centimeters) for 50 individuals. Not surprisingly, the test of H_0: $r=0$ cannot be rejected.

Zeligs, R. (1938). Tracing racial attitudes through adolescence. *Sociology and Social Research*, **23**, 45-54.

> Chronological, mental age, social distance scores, and gender for twelve children tested three times during adolescence. Also indices for racial tolerance over three occasions of measurement for 39 racial groups.

Zimmerman, J. (1917). The Binet-Simon Scale and Yerkes Point Scale: A comparative examination of 100 cases. *Journal of Educational Psychology*, **8**, 551-558.

> Individual data for 100 students on these two IQ tests, with information on student sex, age and native language.

Judith D. Singer and John B. Willett are Associate Professors at the Harvard University Graduate School of Education specializing in quantitative methods. They have written and presented dozens of talks, workshops, and papers on the application of statistical methods in education and the social sciences. Together with other colleagues, they have written two books, *By Design?* and *Who Will Teach?*, both published by Harvard University Press. Their current research interests center on applications of survival analysis in the social sciences. They have recently received the Raymond B. Cattell Award and the Palmer O. Johnson Award from the American Educational Research Association and an NSF Visiting Fellowship from the ASA.

Is There Anyone Out There?:
Electronic Connections
to the Statistics Community

Tim Arnold
North Carolina State University

The Growing Communications Web

Just as the invention of the printing press set in motion forces which continue to have huge consequences for humanity, a new force has gathered that promises to reap as many changes, and with as many undreamed-of ramifications for the future. The development of an international web of communications, quietly begun several years ago, has reached a point of explosive growth. Now it is possible for people the world over to discuss new ideas, share informational tools, ask questions, and seek advice easily, painlessly, and almost instantaneously. The global village is here.

Many universities, companies, colleges, and private individuals world-wide are linked together by various international electronic networks such as BITNET and Internet (collectively called 'the Net'). Each day, more people discover the Net, and each day more sites are added, from community colleges to high schools, grade schools, and even businesses. These inter-connections lend reality to the phrase *global village*. This connectivity, and the ability to access virtual libraries of information located worldwide, promises to change the way we work, communicate, even the way we think. While there is something on the Net for almost everyone, some of the resources available are of special use to people involved in statistics, mathematics and education. The following resources are covered below:

Electronic Forums related to statistics and/or education,
Software Libraries with free software useful in statistics,
Public Access Machines with information on funding and other
government sponsored educational programs.

What Kind of Access Do You Have?

If you have an e-mail address, it is likely that you have a connection to the Net. If you don't have an e-mail address or you're not sure, talk to someone at your site in computer support or your local network administrator. If your own institution does not provide e-mail, contact the computer support department at a local university. They are often happy to provide guest accounts to faculty at local schools so that you can then access your e-mail account from your office or home.

There are two types of connections to the Net. The more restrictive type is a BITNET connection. If you have an electronic mail address (e-mail address) and it looks something like:

username@someplace you probably have a BITNET connection. In that case, your primary access to information will be through mail (to other computers and people on the net).

However, if your address looks like: *username@someplace.something.else* that is, if part of your address contains periods, you probably have Internet type access. If you're not sure what type of access you have, the best thing to do is to get help from a friend knowledgeable about your local computer network or a system administrator on your campus.

In general, the Internet type of access gives you more access to informational resources than the BITNET type. With either type of access, you can send messages and files to other people (if you know their e-mail address), or participate in electronic discussion forums.

A Stream of Voices: People helping people

The Net provides a connection to other people, allowing you to share advice, anecdotes, caveats, and ideas with other people on the net. One way this is done is through the use of electronic forums.

There are over 2,000 active forums you may participate in or *subscribe* to. To subscribe to a forum, you just send a special message to the computer that controls the forum. You only need to know two things: the name of the forum and the name of the computer that controls the forum. Some forums you may be interested in and the procedure for subscribing are listed below.

Once you have subscribed to a forum, you will receive messages from other members of the forum group. These messages may be questions, comments, advice, etc., concerning the forum topic. In this way you receive a stream of messages about topics you care about from people all over the world. You can reply privately (via your regular e-mail) to the person who sent the message, you can respond to the forum (by sending the message to the forum's e-mail address), or you can just watch. Most of the time subscribers simply read the messages.

One of the most interesting characteristics of an electronic discussion is the built-in absence of prejudice. Of couse when subscribers read a messsage, they may also see the writer's e-mail address and name, but they cannot see his or her appearance, religion, race, or color. *Content* of messages is the quantity judged, not the personal characteristics of the writer.

It is not a requirement of subscribers to actively participate, although participation is encouraged. The main rule for any forum is that members should send messages that are of interest to other members of the forum, and that concern the forum topic.

One such forum is EdStat-L, the Statistics Education Discussion Forum. The forum was begun October 23, 1991 and has over 500 members world-wide. The purpose of the forum is to provide a vehicle for comments, techniques and philosophies of teaching statistics. The primary focus is on college level statistics education, both undergraduate and graduate studies.

Some topics which have been (and are still being) discussed through EdStat-L include:

- ▸ techniques currently used in statistical instruction,
- ▸ strategies to prepare students for the future,
- ▸ the part computers should play in instruction,
- ▸ debate about whether computers are a new tool or a new subject for students,
- ▸ tips in presenting certain statistical topics,
- ▸ pointers to books and software useful in teaching statistics,
- ▸ pointers to public-domain software programs and datasets.

This forum attempts to bring together every teacher, student, researcher, and specialist interested in improving statistical instruction.

Subscribe to EDSTAT-L by sending the one-line e-mail message (fill in your real name, of course):

SUBSCRIBE EDSTAT-L *YourFirstName   YourLastName*

to the address:

LISTSERV@NCSUVM

If you have Internet-type access, use the address:

LISTSERV@NCSUVM.CC.NCSU.EDU

Other lists which may be of interest to statistical educators are:

listaddress	*listname*	*Description*
MCGILL1	*Stat-L*	*Statistical Consulting*
OHSTVMA	*SPSSx-L*	*SPSSx(r) Discussion*
OHSTVMA	*SAS-L*	*SAS(r) Discussion*
OHSTVMA	*EdTech*	*Educational Technology*
WUVMD	*AcSoft-L*	*Academic Software Develop.*
IRLEARN	*PStat-L*	*Discussion relating to Pstat*

To subscribe to these lists, send the message:

SUBSCRIBE *listname   YourFirstName   YourLastName*

to the address:

LISTSERV@*listaddress*

If you have Internet-type access, try the address

LISTSERV@*listaddress.bitnet*

When you subscribe to a forum, you will get a mail message informing you that you've been added to the list. You will then start getting mail from the forum. You should read your introductory message carefully since usually the rules of the forum and commands for changing certain of your individual forum-related settings will be discussed in that message.

If you would like to learn more about forums and the computers that manage them, send the command (the one-line message) to the address given in the table below. If you have bitnet-type access, use the address exactly as given—for Internet users, add the characters **.bitnet** to the address in the table—this should work for most users.

Address	*Message (command)*	*Description of help*
netserv@bitnic	*GET BITNET USERHELP*	*Introduction to BITNET*
listserv@ncsuvm	*INFO GENINTRO*	*Introduction to Forums*
listserv@ncsuvm	*LIST GLOBAL*	*(large) list of forum names*

When you send one of these commands, the message will be received by a computer program which processes your commands (that's the one-line message). The program will send you a message containing the information you requested.

Other Tools You Can Use: Software and Information Sites

There is a wealth of information and software available to you if you have Internet access. Four locations are covered here.

1. Statistical Software/Datasets

StatLib is a system for distribution of statistical software by electronic mail. StatLib resources include public-domain (free) software and algorithms, a directory of addresses and e-mail addresses for statisticians, many interesting datasets, and functions for the S language. For more information, send the one-line mail message **send index** to the address **statlib@lib.stat.cmu.edu**

2. Mathematical Software

NetLib is a library of public domain mathematical software. The software includes routines for linear algebra, benchmarking, curve fitting, Fast Fourier Transforms, linear programming, differential equation solvers, nonlinear optimization, and matrices. For more information, send the one-line mail message **send index** to the address **netlib@ornl.gov**

3. NSF Information (including grants and program announcements)

The Science and Technology Information System (STIS) is an electronic information dissemination system which provides easy (free) access to the National Science Foundation publications. Publications include the *NSF Bulletin*, the *Guide to Programs*, grants booklet, forms, program announcements, and more. To use this public-access machine, use the telnet command:

telnet stis.nsf.gov

and login with the username *public*. You will be presented with a simple menu to help in navigating the system.

4. Information on U.S. funding programs, surplus equipment, etc.

FEDIX is a free information service linking the education community and the federal government to facilitate research, education and services. Information provided includes federal education programs (with descriptions of eligibility, funding, and grants), used government research equipment, minority assistance education programs, and more. To use this public-access machine, use the telnet command:

telnet fedix.fie.com

and login with the username *fedix*. You will first see new announcements of the day, then you will be presented with a simple menu to help in navigating the system.

There are many other sites on the Net, each with a plethora of resources. Experiment with the access you have--you will find other pointers to more information. Other resources available on the Net include access to supercomputers, library catalogs, reports from government agencies, various newsletters, full text of some literature classics, and huge software libraries.

Access to the Internet:

Many universities are connected to the Internet by way of various regional networks. Many companies have gained access to the Internet in the same fashion. If you are looking for an Internet connection, the first thing to do is check with your local network administrator. The following is a list of some network access providers:

UUNet Technologies, Inc.
3110 Fairview Park Drive
Suite 570
Falls Church, VA 22042
(703) 876-5050
info@uunet.uu.net

Performance Systems International (PSI)
11800 Sunrise Valley Drive
Suite 1100
Reston VA 22091
(800) 827-7482 (703) 620-6651
info@psi.com

ANS Inc.
100 Clearbrook Road
Elmsford, NY 10523
(800) 456-8267
(914) 789-5300
info@nis.ans.net

California Education & Research Federation Network
c/o San Diego Supercomputer Center
P.O. Box 85608
San Diego, CA 92186-9784
(800)876-2373
help@cerf.net

Tim Arnold is director of instructional computing in the department of statistics at North Carolina State University. His current interests include development of graphical interactive tutorial software, use of electronic communications in the classroom, and the development of an international electronic library of resourses for statistics education. He is also interested in the development of innovative teaching techniques which use the computer.